Aspects of Micropalaeontology

Professor Tom Barnard

Aspects of Micropalaeontology

Papers presented to Professor Tom Barnard

Edited by
F.T.Banner and A.R.Lord

Department of Oceanography *Department of Geology*
University College Swansea   *University College London*

London
GEORGE ALLEN & UNWIN
Boston   Sydney

© F. T. Banner, A. R. Lord and contributors, 1982
This book is copyright under the Berne Convention.
No reproduction without permission. All rights reserved.

**George Allen & Unwin (Publishers) Ltd,
40 Museum Street, London WC1A 1LU, UK**

George Allen & Unwin (Publishers) Ltd,
Park Lane, Hemel Hempstead, Herts HP2 4TE, UK

Allen & Unwin Inc.,
9 Winchester Terrace, Winchester, Mass 01890, USA

George Allen & Unwin Australia Pty Ltd
8 Napier Street, North Sydney, NSW 2060, Australia

First published in 1982

British Library Cataloguing in Publication Data

Aspects of micropaleontology.
1. Barnard, Tom
2. Micropaleontology – Addresses, essays, lectures
I. Banner, F. T.   II. Lord, A. R.
III. Barnard, Tom
560   QE719
ISBN 0–04–563003–2

Library of Congress Cataloging in Publication Data

Main entry under title:
  Aspects of micropalaeontology.
Includes bibliographies and index.
1. Micropaleontology – Addresses, essays, lectures.
2. Barnard, Tom.
3. Paleontologists – England – Biography.
I. Barnard, Tom.   II. Banner, F. T. (Frederick Thomas), 1930–
III. Lord, A. R. (Alan Richard), 1942–
QE719.A84   1982   560   82–8735
ISBN 0–04–562003–2   AACR2

Set in 10 on 11 point Times by Bedford Typesetters Ltd,
and printed in Great Britain
by Mackays of Chatham

Preface

This volume is a collection of papers presented to Professor Tom Barnard by former students, colleagues and friends to mark thirty-two years of teaching and research in micropalaeontology at University College London. This period represents the major part of Tom Barnard's career with microfossils, which actually began rather earlier, but in 1949 his first postgraduate students were registered. Since then some 150 students have worked for higher degrees studying foraminifera, ostracods, calcareous nannofossils, dinoflagellates and palynomorphs, in company with a series of Research Assistants and Visiting Scientists.

The nature of micropalaeontology at 'UC' under Tom Barnard has always been unashamedly biostratigraphical. As a result many students have entered and continue to enter the petroleum industry, not least of all because their mentor has always had a pragmatic view of academic research and its direction. Despite this emphasis, with a particular attention to Mesozoic foraminifera, a major investigation of Recent Caribbean foraminiferal faunas has been carried out and most recently MSc classes have worked with material from the continental shelf of southern Africa. Work with Mesozoic ostracods was initiated in 1956 and during the past decade a growing number of students have concentrated on calcareous nannofossils. A book summarising the results of biostratigraphical work with nannofossils is at present in the press (Lord, A. R. (ed.) *A stratigraphical index of calcareous nannofossils*. Chichester: Ellis Horwood).

Tom Barnard's career spans a period of quite astonishing change in geology, from a time when micropalaeontologists were few in number to the present day when most palaeontologists are concerned with microfossils. Disparate groups of small fossil animals and plants have provided the key for re-interpretation of classic stratigraphies, for the exploration and understanding of the nature of the ocean basins and, not least, for subsurface hydrocarbon discovery. As Dr Field shows in his introduction, this change in emphasis has been closely linked with industrial needs. In this respect Tom Barnard has always paid close attention to the employment prospects of his students, attempting to combine as far as possible good research with practical utility. Tom Barnard's own published research has contributed very significantly to micropalaeontology, especially towards our understanding of the foraminiferal and nannofossil stratigraphy of the northwestern European Mesozoic succession. His great influence as a teacher of others has not been so easy for many to appreciate. We hope that this volume will be a recognised tribute to his achievement.

31 July 1981

F. T. BANNER
A. R. LORD

Acknowledgements

Mrs Mary Barnard helped us in a number of ways, not least of all in our attempt to keep this project a secret from Tom Barnard himself. Mr J. W. Arterton (Registrar), Mrs Y. D. Barrett, Mr E. W. Bridger, Dr J. E. Robinson (University College London) and Professor R. A. Howie (King's College London) kindly helped us with background research. Miss M. Norton, Mr M. M. Gay and Dr A. J. J. Rees (University College London) helped with production work.

We are indebted to Mr Roger Jones and Mr Geoffrey Palmer of George Allen and Unwin for their helpful and constructive advice.

F.T.B.
A.R.L.

Contents

Preface	*page* vii
Acknowledgements	viii
List of contributors	xii
Introduction *R. A. Field*	xiii
Publications by Professor T. Barnard	xv

1 Architecture and evolution of the foraminiferid test – a theoretical approach *M. D. Brasier*   1

Introduction	1
The models	2
Non-septate contained growth	3
Non-septate continuous growth	8
Septate growth	14
Conclusions	37
References	39

2 The subgeneric classification of *Arenobulimina* *C. Frieg and R. J. Price*   42

Introduction	43
Systematics	45
Evolutionary lineages	71
Acknowledgements	72
References	73
Plates	77

3 Middle Eocene assilinid foraminifera from Iran *F. Mojab*   81

Introduction	81
Descriptive micropalaeontology	84
Acknowledgements	98
References	98
Plates	100

4 Ecology and distribution of selected foraminiferal species in the North Minch Channel, northwestern Scotland *P. G. Edwards* page 111

Introduction	111
Geographic position and regional climate	113
Hydrographic conditions and their effect on the distribution of foraminifera	113
Distribution of sediment types in the North Minch Channel	122
The relationship between foraminiferal shell form and substrate type	125
Distribution of selected foraminiferal species	125
Generic distributions and dominance	132
Conclusions	135
Acknowledgements	136
References	136
Appendix: Total benthonic foraminiferal species recorded in the North Minch Channel	137
Plates	139

5 A classification and introduction to the Globigerinacea *F. T. Banner* 142

Introduction	142
The genera and subgenera of the Globigerinacea	147
Nominally available names of the genus-group	148
Appended notes: other generic names which have been applied	196
Key to the discrimination of globigerinacean genera	199
Review and discussion	212
Acknowledgements	225
References	225

6 Differential preservation of foraminiferids in the English Upper Cretaceous – consequential observations *D. Curry* 240

Introduction	240
Collection and preparation of materials	241
Comparative analysis of chalk and flint meal preparations	242
Differential loss of planktonic foraminiferids	246
Differential loss of benthonic foraminiferids	247
Other evidence in relation to selective preservation	247
Discussion	254
Timing of dissolution	254
Selective destruction of planktonic foraminiferids	256
Selective survival of the Textulariina	257
Depth of the Chalk sea	257

Interpretative use of fossil foraminiferid assemblages	*page* 259
Conclusions	259
Acknowledgements	260
References	260

7 Metacopine ostracods in the Lower Jurassic A. R. Lord 262

Introduction	262
Zonal biostratigraphy	266
The extinction of the Metacopina	267
Discussion	272
Acknowledgements	273
References	274
Appendix: Bilecik, Turkey	276

8 Palynofacies and salinity in the Purbeck and Wealden of southern England D. J. Batten 278

Introduction	278
Geology	280
Purbeck Beds palynofacies	283
Hastings Beds palynofacies	284
Weald Clay palynofacies	287
Wealden Marls and Wealden Shales palynofacies, Isle of Wight	288
Conclusion	289
Acknowledgements	289
References	290
Plates	293

Appendix: Past and present students of micropalaeontology at University College London 309

Index 314

General index	314
Taxonomic index	316

List of contributors

Professor F. T. Banner,
Department of Oceanography, University College Swansea, Singleton Park, Swansea, West Glamorgan SA2 8PP, Wales

Dr D. J. Batten,
Department of Geology and Mineralogy, Marischal College, University of Aberdeen, Aberdeen AB9 1AS, Scotland

Dr M. D. Brasier,
Department of Geology, The University, Kingston upon Hull HU6 7RX, England

Professor D. Curry,
Department of Geology, University College London, Gower Street, London WC1E 6BT, England

Dr P. G. Edwards,
Robertson Research International Ltd, 'Ty'n-y-Coed', Llanrhos, Llandudno, Gwynedd LL30 1SA, Wales

Dr R. A. Field,
British Petroleum Research Centre, Chertsey Road, Sunbury-on-Thames, Middlesex TW16 7LN, England

Dr C. Frieg,
Laarmannstrasse 29, D-4630 Bochum 5, West Germany

Dr A. R. Lord,
Postgraduate Unit of Micropalaeontology, Department of Geology, University College London, Gower Street, London WC1E 6BT, England

Dr F. Mojab,
27 Koucheh Karimi Zand, Palestine Street, Shiraz, Iran

Dr R. J. Price,
Amoco Canada Petroleum Company Ltd, 444 Seventh Avenue SW, Calgary, Alberta T2P 0Y2, Canada

Introduction

R. A. Field
Chief Palaeontologist, British Petroleum

Professor Tom Barnard has been a strong force in European micropalaeontology since the early 1950s. He was the first British Professor of Micropalaeontology and under his leadership the Micropalaeontology Unit of the Geology Department at University College London has flourished and continues to grow.

Tom Barnard began his long association with university life in 1941 when he graduated with a First in Geology at University College London, followed, in 1947, with a PhD on the foraminifera of the Middle and Upper Lias of the Midlands. He had already been appointed to the staff of University College London, becoming an Assistant Lecturer in 1946. He was promoted to Lecturer in 1948, Reader in 1953 and was appointed to a Personal Chair in 1963. His academic achievements reached a zenith in 1976 when he was awarded a DSc.

His career has shown a diversification in keeping with trends in the broader micropalaeontological sphere of industry and it is this association, both personally and through the medium of the majority of his students, with the world of industrial micropalaeontology that is one of the most notable aspects of his life's work. His early works followed on from his PhD in concentrating on Jurassic foraminifera and in particular the lagenids, to which he contributed some fundamental studies, but in the early 1970s his attentions turned to the calcareous nannofossils and, in company with W. W. Hay and others, he contributed a number of fundamental papers to this developing subject. His transfer of energy to the nannofossils was coincident with the earliest realisations of the significance of this group as stratigraphical indicators and his work on these fossils, particularly in the Jurassic, was a pioneer effort, as that on the lagenids had been twenty years earlier.

Perhaps Tom Barnard's greatest contribution to micropalaeontology, however, has been through the medium of his teaching. His contribution to the education of the current worldwide micropalaeontological working community has been substantial. Between the years 1950 and 1981, micropalaeontological students qualifying from University College London have numbered over 150, of which some forty-nine comprised a truly cosmopolitan overseas contingent hailing from a total of twenty-one countries from Europe, the Middle East, the Far East, the Caribbean and North America. His sphere of influence, via the offices of his students, has been thus considerably extended. Over half of them have entered the oil industry, including such organisations as BP, BNOC, NIOC and Shell, whilst a large proportion of the remainder have themselves entered university teaching; others occupy posts in mining, the Civil Service and geological consultancy, the latter including Robertson Research International and Paleoservices. The majority of these placings reflect a personal interest in the welfare and future careers of his pupils, and the enduring contacts with individuals in industry that Professor Barnard formed during his long career assured many a scion of a foot in the oil company door.

Many of the research projects carried out at University College reflect Tom Barnard's interest in the application as well as the purely taxonomic or nomenclatural side of micropalaeontology. He has aptly bridged the gulf between the academic and industrial worlds and has contributed to a breaking down of the resentments often engendered by the necessary secrecy attached to much oil company work. His facility for friendship, allied with an embracing knowledge of stratigraphic concepts and principles, assured a mutual understanding with the industrial world, and the interdependence of industry and education has been epitomised in his working relationships. These relationships, and his co-operation and involvement with the world of industry, have not only mapped out the careers of many of his students, but have also frequently tapped a source of funds which has proved vital to the steady growth of his group.

The science of industrial micropalaeontology, as we know it today, began to bear fruit around 1925 when microfossils first began to be used on a regular basis for subsurface biostratigraphic interpretation. It has matured from the almost exclusive use of benthic foraminifera in the Gulf Coast nearshore environments, to the routine use of both benthic and planktic foraminifera, palynomorphs and calcareous nannoplankton, separately or together, according to the dictates of facies and stratigraphy and the degree of detail required. In latter years the trends have enlarged the science towards not only vastly improved detail in stratigraphy but into the more subjective realms of palaeoecology and palaeoenvironmental interpretations. Palaeobathymetry and palaeotemperature work, along with the more conventional time lines, unconformities, facies changes and structural interpretations, all complicated by caving and reworking, now stem from the micropalaeontologists' studies. The biostratigraphic approach itself, however, having identified the ideal micropalaeontological indices of evolutionary change, is now, in the less than ideal circumstances of the subsurface, capable of correlating to a thickness of a few metres over hundreds of miles.

To this rapid scientific evolution, the contributions of Professor Tom Barnard have been numerous, varied and frequently fundamental. He has always been a controversial figure and his personality has either endeared him (to many) or, inevitably, infuriated a few, but his outspoken and light-hearted approach to our business will be sadly missed on his retirement. There is no ready replacement to occupy his unique niche.

Publications by Professor T. Barnard

Ophthalmidium: a study of nomenclature, variation, and evolution in the foraminifera. *Q. J. Geol Soc. Lond.* **102**, 77–113, 1946 (with A. Wood).

The uses of foraminifera in Lower Jurassic stratigraphy. *International Geological Congress, Part XV*, 1948.

An abnormal chalk foraminifer. *Proc. Geol. Assoc.* **60**, 284–7, 1949.

Foraminifera from the Lower Lias of the Dorset coast. *Q. J. Geol Soc. Lond.* **105**, 347–91, 1950.

Foraminifera from the Upper Lias of Byfield, Northamptonshire. *Q. J. Geol Soc. Lond.* **106**, 1–36, 1950.

Notes on *Spirillina infima* (Strickland) foraminifera. *Ann. Mag. Nat. Hist.* **5**, 905–9, 1952.

Foraminifera from the Upper Oxford Clay (Jurassic) of Warboys, Huntingdonshire. *Proc. Geol. Assoc.* **63**, 336–50, 1952.

Foraminifera from the Upper Oxford Clay (Jurassic) of Redcliff Point, near Weymouth, England. *Proc. Geol. Assoc.* **64**, 183–97, 1953.

Arenaceous foraminifera from the Upper Cretaceous of England. *Q. J. Geol Soc. Lond.* **109**, 173–216, 1953 (with F. T. Banner).

Hantkenina alabamensis Cushman and some related forms. *Geol Mag.* **91**, 384–90, 1954.

An unusual worm tube from the Lower Lias. *J. Paleont.* **30**, 1273–4, 1956.

Some Lingulininae from the Lias of England. *Micropaleontology* **2**, 271–82, 1956.

Frondicularia from the Lower Lias of England. *Micropaleontology* **3**, 171–81, 1957.

Some Mesozoic adherent foraminifera. *Palaeontology* **1**, 116–24, 1958.

Some arenaceous foraminifera from the Lias of England. *Contrib. Cushman Foundn Foramin. Res.* **10**, 132–6, 1959.

Some species of *Lenticulina* and associated genera from the Lias of England. *Micropaleontology* **6**, 41–55, 1960.

Note on *Spirillina arenacea, in Ammodiscus* Reuss 1862 (Foraminifera); proposed designation of a type-species under the plenary powers. *Bull. Zool Nomen.* **19**, 27–34 (by W. A. Macfadyen).

Evolution in certain biocharacters of selected Jurassic Lagenidae. In *Evolutionary trends in foraminifera*, G. H. R. von Koeningswald, J. D. Emeis, W. L. Buning & C. W. Wagner (eds), 79–92. Amsterdam: Elsevier, 1963.

Polymorphinidae from the Upper Cretaceous of England. *Palaeontology* **5**, 712–26, 1963.

The morphology and development of species of *Marssonella* and *Pseudotextulariella* from the Chalk of England. *Palaeontology* **6**, 41–54, 1963.

Electron microscope studies of Oxford Clay coccoliths. *Eclog. Geol. Helv.* **64**, 245–71, 1971 (with A. P. Rood and W. W. Hay).

On Jurassic coccoliths: *Stephanolithion, Diadozygus* and related genera. *Eclog. Geol. Helv.* **65**, 327–42, 1972 (with A. P. Rood).

Aberrant genera of foraminifera from the Mesozoic (sub-family Ramulininae Brady 1884). *Rev. Esp. Micropaleont.* **4**, 387–402, 1972.

Electron microscope studies of Lower and Middle Jurassic coccoliths. *Eclog. Geol. Helv.* **66**, 365–82, 1973 (with A. P. Rood and W. W. Hay).

On Jurassic coccoliths: a tentative zonation of the Jurassic of southern England and north France. *Eclog. Geol. Helv.* **67**, 563–85, 1974 (with W. W. Hay).

The Ataxophragmiidae of England: Part I, Albian–Cenomanian *Arenobulimina* and *Crenaverneuilina*. *Rev. Esp. Micropaleont.* **12**, 383–430, 1980 (with F. T. Banner).

Kimmeridgian foraminifera from the Boulonnais. *Rev. Micropaléont.* **24**, 3–26 (with D. J. Shipp).

Foraminifera from the Oxford Clay (Callovian–Oxfordian of England). *Rev. Esp. Micropaleont.* **13**, 383–462 (with W. G. Cordey and D. J. Shipp).

1 Architecture and evolution of the foraminiferid test – a theoretical approach

M. D. Brasier

Three main growth modes occur in the tests of Foraminiferida and related rhizopods: non-septate contained growth, non-septate continuous growth and septate growth. Within each of these groups, geometrical models of unit volume are used to examine the effects of varying the rate of growth translation, rate of chamber volume expansion, of chamber shape and apertural form upon a number of skeletal parameters. Chief among these are the minimum line of communication (MinLOC) from the back of the proloculus to the nearest aperture and the maximum line of communication (MaxLOC) from the most remote point of a distal chamber to its nearest aperture. Comparison of these models with fossil foraminifera allows some general comments about architectural evolution in the group; e.g. those from shallow-water facies show a tendency to 'shortened MinLOC' for a unit volume through the Palaeozoic and again in the Mesozoic, perhaps in response to increased competition.

Introduction

The tests of foraminifera exceed in architectural variety those of any other invertebrate group, including those of their kindred sarcodines, the Radiolaria. And a question arises as to whether these different growth plans have any functional significance? An affirmative answer would have considerable implications for systematics, ecology and evolution of the Foraminiferida. The aim of this paper is to probe this possibility by a study of some geometrical models.

Rhumbler (1895) believed foraminiferid evolution was a quest for increased constructional strength from feeble non-septate tubes to compact multiocular coils. Conversely, Galloway (1933) thought foraminifera were progressing from a low to a high surface area:volume ratio and from symmetry to asymmetry, with specialised and degenerate agglutinated forms at the apex of evolution. But much of this was speculation without reference to ecology, to the fossil record or to functional models. Scepticism about adaptive morphology was therefore voiced in the works of Thompson (1942), Glaessner (1945) and Smout (1954).

More recent biological and ecological research has again raised the question of test function (e.g. Haynes 1965, Marszalek *et al.* 1969, Banner *et al.* 1973, Brasier 1975a, Chamney 1976, Hottinger 1978) from which it seems that the test serves to reduce biological, physical and chemical stress (e.g. predation, parasitism, radiation, abrasion, turbulence, transport, salinity, pH, CO_2, O_2), to enhance feeding in particular habitats and to maintain negative buoyancy. When subjected to stress the protoplasm can withdraw into earlier chambers until conditions improve. In planktonic marine protists, however, where such stresses are few, the test serves to assist regulation of near-neutral buoyancy and for internal support. Frothy ectoplasm, supportive spines and stiff axopodia are functional adaptations of the planktonic sarcodines (Radiolaria, Acantharia, Heliozoa, globigerinacean Foraminiferida; e.g. Anderson & Bé 1978).

While the stresses of a benthic life may be great, they afford an opportunity to increase individual size beyond that possible in buoyancy-conscious planktonics. A gradual increase in size with time is traceable in many lineages of larger and smaller foraminifera (e.g. Hottinger 1978, Hofker 1959). This may seem paradoxical as it has been argued that larger size in foraminifera indicates delayed reproduction under unfavourable conditions (e.g. Sliter 1970). But in general terms there are two size-related trends: r-selection under unstable conditions, which favours small size as a consequence of high reproduction rates and precocious maturation; K-selection under predictable conditions, which favours large size as a consequence of low reproduction rates and delayed maturation (e.g. Tappan 1976; Hallam 1978). In foraminifera, of course, the enlargement of a single cell poses special problems of construction and internal organisation. Chamney (1976) and Hottinger (1978) suggested that mobility of the nucleus was one of these; there is apparently a definite ratio between protoplasmic volume and the number (or surface area) of nuclei (Myers 1936). Other size-related problems probably include vacuolar movement, economy of osmoregulation, skeletal material, life attitude and pseudopodial feeding area. The models outlined below are used to illustrate the possibility that different architectural modes have evolved as a means of modulating the internal–external lines of communication.

The models

The value of models as a guide to what *may* be the function of a skeleton has been demonstrated in many macrofossil groups (e.g. Raup & Stanley 1978) but less so in microfossils. Berger (1969) applied this paradigm principle to problems of buoyancy and ontogeny in planispiral planktonic foraminifera, while Scott (1974) modelled the *Orbulina* lineage. The writer briefly illustrated how multilocular tests may arise as the result of interaction between three variables during growth: the rate of translation (i.e. the net rate of movement along the growth axis to the net movement away from the growth axis), the rate of chamber expansion and the chamber shape (Brasier 1980, Figs 13.6 & 7, p. 95).

The models adopted below are two-dimensional drawings based on three-dimensional figures. The arbitrary unit of measurement taken here is the diameter of the proloculus in Figures 1.7–14. Most of the septate models have a spherical proloculus with a single aperture, followed by six chambers with a relatively high rate of chamber volume expansion of $E=2.9$, where $E=V_n/V_{n-1}$ and V represents the volume of the nth or $(n-1)$th chamber; the rate was based on that of *Hormosina globulifera*, a spheroidal chambered uniserial test figured by Loeblich and Tappan (1964, Fig. 128.4a). This gives an overall volume of 52 283 cubic units, which is standard for all other models and allows a direct comparison between other parameters. Four of these parameters are considered:

(a) The surface area:volume ratio; this is related to the economy of shell material.
(b) The minimum line of communication ('MinLOC') from the most remote point of the proloculus to the nearest aperture in contact with the external milieu. This can be either an absolute measurement or a relative one (i.e. relative to a standard volume); the latter is the case here. This parameter is also an index of test compactness and, indirectly, of test strength. Biological interpretations presume that the proloculus is occupied by cytoplasm throughout life and that the aperture of the final chamber is the main opening for pseudopodia and feeding. This may not be so in some perforate foraminifera but it holds true for most taxa.
(c) The maximum line of communication ('MaxLOC') from the most remote point of a distal chamber to its nearest aperture in contact with the external milieu. Again this is taken as a relative measurement. Biological interpretation should bear in mind that in many living foraminifera the penultimate, rather than the final chamber, is filled with cytoplasm.
(d) The maximum ambit of a miliolid chamber, which is the maximum circumference of a chamber from one foramina to the next.

Non-septate contained growth

Tests comprising a single chamber that cannot be much enlarged except by redistribution of test material ('contained growth') occur in both primitive and degenerate sarcodines of the Class Rhizopoda. They predominate in the mainly freshwater Orders Arcellinida (Lower Carboniferous–Recent) and Gromiida (Middle Eocene–Recent) and some of the fresh to marine Order Foraminiferida. The hyaline tests of *Lagena*, *Oolina* and *Fissurina* are post-Triassic degenerates, perhaps associated with a parasitic life (e.g. Haward & Haynes 1976) while those of the tectinous Allogromiina and

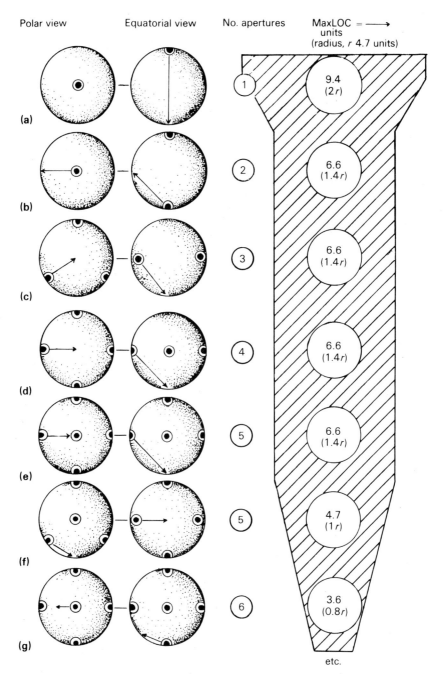

Figure 1.1 Model spherical tests with non-septate contained growth and of unit volume, showing a decrease in the MaxLOC with an increase in the number of apertures. See text for explanation of MaxLOC and MinLOC.

agglutinated Textulariina are generally thought to be primitive (Lower Cambrian–Recent). The protoplasmic organisation of these groups is varied, with lobopodia (Arcellinida, Trichosida) or filopodia (Gromiida) or reticulopodia (Foraminiferida), but on present evidence it seems that the primitive unilocular forms may share one important characteristic: the nucleus (or nuclei) is made rigid by a thick nuclear membrane and may be too large to pass through the aperture, if there is one (Loeblich & Tappan 1964, p. C66), though *Iridia* is capable of leaving its multipored test (Marszalek *et al.* 1969). Generally, the nucleus is confined to the back of the chamber until reproduction allows the formation of daughter cells small enough to pass the barrier. This differs significantly from septate foraminifera in which a small and deformable nucleus (or nuclei) tend(s) to occur near the distal end. If mobility of the nucleus is an important factor in development, then both a large rigid nucleus and a small confining aperture may constrain growth.

Form. The main geometrical figures adopted by contained-growth rhizopods include the sphere, hemisphere, flask, spindle, cylinder, cone and star. For a given volume and radius, there are considerable differences in surface area between these (Fig. 1.3). The spindle with a 60° apical angle (which can hardly be obtained precisely) has less surface area than a sphere. If the flat, unexposed bases of adherent hemispheres, cylinders and cones are excluded, then these figures can also have low surface area: volume ratios. Elongation along the symmetry axis, of course, leads to a corresponding increase in surface area (cf. Fig. 1.7).

The MaxLOC parameter also varies with the number and position of apertures. With a single polar aperture, and a fixed radius and volume, the flat cylinder, hemisphere (cf. *Colonammina*) and sphere (cf. *Saccammina*) have the shortest MaxLOC while the addition of more apertures serves to reduce this parameter further (Figs 1.1 & 1.2). Hence spheres with multiple apertures (e.g. *Echinogromia, Thurammina*) and hemispheres with a basal aperture (e.g. many Arcellinida) have among the best lines of communication for any test with the same volume, while elongate cones, spindles (e.g. *Technitella*), cylinders and flasks (e.g. *Lagenammina*, many Arcellinida and Gromiida) have longer lines of communication.

Ecology. Those with a single aperture are mainly mobile and benthic in habit, often with the axis held nearly vertical (e.g. most Arcellinida and Gromiida). Elongate forms with several apertures adopt a prostrate posture and seem to be less active (Loeblich & Tappan 1964, p. C67). Hemispherical forms with a flat basal surface and lateral or terminal apertures (if any) are generally adherent (e.g. Hemisphaeramminae, *Iridia*) and so are some spherical forms that occur in clusters (e.g. *Sorosphaera*). Other spherical forms may be reclining or mobile and one (*Orbulinoides*) with multiple apertures is planktonic (Boltovskoy & Wright 1976).

A survey of size, shape and distribution allows several generalisations:

(a) Compact, apertured tests less than 1 mm in diameter are favoured by freshwater rhizopods but occur widely in marine ones.
(b) Tubular tests, those without apertures or all tests larger than 1 mm in diameter are more typical of estuarine and marine than non-marine rhizopods.
(c) Non-testate rhizopods are mostly non-marine.

It may also be significant that the predominantly freshwater Arcellinida and Amoebida have lobopodia with a low surface area:volume ratio; fresh to marine Gromiida have filopodia of greater surface area but reticulopodia of marine Foraminiferida have the highest surface area. Considered with the foregoing this suggests a model in which osmotic pressure may be a prime control of the form of protoplasm and test alike. Thus, in fresh water, where osmosis is greatest, a compact test and lobopodia keep the surface area:volume ratio low and the aperture allows egress of water in contractile vacuoles. But the greater adhesive potential of lobopodia may also reduce the need for 'ballast' in the form of a test, hence the naked state of Amoebida (Marszalek *et al.* 1969). In sea water, the osmotic pressure of the protoplasm approximates to that of the surrounding medium and contractile vacuoles

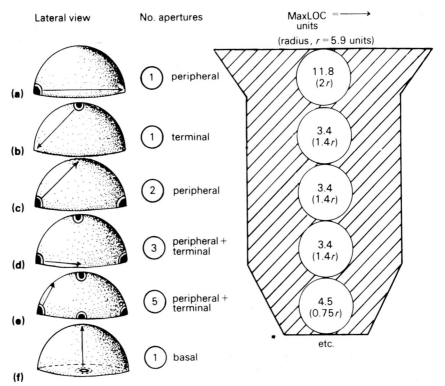

Figure 1.2 Model hemispherical tests with non-septate contained growth and of unit volume, showing a decrease in the MaxLOC with variation in the number and position of apertures.

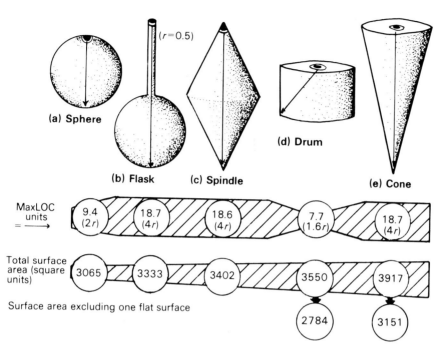

Figure 1.3 Models of spherical, flask-shaped, spindle-shaped, cylindrical and conical tests (non-septate contained growth) with unit volume, radius ($r=4.7$) and a single terminal aperture. This shows how geometry affects MaxLOC and surface area.

are absent. Thus marine rhizopods can assume elongate forms and dispense with proper apertures but the higher surface area:volume ratio of reticulopodia is less effective in the buoyant water (hence the weight of a test) but more efficient in gathering food (hence one reason for the larger size range).

Evolution. Galloway (1933) and Boltovskoy and Wright (1976) believed that the most primitive foraminifera were globular, with a single aperture. Inspection of the fossil record shows that the earliest contained-growth foraminifera may have been agglutinated hemispheres found in the sub-*Fallotaspis* beds in Canada (Conway Morris & Fritz 1980) and spheres of late Precambrian–early Cambrian age from Wales (J. C. W. Cope, personal communication). Simple flask and spindle shapes appeared later in the Cambrian and early Ordovician while most other shapes occurred by early Silurian times (Fig. 1.15). Morphological innovation in the unilocular Textulariina has been negligible since the Silurian, when their acme of generic diversity was reached. However, the doubled diversity of Recent textulariine examples indicates that the fossil record has only preserved the more robust forms, which may explain the scarcity of fossil Allogromiina (Upper Cambrian–Recent), Arcellinida (Lower Carboniferous–Recent) and Gromiida (Middle Eocene–Recent). It is possible that the Arcellinida and Gromiida colonised freshwater habitats successively and independently from a marine reticulopod stock such as the Allogromiina.

The microgranular calcite test of *Saccamminopsis* (Upper Ordovician–Carboniferous) deserves further study since it provides one of several possible ancestors to the septate Fusulinina. Most other Parathuramminacea have been declared inorganic or algal (see Toomey & Mamet 1979).

Non-septate continuous growth

Attempts to increase biomass by growth of a nearly continuous tubular chamber represented the first step towards more advanced architecture. And while a few Gromiida and Allogromiina have adopted this growth mode, examples come mainly from the textulariine Ammodiscacea and the most primitive Miliolina and Rotaliina. Quite why the tubular growth mode has been favoured by primitive marine rhizopods deserves some thought. There is, first, the likelihood that, in the absence of a septum, tubular chambers afford more protection than broad ones if the cytoplasm is withdrawn (Marszalek *et al.* 1969) since chemical diffusion takes longer down a narrow tube. But this does not account for the lack of septa.

The acquisition of septa was undoubtedly a later trend (Fig. 1.15). It required periodic changes in behaviour that, one could argue, were difficult to 'program' successfully. However, chains and clusters of successive tests without connecting foramina were common in adherent Hemisphaeramminae from early Silurian times onward (Conkin & Conkin 1977) while Radiolaria had 'programmed' polycystine skeletons by mid-Cambrian times (Nazarov 1975). Another possible reason for delayed septation may have been the form of the nucleus.

The nucelus (or nuclei) of simple agglutinated forms may be relatively large and rigid (up to 0.6 mm across in *Bathysiphon*) and either remains in a roomy proloculus (e.g. *Vanhoeffenella*) or migrates freely along the tube (e.g. *Spirillina*). In the case of a migrating nucleus, the chamber must be wide enough to allow the ever-growing nucleus to move forward or backward without obstruction. But while a widening tube allows the nucleus to move forward it may prevent a large rigid nucleus from retiring backwards at times of stress or test damage; a non-tapered tube avoids this circumstance. Even were this not so, the advantages of a rapidly widening test may be offset by the increasing area of unprotected aperture (Fig. 1.5), leaving the cytoplasm vulnerable. If the form of the nucleus was truly a constraint on growth and septation, the following evolutionary steps may have occurred: (a) non-septate tubes with a non-deformable, large nucleus; (b) non-septate cones with a nucleus of restricted growth or with slight elasticity; (c) expanding cones with resorbable septa or constrictions and a small, or more elastic, nucleus; (d) greatly widening growth forms with permanent septa and small deformable nuclei. Aside from these speculations it is possible to model more obvious aspects of non-septate growth, such as lines of communication and surface area.

Form

Simple and branched tubes. Non-coiled tubes may be simple and closed at one end (e.g. by a proloculus as in *Botellina*), open at both ends (e.g. *Marsipella* and *Bathysiphon*, though some of the latter are blocked at one end) or they may be branched into three or more arms with apertures at the ends (e.g. *Rhabdammina, Astrorhiza*).

For a given volume, the surface area:volume ratio of tubular tests is enlarged by increasing the ratio of chamber length to width (cf. Figs 1.4a & 1.6a) and for a given diameter, by increasing the number of branches (Figs 1.4a–c). Internal–external lines of communication for a given volume can be shortened by decreasing the ratio of chamber length to width and for a given diameter by providing an aperture at both ends of a simple tube (Fig. 1.4b, cf. *Bathysiphon*) or by increasing the number of branches, each provided with an aperture (Fig. 1.4c, cf. *Rhabdammina*).

Straight cones. In a series of cones of equal volume (Fig. 1.5) a reduction in the length to width ratio results in a gradual decrease in the MinLOC

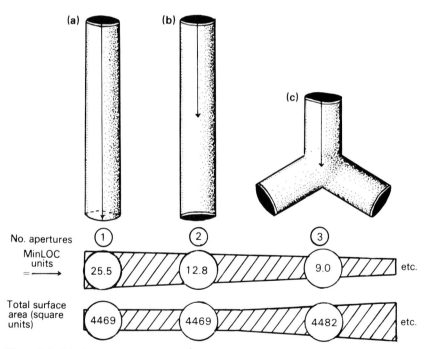

Figure 1.4 Models of simple and branched tubular tests (non-septate continuous growth) with unit volume and radius ($r=2.3$). This shows how branching, and the multiplication of apertures, reduces MinLOC and MaxLOC and increases surface area.

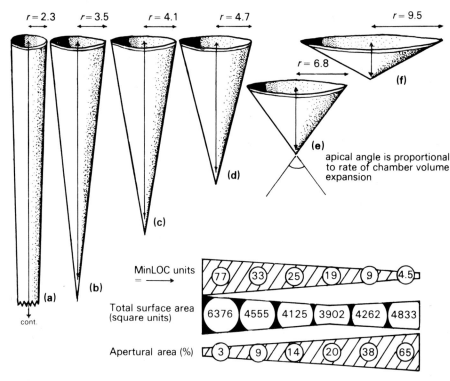

Figure 1.5 Model conical non-septate tests with continuous growth and unit volume, showing how an increased rate of chamber volume expansion reduces MinLOC, increases apertural area and varies the total surface area. Models like (c) and (d) seldom occur and (e) and (f) never occur in non-septate foraminifera but are typical of some septate forms.

parameter. The same series also results in a decrease in surface area up to the point at which the apical angle is exactly 60°. Beyond this, the total surface area increases again. The surface area of the aperture, which may be critical, continues to increase through the series. From the structural point of view the most compact cone (that with an apical angle of 60°) is likely to be stronger and comparable in strength with a compact spindle, sphere or hemisphere.

Most genera that build non-septate cones have an average apical angle of about 10° or less and therefore do not take full advantage of these trends. Only a few (e.g. *Hippocrepina*) reach up to 40°. Septum-like apertural closures are evident at the broad end of the latter, as might be expected.

Curved and coiled tubes and cones. Most curved and coiled tests with continuous growth are of the slowly expanding kind, with a small spherical proloculus (e.g. Ammodiscidae). Their tests are variously meandering in wide loops (e.g. *Tolypammina*), in closer zig-zags (e.g. *Ammovertella*), in planispirals (e.g. *Ammodiscus*), in open trochospirals (e.g. *Ammodiscoides*),

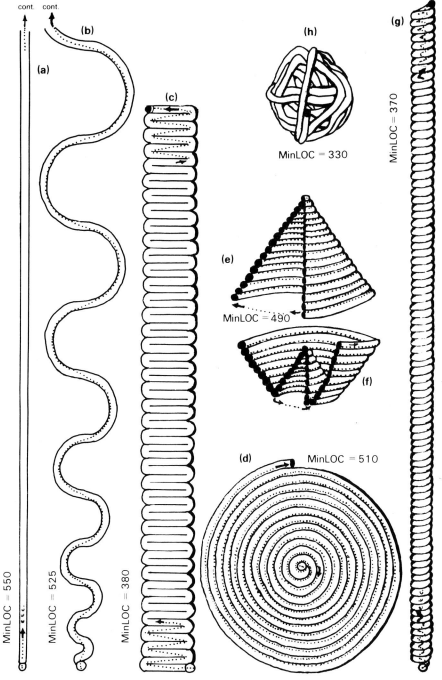

Figure 1.6 Model coiled non-septate tests with continuous growth and unit volume and radius, showing how tighter coiling reduces the MinLOC.

in close-coiled high trochospirals (e.g. *Turitellella*) and in ball-of-wool glomospirals (e.g. *Glomospira*).

The models in Figure 1.6 are untapered (since these are more easily constructed) but give an indication of the importance of curving and coiling to both untapered tubes and slowly expanding cones. For a given rate of chamber volume expansion the MinLOC decreases with the tightness of coiling. Hence a tight zig-zag (Fig. 1.6c) may provide a shorter MinLOC than a series of wide meanders (Fig. 1.6b), though the saving is dependent on the loops of the arms being closely adpressed and of little length because shortening depends on 'cutting the corners', as in Grand Prix motor racing. If the zig-zagging keeps to a constant width, the shortening of MinLOC keeps to a steady rate with growth, but this is not the case with a planispiral tube (Fig. 1.6d). Here the shortening is best on the inner circuit and gets proportionally less with growth. This situation may be redeemed a little by a higher rate of chamber volume expansion or by keeping later whorls as close as possible to the growth axis, i.e. by trochospiral or glomospiral coiling. Tight trochospirals (Fig. 1.6g) provide a shorter MinLOC than more open ones (Fig. 1.6e & f) but their more elongate form may reduce test strength and mobility. Glomospiral coils (Fig. 1.6h) are more compact and, by altering their axis of coiling, keep the coils as close to the centre of gravity as possible, thereby achieving relatively short lines of internal–external communication.

Ecology. In general terms the following growth forms are sessile and adherent: meandering and zig-zag (e.g. Tolypamminae, Calcivertellinae), planispiral (e.g. *Ammodiscus, Cyclogyra, Spirillina*), some high trochospiral forms (e.g. *Trepeilopsis*) and branched and dendritic forms (e.g. *Dendrophrya*). Other simple and branched tubes recline or creep over the sea floor, especially those with fragile pseudopodia and multiple apertures (e.g. *Shepheardella, Nemogullmia, Astrorhiza*). Straight tubes and cones with a proloculus and single aperture are more usually mobile and, if the pseudopodia are relatively rigid, may carry the axis upright (e.g. *Hippocrepina*).

As noted above, straight and branched tubes are of mostly marine origin and may reach a considerable size (<50 mm for *Bathysiphon*). Coiled tests, if 'uncoiled' may be even longer (*c.* 80 mm for *Ammodiscoides turbinatus*; Loeblich & Tappan 1964, Fig. 122.4). But when their volumes are computed, they are seldom more than twice the volume of marine forms with spherical tests (e.g. *Psammosphaera*) and are probably much the same. Few non-septate tubular foraminifera thrive in brackish or estuarine waters (e.g. *Spirillina*; Scott 1976) and most are found at depths greater than 200 m (e.g. Culver & Buzas 1980). However, it is likely that the habitat of many genera has become progressively deeper with time because this group was typical of shallow-water limestones and shales in the early Palaeozoic (Conkin & Conkin 1977). Evolution of more 'efficient' test forms from the Devonian onwards may be responsible for this displacement.

In the Cretaceous, Chamney (1976) noted some interesting trends. Straight tubular *Bathysiphon* and *Hyperammina* and branched *Schizammina* are abundant and persistent in deeper marine facies (flysch of foredeeps and deep slopes). Planispiral *Ammodiscus* is more usual in deep trough to shallow shelf facies deposited under stable marine conditions while glomospiral–planispiral *Glomospirella* tolerated adverse marine and subsaline conditions (cf. the hypersaline marsh occurrence of Recent *Glomospira*; Murray 1973). The pseudochambered conical *Hippocrepina* indicated shallow marine conditions and, possibly, nearby subsaline conditions. Hence morphologies which allow shorter lines of internal–external communication and a lower surface area:volume ratio thrive in shallow and less stable marine facies. But the trend needs further documentation.

Evolution. Both Rhumbler and Cushman believed that non-septate tubes with multiple branches were the most primitive kind of test, but this is not supported by the fossil record. *Platysolenites* and *Spirosolenites* are straight and planispiral tubes of agglutinated matter from beds in the Precambrian–Cambrian transition in northwestern Europe. They closely resemble *Bathysiphon* and *Ammodiscus* though *Spirosolenites* lacks a preserved proloculus (Føyn & Glaessner 1979). Adherent coils resembling *Lituotuba* but of calcareous composition ('*Wetheredella*') are reported from high Lower Cambrian reefs in Labrador (Kobluk & James 1979). The branched tectinous tube of *Chitinodendron* (Upper Cambrian) is the only other record until the Lower Ordovician where straight tubes with a proloculus appear (adherent *Ammolagena* and free *Hyperammina*). Agglutinated tubes with branches appeared in the Middle Ordovician and meandering *Tolypammina* in the Upper Ordovician. A radiation of coiled forms (*Ammodiscus, Turitellella, Glomospira*) took place in the early Silurian, since when there have been no important architectural innovations except, perhaps, the tight zig-zags of adherent *Ammovertella* (Lower Mississipian–Recent) and the trochospiral reversals of *Ammodiscoides* (Middle Pennsylvanian–Recent). Among the Textulariina the fossil record reveals little change in generic diversity from the Carboniferous onwards, though a marked increase in the number of Recent non-coiled tubes suggests that nearly half of these have a poor preservation potential.

The apparent evolutionary stasis, however, is misleading since several calcareous stocks may have originated in the Ammodiscidae, namely the porcelaneous Miliolina, some hyaline forms (Spirillinacea, Involutinidae, e.g. Loeblich & Tappan 1974) and some microgranular forms (Tournayellidae, Archaediscidae, Lasiodiscidae, e.g. Glaessner 1963). One might go further and unite these tests in a suborder Biloculinidea (Sigal 1952), a possibility which downgrades the importance of test ultrastructure and composition. The morphological convergence between genera of different test composition is certainly striking but their separation is retrospectively justified by the fact that each stock gave rise independently to septate forms. Shell structure studies by Hohenegger and Piller (1975) have now placed the Tournayellidae in the Fusulinina, and the Spirillinacea, Involutinidae, Archaediscidae and Lasiodiscidae in the Suborder Spirillinina. The latter are derived from the former, and progressed from evolute glomospiral–planispiral forms in the late Palaeozoic to trochospiral forms in the Triassic and from evolute to involute forms in the late Triassic (Hohenegger & Piller 1975).

Septate growth

Septate growth involves the periodic construction of distinct chambers separated by septa but with connecting foramina. Such growth represents a major development in protistan architecture and appeared independently in five or more lineages of the Foraminiferida during the mid–late Palaeozoic. Evolution of a small and deformable nucleus may have been an important step towards septate growth.

What, though, is the advantage of serial growth with septa? It is not for an increase in biomass alone, since this is possible with non-septate tubes or slightly tapered cones. More likely, it seems that the septum allows a greater rate of volumetric expansion of the test by shielding the otherwise exposed area. And a greater rate of expansion allows a marked improvement in internal–external lines of communication and surface area:volume ratio compared with non-septate tubular chambers. Furthermore, the presence of septa, when combined with the advantageous placement of apertures and foramina, allows the development of novel architectural forms (e.g. broad cones, discs and complex spindle shapes).

Since variation in chamber shape and apertural form may be interpreted as specific solutions to problems set by growth plan, it is these plans that will be taken as the main headings, i.e. uniserial, biserial, planispiral, trochospiral, miliolid, annular, multiform. Except for the last, this sequence is approximately that of their geological order of appearance. The significance of the rate of chamber volume expansion and of chamber shape, however, is mainly outlined under uniserial growth.

Uniserial form

A uniserial arrangement of chambers in a straight or slightly arched chain is the simplest and probably the oldest septate growth plan (?Upper Ordovician, Middle Devonian–Recent). The writer has previously described it as the end member of a series with increasing growth translation, with one chamber per whorl (Brasier 1980, pp. 94–5) but uniserial tests have dual origin: primary ones with no known spiral or biserial ancestry (e.g. *Reophax*) and secondary ones with such an ancestry (e.g. *Nodosarella*). However, the principles of variation in these and all other septate forms are the same: (a) variation in the rate of chamber volume expansion; (b) variation in the shape of chambers; (c) variation in the number and position of apertures.

The rate of chamber volume expansion. This is the rate of volume expansion (E) from one chamber (V_{n-1}) to the next (V_n) calculated from $E = V_n/V_{n-1}$. Berger (1969) and Scott (1974) used a similar ratio between chamber diameters, which is a simple measure of volume for spherical chambers but cannot be so easily applied to other chamber shapes.

The rate of chamber volume expansion is unlikely to remain constant with growth and probably tends to decelerate in later life. The ways in which rates of growth and chamber addition vary have already been modelled by Berger (1969). In the following figures, however, a constant rate of increase is assumed unless otherwise stated. From the models in Figure 1.7 it is clear that, for a given chamber shape and total volume, an increase in the rate of chamber volume expansion reduces the MinLOC and decreases the surface area.

The MinLOC of megalospheric tests is likely to be less than of microspheric ones of the same species because the size of the proloculus tends to affect the rate of chamber volume expansion directly (Redmond 1953; Hofker 1954) and may be inversely proportional to the number of chambers and length of the adult test (e.g. Sliter 1970).

Chamber shape. The writer has already suggested how, for a fixed volume, variation in the chamber co-ordinates l and w may dramatically modify test shape (Brasier 1980, pp. 94–6). Only a few of the basic possibilities are considered here (Fig. 1.8).

In spherical chambers, the maximum axial diameter (l) and co-ordinates normal to it (w_1, w_2) expand uniformly with growth; in the *Hormosina*-like example (Fig. 1.8a) this growth is exponential, producing an evenly tapered profile. Slight modification of chamber shape with little change in profile is possible by flattening the sides so that each chamber resembles a truncated cone (a **frustum**, Fig. 1.8b) or a squat cylinder (a **drum**, Fig. 1.8c) while keeping the maximum width (w) co-ordinates the same. The drum-shaped chambers effect a considerable reduction in MinLOC and MaxLOC, which may be one reason why they are particularly common in uniserial forms (e.g. *Colomia*). Further reduction in length (l) and increase in width (w) of the chambers continues the trend of reduced MinLOC almost indefinitely. The advantages of a low conical shape and smooth profile, as in *Orbitolina*, become apparent (cf. Fig. 1.8f). But there are several turning points. The surface area:volume ratio will decrease with flattening of the cone until the apical angle is 60°, but continued flattening increases it again. Furthermore, it is at about this point that the shortening of the MinLOC loses any advantage because it is exceeded by the MaxLOC. The solution of this problem is apertural modification (see below).

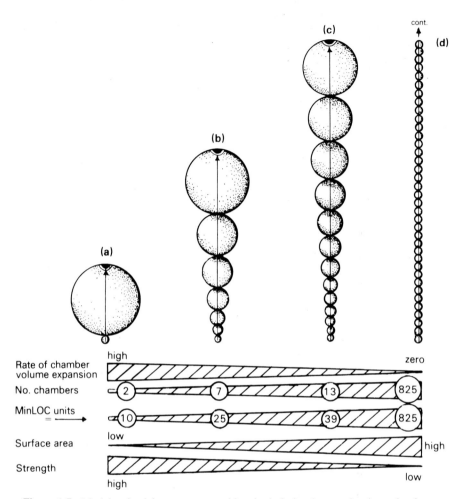

Figure 1.7 Model uniserial septate tests with spherical chambers and unit total volume, showing how a reduced rate of chamber volume expansion increases the number of chambers, the MinLOC and the surface area.

In Figures 1.8d and f, length is held constant at 1.0 and w_1 and w_2 increase with growth. But chamber shape may also be modified by flattening parallel to the growth axis, so that one of the width parameters remains unchanged or only increases slightly with growth (Fig. 1.8i). Here w_2 is held constant at 1.0 and w_1 increases so that the MinLOC is about the same as for spherical chambers (Fig. 1.8a) but the MaxLOC is much greater. For the same volume, MinLOC and MaxLOC could be further reduced by either a higher fixed value or an increasing value for w_2 or by apertural modification. This kind of geometry, with flaring compressed chambers but a single terminal aperture, is largely confined to nodosariaceans and their ancestors (e.g. *Pachyphloia, Flabellamina, Frondicularia*).

A further change in chamber shape is arrived at by holding the width

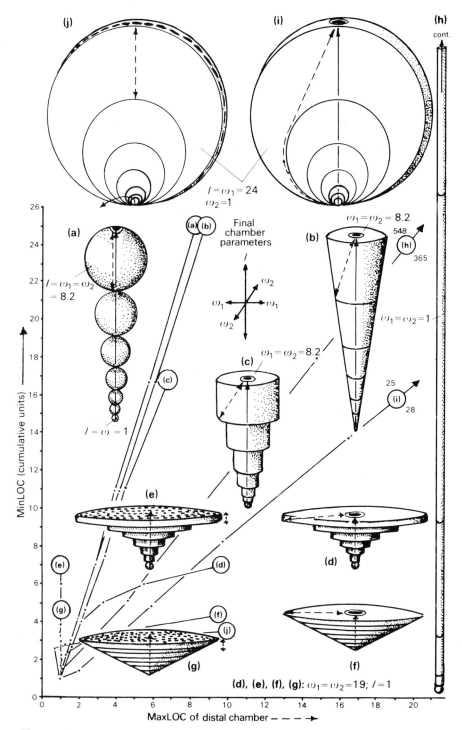

Figure 1.8 Model uniserial tests with unit volume and volumetric expansion. This shows how variation in chamber shape and in apertures may shorten or extend the MinLOC and MaxLOC. In most of the following figures, model (a) is taken as a standard for comparison.

parameters constant and allowing the length to increase, i.e. by building tubular chambers (Fig. 1.8h). The resultant increase in MinLOC may be considerable and very few septate tests adopt this lengthening formula. It is seen in some 'uncoiled' miliolines (e.g. *Tubinella*) but tubular chambers are standard in that group.

Apertures. It is assumed in most models that each primary aperture is covered by the next chamber, since this is usual in foraminifera. And in the above account, a single terminal aperture placed over the growth axis has been adopted as this is also quite normal. But in a sequence of increasingly compressed chambers there are two points at which multiple apertures would improve internal communications: first, where the width (w) exceeds the length (l) of the chamber and, secondly, where the MaxLOC of the largest chamber exceeds the MinLOC of the whole test.

In the case of chambers of shortened length (i.e. with a low conical test shape) the addition of scattered cribrate apertures on the distal septal face has no effect on the MinLOC but it significantly reduces the MaxLOC (Figs 1.8d–e & f–g). The latter closely resemble the Pavonitinae, Orbitolinidae and Rhapydionininae. And in the case with chambers of flattened width (i.e. with a discoidal, flabelliform or palmate test shape) the addition of scattered apertures around the peripheral septal face shortens both MinLOC and MaxLOC (Figs 1.8i–j). This is exemplified by the uniserial portions of multiform *Arenonina, Semitextularia, Peneroplis* and *Pavonina* as well as in similar planispiral and annual growth plans (e.g. Spirocyclininae, Dicyclinidae, Soritidae). It does not seem to occur in wholly uniserial tests.

A conspectus of lituolacean genera also supports the relationship between chamber shape, MaxLOC and cribrate apertures in uniserial forms. While only 45% of uniserial chambers have cribrate apertures, these always occur in compressed chambers ($w>l$). Conversely, all equant ($l=w$) and elongate ($l>w$) chambers have a single terminal aperture. Where larger foraminifera employ uniserial chambers they generally do so in a way that minimises internal lines of communication. The internal partitions typical of larger foraminifera can be interpreted as buttresses to strengthen the highly flattened chambers (for low MinLOC) against compression, since they are vulnerable in direct proportion to their $l:w$ ratio.

Ecology. The ecological range of uniserial tests is very wide. However, they do not typically occur in fresh water (except *Reophax arctica*; Boltovskoy & Lena 1971) and are not abundant in brackish and estuarine facies. Chamney (1976) regarded Cretaceous uniserial forms as less tolerant than most other growth plans to environmental change, being more typical of shelf and slope facies. The structural weakness of spherical chambered forms might be one reason why they mostly occur in outer shelf and deeper facies, though fragile species of *Reophax* live in shallow-water subsaline bays and estuaries (e.g. Murray 1973).

Multiform tests with a uniserial end stage also occur in a wide range of habitats. The suggestion of Chamney that they are typically of shelf or bathyal facies is a generalisation that has many exceptions. Those that show a tendency to a pronounced shortening of MinLOC and MaxLOC (e.g. the large orbitolinids and soritids) have mainly evolved in warm shallow carbonate facies of normal or slightly hypersalinity. Those with a tendency to increase the MinLOC by 'uncoiling' (e.g. *Ammobaculites, Ammomarginula, Ammoscalaria, Ammotium, Pseudoclavulina*) may abound in brackish tidal marshes or in hypersaline lagoons (e.g. *Clavulina, Nodobaculariella, Articulina, Spirolina, Monalysidium*). It is also possible that competition has caused the ecological range of uniserial tests to change with time. In mid- and late-Devonian times, uniserial tests dominated carbonate facies (e.g. Toomey 1972), but these gave way to the newer coiled and multiform tests, except in the more marginal environments, during the Carboniferous and Permian (e.g. Toomey & Winland 1973). A similar story of uniserial radiation and progressive displacement can be read in the history of Jurassic to Cretaceous uniserial Nodosariacea (e.g. Barnard 1963, Brotzen 1963).

Evolution. If the microgranular spheroid *Saccamminopsis* (Upper Ordovician–Carboniferous) was uniserial in the Upper Ordovician (which is uncertain), then this was the first septate growth plan and a *Saccammina*-like ancestor is likely. A similar origin seems probable for the agglutinated, uniserial, adherent *Oxinoxis* (Middle Devonian–Recent), though free *Reophax* (Upper Devonian–Recent) may have an origin in *Hyperammina*.

Calcareous *Paratikhinella* (Middle–Upper Devonian) may have developed from a *Saccamminopsis*-like ancestor or from a non-septate tube like *Earlandia* (Upper Silurian–Pennsylvanian), and probably spawned the line leading to the extant hyaline Nodosariacea; Hohenegger and Piller (1975) acknowledge this lineage in the Suborder Lagenina.

In many other lineages the uniserial arrangement terminates a multiform test and poses the question as to which character is more ancient? The wholly uniserial arrangement is secondary in the Rhapydionininae (Jurassic–Recent), Orbitolinidae (Lower Cretaceous–Eocene) and some Pleurostomellidae (Lower Cretaceous–Recent) for example, and in some dimorphic generations of multiform taxa. But the partly uniserial arrangement of *Ammobaculites* (Upper Devonian–Recent) antedates that of *Haplophragmoides* (Lower Pennsylvanian–Recent), in which the uniserial portion perhaps has been 'retarded out' with time.

In *Orbitolina*, an evolutionary trend from small high cones to large flattened cones and discs (Rat 1963) may be interpreted as a MinLOC-shortening strategy.

Biserial form

Like the uniserial arrangement, biserial growth was one of the first

20 FORAMINIFERID ARCHITECTURE AND EVOLUTION

multilocular growth plans (Middle Devonian–Recent). It may be regarded as a form of high trochospiral growth (i.e. with a high rate of growth translation) in which only two chambers are added per whorl, each at 180° to its neighbour. The wide distribution of the biserial plan in the Foraminiferida may be due to its inherently short lines of communication.

As with uniserial tests, changes in the rate of chamber volume expansion have a marked effect on the MinLOC and surface area. These aspects will not be repeated here. The advantage of a biserial arrangement depends as much on the placement of the aperture. In Figure 1.9 are some biserial

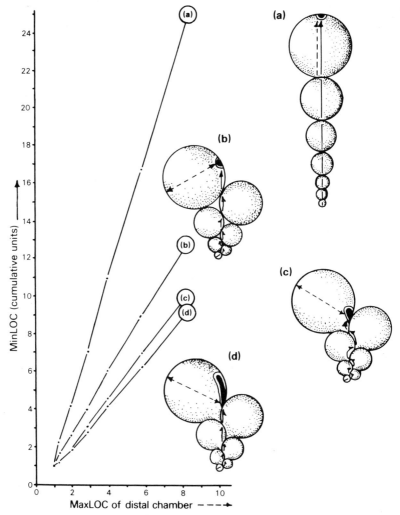

Figure 1.9 Model biserial tests with unit volume and volumetric expansion. This shows how slight variation in chamber shape, apertural form and position may shorten the MinLOC with little effect on MaxLOC.

models with spherical and near-spherical chambers, compared with a uniserial model of identical chamber parameters. The simplest biserial arrangement (Fig. 1.9b) requires no chamber modification; the aperture is situated on the growth axis at the impending point of tangent with the next chamber (i.e. areal), leaving a straight foraminal canal linking proloculus with aperture. The MinLOC of this form is about 50% shorter than that of the uniserial form. This stage is found in *Paratextularia*.

A further shortening of the MinLOC is achieved by placing the aperture at the base of the chamber (i.e. basal), adjacent to the preceding chamber and again lying over the growth axis (Fig. 1.9c, cf *Textularia*). Such an arrangement demands a greater degree of chamber overlap in the axial region to provide a passage for a shorter but more sinuous to zig-zagging foraminal canal.

Superior to both the above is a broad or slit-like aperture, running from the areal to basal positions of the former, since this removes the need for a zig-zag canal around the chambers (Fig. 1.9d, cf. *Semivulvulina, Brizalina*). In this model the MinLOC distance is only about 40% of that in the uniserial model, which may explain why many biserial tests have a 'basal slit' aperture.

Compression of the chamber shape in biserial tests carries with it the same advantages and problems as in uniserial tests (q.v.). In chambers of shortened length (with a low conical test shape) the MinLOC is much reduced (e.g. *Textularia conica*), while cribrate pores on the septal face would reduce the MaxLOC. In the more widespread case where one width parameter is held (giving a discoidal flabelliform or palmate test shape) the addition of peripheral apertures reduces both MinLOC and MaxLOC (e.g. *Cuneolina*). But the advantage of the biserial arrangement lies mainly with the shortening of the axial canal between relatively equant chambers. With increased chamber compression the difference between biserial and uniserial arrangements becomes progressively less and ultimately non-existent (except for the support an additional septum may provide). Thus many biserial foraminifera with compressed chambers revert to a simpler uniserial plan with growth (e.g. *Semitextularia*).

Ecology. Biserial forms are largely mobile and benthic in habit, with their test axis held upright (e.g. Haward & Haynes 1976) but *Heterohelix* and related Heterohelicacea were planktonic in Mesozoic and Palaeogene times (e.g. Hart & Bailey 1979). At present they do not live in fresh water but *Textularia earlandi* dwells in brackish water (Murray 1973). Simple biserial forms like *Textularia* were regarded by Chamney (1976) as adapted to normal marine conditions with moderate energy while multiform taxa such as *Dorothia* were more sensitive and restricted in distribution. This may be so, but more documentation and analysis is needed.

Evolution. The biserial tests *Pseudopalmula, Semitextularia* and *Paratextularia* first appeared during the Middle Devonian (Eifelian–Givetian) radiation of multilocular types with microgranular calcareous walls (e.g. Toomey & Mamet 1979, Poyarkov 1979) but their origin is uncertain. Possible Lower Devonian ancestors include fixed-form and non-septate spheres and tubes (e.g. *Saccammina, Saccamminopsis, Earlandia, Pseudoglomospira*), septate tubes (e.g. *Moravammina, Paratikhinella*) or, of course, the contemporary nodosinellids or endothyrids. Since there is no trace of spiral growth or lamellar walls in any of these Semitextulariidae, uniserial ancestry via *Saccamminopsis* or *Paratikhinella* might be favoured. Poyarkov (1979), however, derives his Order Semitextulariida perhaps from *Moravammina*. It is most unlikely that these earliest biserial forms originated from an increased rate of growth translation in high trochospiral tests, as did the hyaline Bolivinitidae (Upper Triassic–Recent, e.g. Loeblich & Tappan 1974) since septate trochospiral growth is not known from the Devonian. The lamellar microgranular Palaeotextulariina (Upper Mississipian–Permian), however, may derive from coiled tournayellids (Hohenegger & Piller 1975).

While biserial growth is common in the Textulariina, it seems to have appeared later than uniserial growth and is first seen at the distal end of planispiral *Spiroplectammina* (Lower Carboniferous–Recent), at the proximal end of *Bigenerina* (Lower Pennsylvanian–Recent) and ultimately throughout *Textularia* (Middle Pennsylvanian–Recent). Thus it has a secondary origin from retardation of the juvenile stage and acceleration of the ephebic stage in multiform tests.

In the hyaline biserial test of *Patellina* (?Permian, Jurassic–Recent), ontogeny suggests an origin from a trochospiral tube such as *Conicospirillina* (Jurassic–Recent). The Spirillinina were apparently the last group to achieve septation.

Planispiral form

In planispiral tests growth translation is zero, with chambers arranged symmetrically in a plane coil about the growth axis. While some of the advantages of such coiling had been exploited by non-septate Ammodiscidae in the early Palaeozoic, the higher rate of chamber volume expansion made possible by septation soon led to the rapid evolution of endothyrids and fusulinids (Upper Devonian–Upper Permian).

Like other growth plans, the rate of chamber volume expansion and chamber shape affect the MinLOC, but attention is mainly focused on the importance of apertures.

The simplest planispiral model resembles a uniserial test with spherical chambers wrapped around the growth axis (Fig. 1.10b). In this, the aperture is situated at the impending point of tangent with the next chamber, in the plane of symmetry (i.e. areal). The MinLOC therefore describes a polygonal helix from the proloculus through the areal foramina to the aperture. Such an arrangement is uncommon but occurs in *Paraendothyra*

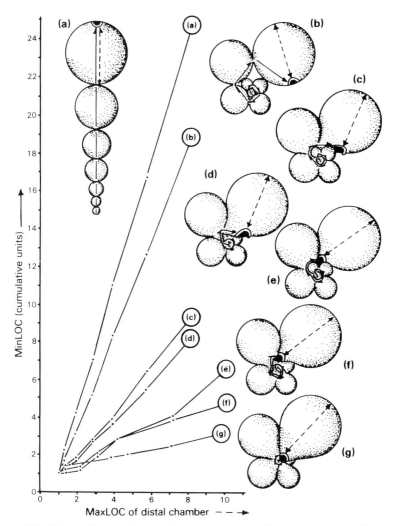

Figure 1.10 Model planispiral tests with unit volume and volumetric expansion. This shows how slight variation in chamber shape and apertural position may shorten the MinLOC but extend the MaxLOC.

and *Trochamminita*, for example. The MinLOC can be shortened by slight modification of chamber shape from spherical to pyriform or cuneiform, and a more basal placement of the aperture, which lies above the point of contact with the previous whorl (Fig. 1.10c). This arrangement is found in many evolute tests (e.g. *Trochamminoides, Loeblichia*). And without changing the aperture, a further shortening of the MinLOC is possible by allowing the whorls to overlap slightly ('semi-involute'), thereby forming a shorter spiral canal (Fig. 1.10d, cf. *Nanicella*). The greater the overlap, the more is the saving in MinLOC, hence the frequency of involute forms with basal equatorial apertures (e.g. *Haplophragmoides, Endothyranopsis*). Even more parsimonious is a basal umbilical aperture (Fig. 1.10g, cf. *Adercotryma*) since this shortens the canal further, though it is perhaps more usual to find an extended basal umbilical–extraumbilical aperture (e.g. *Melonis*). Thus chamber involution can minimise the length of the spiral canal, but the greater width of such chambers may result in the MaxLOC greatly exceeding the MinLOC parameter (Figs 1.10e–g). This anomaly can be countered by the existence of a long apertural slit up the distal face (e.g. *Daxia*) or by placing cribrate apertures on the otherwise remote area of the face (e.g. *Cyclammina*).

One ultimate paradigm for short MinLOC is reached with an open umbilicus so that every involute chamber is in contact with the exterior. This 'hot line' to the exterior will be most effective in tests with only a short growth axis (i.e. with an 8-shaped profile) and least effective in those with a long fusiform growth axis. Accordingly, it is found in *Haynesina*, for example, but not in any fusuline, loftusiine or alveoline test. Another way to provide each chamber with a 'hot line' to the exterior is by the addition of supplementary apertures along the sutures or in the spiral wall (e.g. *Bradyina, Hastigerinoides*).

Functional aspects of larger fusiform planispiral tests have been examined by Hottinger (1978). He illustrated how the numerous short chambers and the row of frontal apertures stretching from pole to pole may together give short lines of communication between the first and last shell compartments (Hottinger 1978, Fig. 11). But this does not tally with the evidence for progressive elongation of the growth axis in several iterative lineages (e.g. fusulines, alveolinids; see Dunbar 1963 and Hottinger 1963) because such a trend successively increases the MinLOC. If short proloculus-to-exterior lines of communication were important to these groups, they would have tended towards an involute 8-shaped axial section (as in the fusulines *Ozawainella* and *Parastaffella*). But by extending the biomass along the growth axis, the number of whorls for a given biomass is reduced. Hence the length of the spiral canal will be shortened; and since greater shortening occurs in the inner whorls than in the outer ones, a considerable reduction in MinLOC will result. The evolutionary trend towards a long axis therefore makes sense because it allows shorter lines of communication to the greatest proportion of cytoplasm, a factor of more importance than communication of the proloculus alone.

Ecology. The planispiral growth plan in smaller benthic foraminifera may be an adaptation for a mobile epifaunal and infaunal habit (Banner & Culver 1978) though it is also found in planktonic forms. Fusiform tests typically occur in reefal carbonates, especially high energy bioclastic sands in which they recline or tumble with little damage and survive on blooms of endosymbiotic algae (e.g. Ross 1972, Brasier 1975a). Forms with a short growth axis occupy the whole environmental range from freshwater tolerance (e.g.

Haynesina; see Banner & Culver 1978) to abyssal plains (e.g. *Cyclammina*).

Brackish-water forms are common. It may be significant that *H. germanica*, the most euryhaline and eurythermal of living Rotaliina, is involute with an 8-shaped axial section and with supplementary latero-umbilical apertures connecting each chamber with the exterior (Banner & Culver 1978), thereby keeping MinLOC very low. Its descendants *H. albiumbilicata, H. orbiculare* and *H. depressula* are stenohaline with the septal lacunae (in which these apertures lie) partly occluded by dense tubercles or lamellae of calcite (Banner & Culver 1978). Other possible descendants, '*Elphidium*' *incertum* and *E. magellanicum*, which live in rather less subsaline conditions than *H. germanica*, have closed their lacunae more completely to form septal canals; these may serve as cytoplasmic reservoirs to allow some test mobility, even when the cytoplasm has withdrawn from the last chambers of the test (e.g. Hottinger 1978).

Among the Textulariina, planispiral forms are found in brackish tidal marshes (e.g. *Haplophragmoides* and planispiral–uniserial *Ammobaculites, Ammomarginula, Ammoscalaria, Ammotium*). Of these, the former is involute with a basal equatorial slit providing for a fairly short spiral canal, but the multiform taxa are uncoiled, thereby lengthening the MinLOC. Brackish *Trochamminita*, with areal apertures and irregular later chambers, seems intermediate.

Thus it appears that the open umbilical apertures are adaptive in littoral habitats of varying salinity but may be less important in some sublittoral and marsh habitats. In the latter, where long dry periods occur, the tubular rotaliid canals may allow *Elphidium* to continue moving and feeding with a reduced risk of desiccation, while the uncoiled portions of *Ammobaculites* and its relatives may minimise the effects of desiccation or lowered salinity. Apertural variation can be pronounced within species from brackish waters (Boltovskoy & Wright 1976).

Nodosariacea break all the rules: coiled forms (e.g. *Lenticulina*) place their radiate apertures on the remote outer margin; compressed uniserial forms (e.g. *Frondicularia*) do not provide peripheral apertures to reduce the increased MaxLOC. This may be one reason why the uniserial and planispiral Nodosariidae were so thoroughly displaced from shallow-water habitats after the Jurassic.

Evolution. *Nanicella* (Upper Givetian–Frasnian), which is a nearly evolute 'endothyrid', is the oldest septate planispiral test in the fossil record. Its origin is unknown but the two-layered wall structure compares with some contemporary Nodosinellidae. Poyarkov (1979) derives it from biserial semitextulariids of similar age. An ancestry via the non-septate to protoseptate Tournayellidae is unlikely since these did not appear until the late Frasnian and flourished in the Fammenian (Poyarkov 1979, Toomey & Mamet 1979). But it may be from this group that the 'plectogyral' (i.e. glomospiral) to planispiral Endothyridae emerged in late Fammenian times.

Evolutionary trends from *Endothyra* to *Fusulina* and beyond can be interpreted as architectual solutions to problems set by increasing size, particularly the maintenance of short MinLOC (see above).

The origin of the multiform agglutinated *Ammobaculites* (Upper Devonian–Recent) is also uncertain but it was contemporary with *Reophax* and *Oxinoxis* and may share a saccamminid ancestry. *Haplophragmoides* (Lower Pennsylvanian–Recent) may have evolved by retardation of the uniserial stage in *Ammobaculites* or by reduced growth translation in *Trochammina* (Lower Mississipian–Recent).

Fusiform planispiral growth has evolved independently several times since the Palaeozoic, e.g. from discoidal planispiral Lituolidae (Loftusiinae; Jurassic–Cretaceous), Soritidae (*Fusarchaias*; Oligocene–Miocene) and from streptospiral Miliolidae (Alveolinidae; Lower Cretaceous–Recent). Likewise the planispiral rotalines have multiple origins, mainly from trochospiral ancestors (e.g. Banner & Culver 1978), though the Nodosariacea are exceptional since a uniserial origin is possible.

Loeblich and Tappan (1974) derived the coiled Nodosariacea from the Endothyridae and uniserial forms from Nodosinellidae. But this overlooks their common inheritance of a terminal radiate aperture, probably from the radial organisation of the uniserial nodosinellid *Colianella* (Brotzen 1963). Planispiral forms are derived from uniserial forms in the Permian according to Hohenegger and Piller (1975), who place the whole group in the Suborder Lagenina.

Trochospiral forms

Trochospiral arrangements of three or more chambers per whorl (with translation along the growth axis) are found in many Textulariina and Rotaliina but in a few Fusulinina or Miliolina. Non-septate trochospiral arrangements appeared in the Silurian but they were unable to exploit the advantages of this growth plan. The advent of septation in the mid to late Devonian was followed by the first experiments in septate trochospiral growth (*Trochammina*; Lower Carboniferous–Recent).

For each stage in the shortening of the MinLOC by planispiral growth (q.v.) there is an equivalent stage in trochospiral growth. Triserial forms, however, compare also with biserial growth (q.v.). If models of these growth plans in Figures 1.9b and 1.10b are compared with Figure 1.11b, it can be seen that, with a single areal aperture, an increase in the rate of translation shortens the MinLOC in the following sequence: planispiral, low trochospiral, high trochospiral (and triserial), biserial. But with the modification of apertures and chamber shape, the advantage passes to planispiral and low trochospiral forms.

In the low trochospiral models of Figure 1.11, translation of the chambers allows a shortening of the spiral canal (in plan view) with little chamber modification. To this distance, however, must be added the increased length

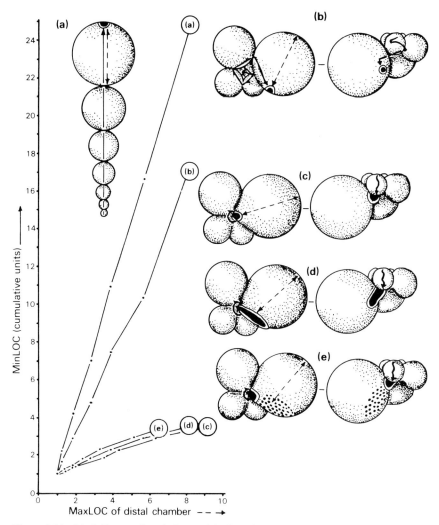

Figure 1.11 Model low trochospiral tests with a flat spiral side, with unit volume and volumetric expansion. (In Raupian terms, the rate of translation $T \simeq 0.7$; see Raup & Stanley 1978, pp. 172–3.) This shows how slight variation in chamber shape and apertural shape, number and position may shorten the MinLOC and MaxLOC.

of the growth axis. A progressive increase in the degree of involution of chambers on the umbilical side combined with a more umbilical placement of the aperture, gives a progressive shortening of the MinLOC but an increase in the MaxLOC (Fig. 1.11c, cf. *Siphotrochammina*). The MaxLOC can be reduced by a slit-shaped umbilical–extraumbilical aperture (Fig. 1.11d, cf. *Trochammina*), by a slit up the apertural face (cf. *Ceratobulimina*), by a slit along the keel (cf. *Hoeglundina*) or by cribrate apertural pores on the distal parts of the face (Fig. 1.11e, cf. *Arenoparrella*).

The ultimate shortening of MinLOC is provided by an open umbilicus, in which each chamber ideally opens directly to the exterior (e.g. *Tiphotrocha, Cymbalopora, Rosalina, Globigerina, Ammonia*) or by supplementary sutural or areal apertures (e.g. *Pseudoeponides, Virgulinella, Cymbaloporetta, Globigerinoides*).

In involute forms with nearly or wholly umbilical apertures, an increase in the rate of translation increases the axial distance with little shortening of the spiral distance of the foraminal canal. Hence the MinLOC increases in the following sequence: high trochospiral, quadriserial, triserial, biserial. It also follows that the MinLOC of umbilical-apertured planispiral and trochospiral forms will be little different, hence the choice of growth plan may depend on the need for mobility or adherence.

Ecology. High trochospiral forms are often mobile benthos with the test apex held upright (e.g. *Bulimina* of Haward & Haynes 1976) though *Guembelitria* and related forms were planktonic in the Mesozoic and Palaeogene. Many have a broad umbilical area well suited for attachment to hard surfaces or vegetation (e.g. *Cibicides, Rosalina, Discorbis*). Low trochospiral tests with a biconvex profile may be secondarily adapted to a mobile epifaunal or infaunal niche (e.g. *Ammonia*), to a life in unstable reef sands (e.g. *Asterigerina, Rotorbinella*) or to a semi-mobile life on hard substances (e.g. *Amphistegina*; Brasier 1975a).

While their environmental range is wide, it is evident that low trochospiral tests abound in brackish-water marshes. *Trochammina inflata* has spheroidal chambers with an involute umbilical side and a basal umbilical–extraumbilical slit aperture. The narrower chambers of *Jadammina macrescens* and *J. polystoma* have the MaxLOC reduced by supplementary cribate apertures on the septal face. The same is true for the elongate chambers of *Arenoparella mexicana*. But the shortest lines of communication in these paralic Trochamminidae are found in *Tiphotrocha comprimata* from brackish marshes and *Discorinopsis aguayoi* from hypersaline marshes (e.g. Murray 1973).

Ammonia beccarii is the most euryhaline of living trochospiral Rotaliina and occurs in a variety of forms, which some call species, that dwell in different habitats. *A. beccarii tepida* is found in brackish (0–35‰) estuaries and lagoons but seldom on exposed marshes. The open umbilicus and sutural apertures ideally provide each chamber with direct access to the exterior (Banner & Williams 1973). In *A. beccarii batava*, which is a neritic species, the open umbilicus has largely been closed by calcite pillars (Banner & Williams 1973). Of the hypersaline Rotaliina, *Glabratella* has an open umbilicus and *Pseudoeponides* has sutural apertures on both the spiral and umbilical side, connecting each chamber with the exterior.

Thus the observations of the planispiral section are repeated. Open umbilical and sutural apertures may be adaptive in littoral habitats of varying salinity but may be less important in sublittoral habitats or in marsh habitats where desiccation is severe.

Evolution. Agglutinated *Trochammina* (Lower Carboniferous–Recent) is the first known trochospiral test, but its ancestry is uncertain. Salami (1976) found the proloculus and cytoplasm comparable with simple allogromiids and saccamminids. Derivation via the curved *Oxinoxis ligula* or planispiral *Ammobaculites* of latest Devonian–Carboniferous times (e.g. Conkin & Conkin 1977) are also possible.

The calcareous two-layered wall of *Tetrataxis* (Upper Mississippian–Triassic) compares with that of coiled endothyrids (ancestral, according to Loeblich & Tappan 1974) but also with the biserial Suborder Palaeotextulariina (Upper Mississippian–Permian) of Hohenegger and Piller (1975). These authors, however, derive from these planispiral, biserial and trochospiral stocks independently from tournayellid ancestors. Other than these few 'experiments', trochospiral growth was little exploited in the Palaeozoic. The richer variety of Mesozoic and Cenozoic Rotaliina may be traced back to the trochospiral Duostominacea, themselves perhaps descendants of the Textrataxidae (Hohenegger & Piller 1975). However, the high trochospiral nodosariaceans, with radiate terminal apertures, are most likely derived from the older uniserial Nodosariidae.

Suggestions that the high trochospiral Buliminacea and Cassidulinacea had an agglutinated origin (see Brotzen 1963) are interesting, but the models presented here suggest that apertural form and growth plan are closely linked, thus architectural convergence is very likely.

Miliolid form

The distinctive miliolid or 'streptospiral' growth mode abounds in septate Miliolina (Middle Carboniferous–Recent) though similar forms occur in hyaline Spirillinina (Middle Jurassic) and agglutinated Rzehakinidae (Lower Cretaceous–Recent). This plan may be seen as a development of glomospiral and ammodiscid coiling in which tubular chambers are added around an elongate growth axis in a more orderly fashion. True septa may be lacking or take the form of 'teeth' or other constrictions. The aperture is terminal and lies at alternate ends of the axis as growth proceeds. This latter arrangement differs from that in streptospiral Rotaliina (e.g. *Guttulina, Globobulimina*) in which the terminal aperture remains at one pole and each chamber is slightly translated along the growth axis; this is a modification of trochospiral growth.

Since the chambers are wrapped around the axis like a ball of wool the same principles and problems apply. Coiling in one plane is simple and natural but variation in the plane of coiling gives greater strength and compactness. For maintenance of symmetry the changes should be more than 90° apart. Thus chambers coiled at 180° apart give good symmetry but little strength, while chambers coiled at, say, 120° and 144° apart give less symmetry but greater strength and compactness.

As with other growth plans, increase in the rate of chamber volume expansion (for a given total volume) will shorten the MinLOC and vice versa. In Figure 1.12, the models demonstrate the significance of coiling mode and chamber shape for a unit volume.

The simplest miliolid arrangement is spiroloculine, with the chambers coiled about the axis in a single plane (i.e. at 180° to each other; Fig. 1.12c, cf. *Spiroloculina*). It resembles that of *Ammodiscus* except that the insertion of constrictions at 180° intervals allows a greater rate of chamber volume expansion and thus a shortening of the MinLOC. Unlike some other miliolid growth forms, spiroloculine coiling is possible with chambers of a simple circular cross section and almost any rate of chamber volume expansion. Shorter lines of communication can be attained by compressing the chambers adaxially so that they overlap more (Fig. 1.12d) and/or by maintaining a short growth axis and a nearly circular plan view.

In quinqueloculine growth, chambers are added at 144° intervals around the growth axis in such a way that five chambers (i.e. two whorls) are always visible (Fig. 1.12f, cf. *Quinqueloculina*). The chambers may be of circular cross section but a strict and fairly high rate of chamber volume expansion is necessary if the pattern is not to be disrupted. The scope for chamber compression is limited because no chamber should extend much beyond 72° of arc. But there is one important advantage; later chambers can be placed very close to the growth axis, keeping the MinLOC relatively low.

In triloculine growth, chambers are added at 120° intervals about the growth axis, leaving a single whorl of three chambers visible from the outside (Fig. 1.12e, cf. *Triloculina*). In this case a compact test is impossible unless the chambers have a compressed cross section. For a given rate of chamber volume expansion, the MinLOC of triloculine tests may be greater than for quinqueloculine ones because each chamber lies upon a larger antecedent, causing the base of each chamber to lie further away from the growth axis. But triloculine coiling brings the bulk of the protoplasm closer to the coiling axis than quinqueloculine growth.

In biloculine growth, involute chambers are added at 180° intervals about the growth axis, leaving a single whorl of two chambers visible from the outside (Fig. 1.12b, cf. *Pyrgo*). Again, this causes the base of each chamber to lie further away from the growth axis than in quinque- or triloculine coiling but a greater proportion of the cytoplasm lies closer to the axis. The ultimate stage in this trend is periloculine growth, with each chamber wholly enveloping the next (Fig. 1.12a, cf. *Periloculina*) and a compact distribution of mass about the axis.

Coiling changes with growth in most miliolids and both 'shortened' and 'extended' strategies may occur. In *Pyrgo* and *Periloculina* the change from quinque- and triloculine growth to bi- or periloculine growth reduces the maximum ambit of the test. But in *Massilina*, the MinLOC is extended by a change from quinqueloculine to spiroloculine growth.

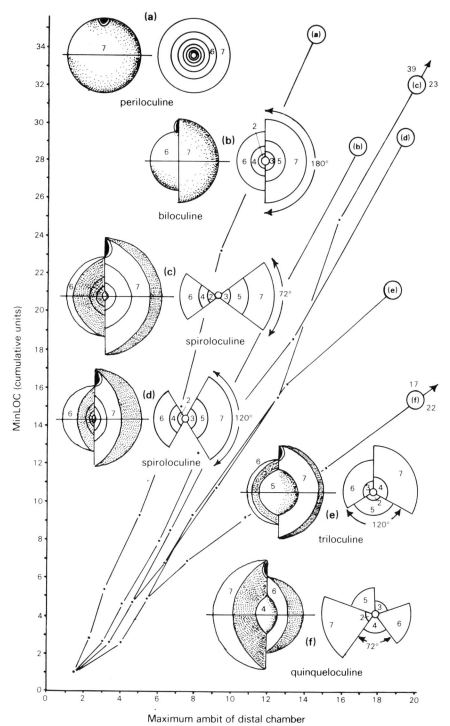

Figure 1.12 Model miliolid tests with unit volume and volumetric expansion. This shows how shortened MinLOC in the series (a), (b), (e), (f) is countered by a shorter maximum ambit in the series (f), (e), (b), (a). The shortening effect of adaxial chamber compression is also shown in the spiroloculine growth of (c) (cf. (f)) and (d) (cf. (e)).

Ecology. Aspects of miliolid morphology and habitat have been outlined by Brasier (1975a). Their abundance in warm, clear and shallow waters may relate more to test composition and ultrastructure (e.g. Greiner 1969) than to architecture. The rzehakinid *Miliammina* can live in fresh water but thrives in subsaline to hypersaline tidal marshes; its chambers are long and tubular with a quinqueloculine arrangement. Porcelaneous forms abound in fine-grained sediments of hypersaline lagoons and marshes (e.g. *Q. subpoeyana, T. oblonga*); these tend to be small with elongate chambers in quinque- or triloculine coils. Aberrant forms in which the final chambers are irregular, planispiral or of reduced length also develop in very marginal habitats. Miliolids from normal marine sediments may be larger and less elongate, with thicker walls (e.g. *Q. lamarckiana*). Those that attach to plant substrates tend to have a rounded or ovate profile and lack a neck or surface sculpture (e.g. *Miliolinella, Massilina*) while those from soft oceanic oozes may be nearly spherical in shape (e.g. *Pyrgo*).

Evolution. The first septate miliolid *Eosigmoilina* (Middle Carboniferous) had tubular chambers with a glomospiral juvenile portion and a spiroloculine–sigmoiline outer portion. It may have developed from *Agathammina* (Lower Pennsylvanian–Triassic, ?Jurassic) a non-septate quinqueloculine with an agglutinated wall, or more likely from *Hemigordius* (Upper Mississippian–Jurassic) a non-septate glomospiral–planispiral miliolid. Purely spiroloculine and quinqueloculine septate growth were achieved by late Triassic times (*Ophthalmidium, Quinqueloculina*) while *Triloculina* and *Pyrgo* appeared in the Jurassic. Late Cretaceous developments included periloculine growth, complex interior structure and cribrate apertures (e.g. *Periloculina*). An early attempt to achieve a *Pyrgo*-like test (*Hemigordiopsis*; Permian) without septation did not survive the end-Palaeozoic extinction event.

The agglutinated Rzehakinidae first appeared in the Lower Cretaceous (spiroloculine *Psamminopelta*, quinqueloculine *Miliammina*). Triloculine *Triloculaena* is a Recent genus. Biological studies might help to determine whether this group is related to uniserial *Hormosina* and its relative (Loeblich & Tappan 1974) or is an independent offshoot of the Ammodiscacea or Miliolina (e.g. from *Agathammina*).

Annular, cyclic and orbuline forms

In annular growth, ring-shaped chambers are added serially to the periphery of the test; commonly they are partitioned into small chamberlets. In cyclic growth, a similar effect is produced by adding a ring of small chambers, each ring laid down in successive cycles. In orbuline growth the chambers are spherical and wholly envelop the earlier ones, much as in periloculine growth described above.

The writer has shown how annular chambers may follow naturally from flabelliform uniserial ones, where the chamber length (l) and width (w_2) are held nearly constant (Brasier 1980, Fig. 13.7). For example, the uniserial growth in Figures 1.8i and j is only a step from the annular growth in Fig. 13b. But one feature of annular growth in foraminifera is the nearly constant length of each successive ring, quite unlike the expanding rings of Figure 1.13b. And while w_2 has been held constant in Figure 1.13c, the resultant figure bears little comparison with any living foraminifera, unless it be a highly exaggerated *Orbitopsella* or *Marginopora*. To produce typical annular

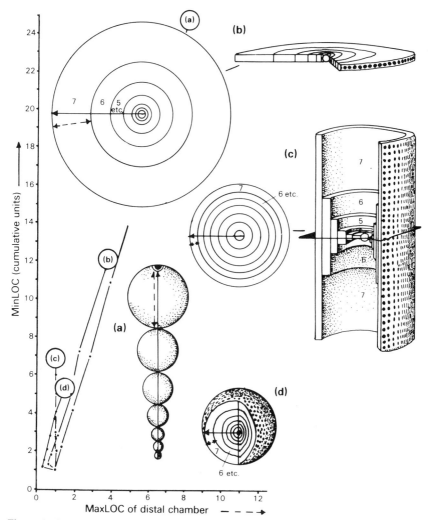

Figure 1.13 Model annular and orbuline tests with unit volume and volumetric expansion. This shows how enveloping, compressed chambers may shorten Min LOC, especially where this compression is adaxial (i.e. towards the axis). The MaxLOC is lowered by numerous areal or peripheral apertures.

chambers it is necessary for the rate of chamber volume to decelerate with time; with chambers of circular axial section, the volumetric decrease is exponential such that $V_2/V_1 = 8$ but $V_7/V_6 = 1.2$ and the outermost chambers differ little from each other in volume (i.e. $V_n/V_{n-1} \simeq 1.0$). Thus growth in *Peneroplis*, which develops from compressed planispiral to annular, is not 'accelerated', as a simple measure of the chamber ambit might suggest (Hottinger 1978, pp. 213–14).

While annular chambers may be constructed with a single peripheral aperture (cf. Fig. 1.8i), the high MaxLOC of such a test may be one reason why this is rare. The more usual scatter of peripheral apertures creates a geometry with relatively short lines of internal–external communication for the greatest proportion of cytoplasm (Fig. 1.13b, cf. Dicyclinidae, Soritidae) though the shortening is greater with a biconcave profile (Fig. 1.13c, cf. *Orbitopsella, Marginopora*).

Superior to either of these arrangements as a means of shortening the MinLOC is the orbuline arrangement. With a single terminal aperture the MaxLOC parameter is high; this is approximately the condition seen in linguline Nodosariidae (e.g. *Daucinoides*) and in the periloculine miliolids described above (Fig. 1.12a). But the planktonic *Orbulina* is the only foraminifer that approaches the ideal arrangement, with scattered pores over the whole surface and a low MinLOC and MaxLOC (Fig. 1.13d), although planktonic radiolarians have built skeletons on this plan since Cambrian and Ordovician times (e.g. *Actinomma*).

Ecology. Annular and cyclic growth is favoured by larger benthic foraminifera living in warm, shallow carbonate habitats. Each chamber is usually divided into chamberlets and the cytoplasm may contain endosymbionts that contribute to the large size of their hosts (e.g. Lee 1974). Their photosynthesis is presumably encouraged by a restriction to clear tropical waters (less than about 90 m deep) and by the high surface area:volume ratio of annular and cyclic growth. Perhaps because this last induces greater friction with movement, the annular Soritidae tend to be immobile, clinging to hard surfaces such as seagrass blades. There is some evidence that this niche developed with the evolution of seagrasses in the Cretaceous (Brasier 1975b).

The multi-apertured arrangement of *Orbulina* and *Actinomma* approaches the optimum paradigm for the collection of food in a uniform medium such as the water column.

Evolution. The final growth stage of the aberrant fusuline *Paradoxiella* (Upper Permian) is cyclic in appearance but such growth is otherwise unknown until the development of annual growth in the Dicyclinidae (Lower Jurassic–Middle Eocene), certain Lituolidae (Upper Jurassic–Lower Cretaceous), Soritidae (Upper Cretaceous–Recent) and of cyclic growth in several lineages of Cretaceous and Cenozoic Rotaliina (e.g. Orbitoidacea). *Orbulina*-like rotaliines have evolved several times since the early Miocene.

Thus these relatively 'sophisticated' forms of architecture were among the last to appear in foraminiferid history.

Multiform growth

Multiform growth, in which several growth plans are combined, generally takes the appearance of uncoiling, with a planispiral, trochospiral or streptospiral initial stage and one or more ephebic stages from the following sequence: triserial, biserial, uniserial, annular or irregular. Biserial enrolled forms such as *Cassidulina* may appear to reverse the trend but even these may uncoil (e.g. *Ehrenbergina*).

While multiform growth is commonplace, its significance is little understood. The initial stage has generally been taken to indicate ancestry and this has led to the view that coiling is primitive while uncoiling is a sign of gerontic character (e.g. Galloway 1933, Cushman 1948). But coiling of the juvenile in biserial *Heterohelix* and *Bolivinopsis* occurred late in lineages of progressively reduced proloculus size (Hofker 1954). The gerontic interpretation also overlooks the fact that multiform growth appeared alongside or even preceded the first uniform septate growth plans of the Palaeozoic. Either way, there is a tendency for the juvenile stage to be more fully developed in microspheric tests than in megalospheric ones where the higher initial volume may 'accelerate it out' (e.g. Hofker 1954).

From the models outlined above it is clear that uncoiling will relatively extend the MinLOC if the rate of chamber volume expansion and chamber shape remain the same and the apertures are optimal for each arrangement. But a change in any of the latter may extend or shorten the MinLOC or MaxLOC.

In Figures 1.14a–b and d–e are two models in which uncoiling has led to shorter MinLOC. The nine chambers of the evolute planispiral or biserial stage are pyriform ($l\pm=w$) with basal apertures. The four chambers of the uniserial stage are compressed and drum-shaped ($w>l$, with both held constant) bearing a single terminal aperture. In this case the rate of volumetric increase is greatly raised between chambers 9 and 10 and thereafter is held constant at $E=1.0$ (thus the cumulative volumes of chambers 10–13 are identical in Figures 1.14a–f but not the volumes of the individual chambers). In Figures 1.14e–f the retention of a medium septum would, of course, keep the growth plan biserial, but there is little to choose between biserial and uniserial arrangements once chamber compression has reached the point where multiple apertures are advantageous (Fig. 1.14c & f). Many uncoiled forms follow this pattern, either with a single aperture (e.g. some *Bigenerina, Pseudospiroplectinata, Palaeobigenerina*) or with cribrate apertures (e.g. *Cribrogoesella, Climacammina*).

Flaring rather than cylindrical growth results if the length of chambers 10–13 in Figures 1.14a and d are fixed at a lower figure and/or with a slight

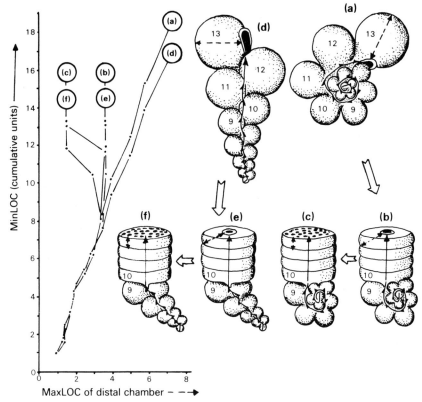

Figure 1.14 Model planispiral (a), biserial (d) and multiform tests (b), (c), (e), (f) with unit volume: (a) and (d) have unit volumetric expansion ($E=1.7$); (b), (c), (e) and (f) have varied volumetric expansion (chambers 1–9, $E=1.7$; chamber 10, $E=2.5$; chambers 11–13, $E=1.0$). This shows that uncoiling of (a) and (d) may shorten MinLOC and MaxLOC if accompanied by shortened chamber length and multiple apertures.

rate of volumetric increase (e.g. *Arenonina*, *Ammospirata*, *Coskinolina*, *Semitextularia*, *Peneroplis*, *Pavonina*).

Another case where uncoiling may shorten the MinLOC is in those tests where the aperture remains areal and terminal. This is because the MinLOC become less in the sequence: planispiral, trochospiral, biserial. Such may be the case in *Siphonides*. In the Nodosariidae, where the aperture lies on the outer ambit of the coil, uncoiling makes little difference but it may shorten the MinLOC if the aperture becomes more medial in the uniserial portion.

The assertion by Redmond (1953) that changes in growth plan are invariably accompanied by an increase in the rate of chamber volume expansion (cf. Figs 1.14b, c, e, f) is not substantiated (Hofker 1954). In many cases the volumetric expansion remains the same, thus uncoiling tends to increase the MinLOC (e.g. *Lituotuba*, some *Ammobaculites*, *Spiroplectinata*, *Clavulina*, *Spirolina*, some *Articulina*, *Rectobolivina*). Extension by a change to more tubular chambers ($l>w$) is rather less frequent (e.g. the last chamber of *Ammoscalaria, Tubinella, Bifarina*).

The ecology and evolution of multiform taxa have been discussed in previous sections.

Conclusions

Foraminiferid architecture may be classified and measured not only by the growth plan but also by a study of minimum and maximum lines of communication as defined above. This method has promising potential for the analysis of both evolution and evolutionary ecology in foraminifera. In such studies the standard of measurement should be a single-apertured sphere. Chambers with MinLOC values each respectively longer than a sphere of equal volume may be said to have 'extended MinLOC' while those shorter than a sphere have 'shortened MinLOC'. Those approximating to a sphere have 'neutral MinLOC'. Ontogenetic changes from a negative to a positive value, or vice versa, may be traced in multiform taxa.

Basic tendencies in the evolution of foraminiferid architecture are indicated in Figure 1.15. Some interesting trends which have emerged from this study deserve much further research but are outlined below.

(a) Early Palaeozoic foraminifera comprised tests with contained growth, short MaxLOC but limited growth, and tubular or conical tests with continuous growth but extended MinLOC. Attempts to minimise these parameters for a unit volume resulted in forms like *Turitellella* and *Glomospira* by early Silurian times.

(b) Septation was achieved independently in at least five lineages during the mid-Devonian to Jurassic: Semitextulariidae, Tournayellidae, Textulariina, Miliolina, Spirillinina. This permitted higher rates of volumetric expansion, varied chamber form and shorter MinLOC. The change from uniserial-dominant to planispiral- and multiform-dominant in shallow-water carbonate habitats during this period was accompanied by a progressive displacement of more primitive agglutinated forms into marginal marine and bathyal habitats. This suggests a trend towards ever more shortened MinLOC values in highly competitive shallow-water habitats.

(c) Following the end-Palaeozoic extinctions, hyaline Spirillinina and Lagenina with negative to nearly neutral MinLOC values dominated Triassic and Jurassic shallow-water facies. They were progressively displaced by trochospiral, planispiral, streptospiral and annular-cyclic tests of Rotaliina, Miliolina and Textulariina with shortened MinLOC parameters, culminating in the complex larger foraminifera of Cretaceous and Palaeogene times. Competition may have been the stimulus to shorten the MinLOC, as in the Palaeozoic.

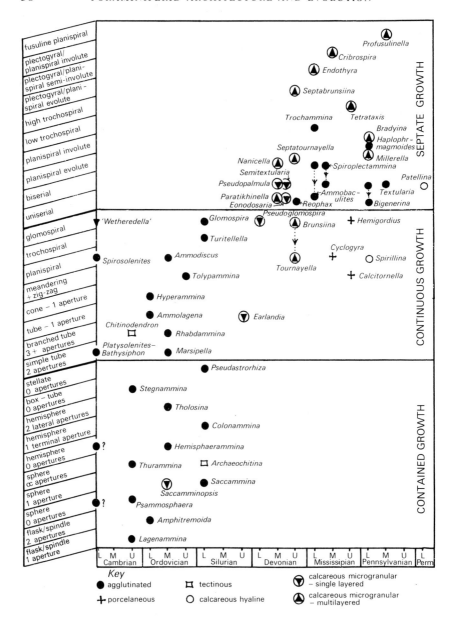

Figure 1.15 The gradual appearance of foraminiferid growth plans through the Palaeozoic, as revealed by the fossil record. Mainly from data in Loeblich and Tappan (1964), Conkin and Conkin (1977), Toomey and Mamet (1979) and Poyarkov (1979).

A further trend which merits careful analysis is the progressive colonisation of brackish-water habitats by septate foraminifera with very low MinLOC values during the Cenozoic.

While the approach outlined above affords a new way of looking at foraminifera, it is still only a series of models at present. Cytological and ecological research are needed to examine the predictions of these models and to help explain the full meaning of lines of communication within the foraminiferid test and with the external milieu.

References

Anderson, O. R. and A. W. H. Bé 1978. Recent advances in foraminiferal fine structure research. In *Foraminifera*, Vol. 3, R. H. Hedley & C. G. Adams (eds), 121–202. London: Academic Press.

Banner, F. T. and S. J. Culver 1978. Quaternary *Haynesina* n. gen. and Paleogene *Protelphidium* Haynes; their morphology, affinities and distribution. *J. Foramin. Res.* **8**, 177–207.
Banner, F. T. and E. Williams 1973. Test structure, organic skeleton and extrathalamous cytoplasm of *Ammonia* Brünnich. *J. Foramin. Res.* **3**, 46–69.
Banner, F. T., R. Sheehan and E. Williams 1973. The organic skeleton of rotaline foraminifera: a review. *J. Foramin. Res.* **3**, 30–42.
Barnard, T. 1963. Evolution in certain biocharacters of selected Jurassic Lagenidae. In *Evolutionary trends in foraminifera*, G. H. R. von Koenigswald, J. D. Emeis, W. L. Bunning & C. W. Wagner (eds), 79–92. Amsterdam: Elsevier.
Berger, W. H. 1969. Planktonic foraminifera: basic morphology and ecologic implications. *J. Paleont.* **43**, 1369–83.
Boltovskoy, E. and H. Lena 1971. The foraminifera (except Allogromidae) which dwell in fresh water. *J. Foramin. Res.* **1**, 71–6.
Boltovskoy, E. and R. Wright 1976. *Recent foraminifera*. The Hague: Junk.
Brasier, M. D. 1975a. Morphology and habit of living benthonic foraminifera from Caribbean carbonate environments. *Rev. Esp. Micropaleont.* **7**, 567–78.
Brasier, M. D. 1975b. An outline history of seagrass communities. *Palaeontology* **18**, 681–702.
Brasier, M. D. 1980. *Microfossils*. London: George Allen & Unwin.
Brotzen, F. 1963. Evolutionary trends in certain calcareous foraminifera on the Palaeozoic/Mesozoic boundary. In *Evolutionary trends in foraminifera*, G. H. R. von Koenigswald, J. D. Emeis, W. L. Bunning & W. C. Wagner (eds), 66–78. Amsterdam: Elsevier.

Chamney, T. P. 1976. Foraminiferal morphology symbol for paleoenvironmental interpretation of drill cutting samples: Arctic America. In *First International symposium on benthonic foraminifera of continental margins*, C. T. Schafer & B. R. Pelletier (eds), 585–624. Halifax, Nova Scotia: *Maritime Sediments*.
Conkin, J. E. and B. M. Conkin 1977. Paleozoic smaller foraminifera of the North Atlantic Borderlands. In *Stratigraphic micropalaeontology of Atlantic Basin and borderlands*, F. M. Swain (ed.), 49–60. Amsterdam: Elsevier.
Conway Morris, S. and W. H. Fritz 1980. Shelly microfossils near the Precambrian/Cambrian boundary, Mackenzie Mountains, north-west Canada. *Nature, Lond.* **286**, 381–4.
Culver, S. J. and M. A. Buzas 1980. *Distribution of Recent benthic foraminifera off the North American coast*. Smithson. Contr. mar. Sci. no. 6.
Cushman, J. A. 1948. *Foraminifera, their classification and economic use*, 4th edn. Cambridge, Mass.: Harvard University Press.

Dunbar, C. O. 1963. Trends of evolution in American fusulines. In *Evolutionary trends in foraminifera*, G. H. R. von Koenigswald, J. D. Emeis, W. L. Bunning & W. C. Wagner (eds), 25–44. Amsterdam: Elsevier.

Føyn, S. and M. F. Glaessner 1979. *Platysolenites*, other animal trace fossils, and the Precambrian–Cambrian transition in Norway. *Nor. Geol Tiddskr.* **59**, 25–46.

Galloway, J. J. 1933. *A manual of foraminifera.* Bloomington, Ind.: Principia.
Glaessner, M. F. 1945. *Principles of micropalaeontology.* New York: Hafner.
Glaessner, M. F. 1963. Major trends in the evolution of the foraminifera. In *Evolutionary trends in foraminifera*, G. H. R. von Koenigswald, J. D. Emeis, W. L. Bunning & C. W. Wagner (eds), 9–24. Amsterdam: Elsevier.
Greiner, G. O. 1969. Recent benthonic foraminifera, environmental factors controlling their distribution. *Nature, Lond.* **223**, 168–70.

Hallam, A. 1978. How rare is phyletic gradualism and what is its evolutionary significance? Evidence from Jurassic bivalves. *Paleobiology* **4**, 16–25.
Hart, M. B. and H. W. Bailey 1979. The distribution of planktonic foraminiferida in the mid-Cretaceous of NW Europe. In *Aspekte der Kreide Europas*, IUGS Series A, no. 6, 527–42. Stuttgart: IUGS.
Haward, N. J. B. and J. R. Haynes 1976. *Chlamys opercularis* (Linnaeus) as a mobile substrate for foraminifera. *J. Foramin. Res.* **6**, 30–8.
Haynes, J. 1965. Symbiosis, wall structure and habitat in foraminifera. *Contr. Cushman Fdn Foramin. Res.* **16**, 40–3.
Hofker, J. 1954. Chamber arrangement in foraminifera. *Micropaleontologist* **8**, 30–2.
Hofker, J. 1959. Orthogenesen von Foraminiferen. *Neues J. Geol Palaeontol. Abh.* **108**, 239–59.
Hohenegger, J. and W. Piller 1975. Wandstrukturen und grossgliederung der foraminiferen. *Öst. Akad. Wiss. Mat.-Naturw. Kl. Abt. I* **184**, 67–95.
Hottinger, L. 1963. Les alvéolines paléogènes, exemple d'un genre polyphylétique. In *Evolutionary trends in foraminifera*, G. H. R. von Koenigswald, J. D. Emeis, W. L. Bunning & C. W. Wagner (eds), 298–314. Amsterdam: Elsevier.
Hottinger, L. 1978. Comparative anatomy of elementary shell structures in selected larger foraminifera. In *Foraminifera*, Vol. 3, R. H. Hedley & C. G. Adams (eds), 203–66. London: Academic Press.

Kobluk, D. R. and N. P. James 1979. Cavity-dwelling organisms in Lower Cambrian patch reefs from southern Labrador. *Lethaia* **12**, 193–218.

Lee, J. J. 1974. Towards understanding the niche of the foraminifera. In *Foraminifera*, Vol. 1, R. H. Hedley & C. G. Adams (eds), 207–60. London: Academic Press.
Loeblich, A. R., Jr and H. Tappan 1964. Protista 2: Sarcodina, chiefly 'Thecamoebians' and Foraminiferida. In *Treatise on invertebrate paleontology*, Part C, Vols 1 and 2, R. C. Moore (ed.). Kansas City: Geological Society of America and University of Kansas Press.
Loeblich, A. R., Jr and H. Tappan 1974. Recent advances in the classification of Foraminiferida. In *Foraminifera*, Vol. 1, R. H. Hedley & C. G. Adams (eds), 1–53. London: Academic Press.

Marszalek, D. S., R. C. Wright and W. W. Hay 1969. Function of the test in the foraminifera. *Trans Gulf Coast Assoc. Geol Socs* **19**, 341–52.
Murray, J. W. 1973. *Distribution and ecology of living benthic foraminiferids.* London: Heinemann.
Myers, E. H. 1936. The life cycle of *Spirillina vivipara* Ehrenberg, with notes on the morphogenesis, systematics and distribution of the foraminifera. *Jl R. Microsc. Soc. Lond.* **56**, 120–46.

Nazarov, B. B. 1975. Lower and Middle Palaeozoic radiolarians of Kazakhstan. *Trudy Geol Inst. Akad. Nauk SSSR*, **275** (in Russian).

Poyarkov, B. V. 1979. *Evolution and distribution of Devonian foraminifera.* Moscow: Izdat. Nauka. (In Russian).

Rat, P. 1963. L'accroissement de taille et les modifications architecturales corrélations chez les orbitolines. In *Evolutionary trends in foraminifera*, G. H. R. von Koenigswald, J. D. Emeis, W. L. Bunning & C. W. Wagner (eds), 93–111. Amsterdam: Elsevier.

Raup, D. M. and S. M. Stanley 1978. *Principles of paleontology.* San Francisco: W. H. Freeman.

Redmond, C. D. 1953. Chamber arrangement in foraminifera. *Micropalaeontologist* **7**, 16–22.

Ross, C. A. 1972. Palaeobiological analysis of fusulinacean (Foraminiferida) shell morphology. *J. Paleont.* **46**, 719–28.

Rhumbler, L. 1895. Entwurf eines natürlichen Systems der Thalamophoren. *Ges. Wiss. Göttingen, Math.-Physik Kl. Nacht* **1**, 51–98.

Salami, H. B. 1976. Biology of *Trochammina* cf. *T. quadrilobia* Hoglund (1947), an agglutinating foraminifer. *J. Foramin. Res.* **6**, 142–53.

Scott, D. B. 1976. Brackish-water foraminifera from Southern California and description of *Polysaccammina ipohalina* n. gen., n. sp. *J. Foramin. Res.* **6**, 312–21.

Scott, G. H. 1974. Biometry of the foraminiferal shell. In *Foraminifera*, Vol. 1, R. H. Hedley & C. G. Adams (eds), 55–151. London: Academic Press.

Sigal, J. 1952. Foraminifera. In *Traité de paléontologie*, Vol. 1, J. Piveteau (ed.), 133–78 and 192–301. Paris: Masson.

Sliter, W. V. 1970. *Bolivina doniezi* Cushman and Wickenden in clone culture. *Contr. Cushman Fdn Foramin. Res.* **21**, 87–99.

Smout, A. H. 1954. *Lower Tertiary foraminifera of the Qatar Peninsula.* London: British Museum (Natural History).

Tappan, H. 1976. Systematics and the species concept in benthonic foraminiferal taxonomy. In *First international symposium on benthonic foraminifera of continental margins*, C. T. Schafer & B. R. Pelletier (eds), 301–13. Halifax, Nova Scotia: *Maritime Sediments*.

Thompson, D'A. W. 1942. *On growth and form.* Cambridge: Cambridge University Press.

Toomey, D. F. 1972. Distribution and paleoecology of Upper Devonian (Frasnian) algae and foraminifera from selected areas in western Canada and the northern United States. *24th Int. Geol Congr., Montreal* **7**, 621–30.

Toomey, D. F. and B. L. Mamet 1979. Devonian Protozoa. In *The Devonian system*, M. R. House, C. T. Scrutton & M. G. Bassett (eds), 189–92. Oxford: Oxford University Press.

Toomey, D. F. and H. D. Winland 1973. Rock and biotic facies associated with Middle Pennsylvanian (Desmoinesian) algal buildup, Nena Lucia field, Nolan County, Texas. *Bull. Am. Assoc. Petrol Geol.* **57**, 1053–74.

2 The subgeneric classification of *Arenobulimina**

Amendments to the subgeneric classification of *Arenobulimina* (after Voloshina 1965) and an interpretation of their evolutionary development in relation to other members of the Family Ataxophragmiidae (Schwager 1877, amended Balahakmatova 1972) during the mid-Cretaceous (Albian and Cenomanian) in northwestern Europe

Clemens Frieg and Roger J. Price

The subgeneric arenobuliminid classification of Voloshina (1965, 1972) is followed, whereby the genus *Vialovella* is recognised (with a new species, *V. praefrankei*, ancestral to *Crenaverneuilina frankei*), and *Arenobulimina (Arenobulimina)*, *Arenobulimina (Harena)* and *Arenobulimina (Pasternakia)* are agreed subgenera. *Arenobulimina (Hagenowella)* is employed as a senior synonym of *A. (Novatrix)*. Within these subgenera, one new species is proposed, namely *A. (P.) barnardi*. The new subgenus *Arenobulimina (Sabulina)* is erected, and a new species *A. (S.) gaworbiedowae* proposed. The subgenus *Arenobulimina (Voloshinoides)* is put in synonymy with *Hagenowina*.

The classification of the Ataxophragmiidae (Schwager 1877, amended Balahakmatova 1972) is followed, whereby the Globotextulariinae (Cushman 1927) and Valvulininae (Berthelin 1880) are incorporated into the Ataxophragmiinae. In addition, the classification and phylogeny proposed by Barnard and Banner (1981) is compared and contrasted with the scheme followed herein wherever possible.

The evolutionary development of the arenobuliminids is traced. *A. (P.) macfadyeni* Cushman is the Early and Middle Albian species from which *A. (P.) chapmani* Cushman arose during the ?latest Middle or Late Albian. During the late Late Albian *A. (A.)* cf. *obliqua* (d'Orbigny), *A. (P.) truncata* (Reuss), *A. (P.) minima* Vasilenko, *A. (P.) bochumensis* Frieg, *A. (P.) barnardi* nov. sp., *H. advena* (Cushman), *H. anglica* (Cushman) and *H. d'orbignyi* (Reuss) appear, probably radiating from *A. (P.) chapmani*. *A. (A.) presli* (Reuss) possibly arose from *A. (A.)* cf. *obliqua* in the Early Cenomanian. *Ataxophragmium* spp. probably arose from *Hagenowina* in the latest Late Albian or Early Cenomanian. The evolutionary lineage of *A. (S.) sabulosa* (Chapman) and *A. (S.) gaworbiedowae* nov. sp. is uncertain. The former is found in

*Contribution to the IUGS (IGCP) project 'Mid-Cretaceous Events', with national promotion from the Deutsche Forschungsgemeinschaft. Paper no. 5 of the working group at Münster, West Germany.

large numbers in the late Late Albian of England, whereas the latter, at present, is found only in northwestern Germany. Similarly *A. (H.) courta* (Marie), *A. (H.) elevata* (d'Orbigny) and *A. (H.) obesa* (Reuss) suddenly appear in the early Cenomanian.

Vialovella praefrankei nov. sp. is considered to belong to the subfamily Verneuilininae, not the Ataxophragmiinae as in the classification proposed by Voloshina (1972). The separate evolutionary development of *Crenaverneuilina intermedia* (ten Dam) from *V. praefrankei* during the latest Albian is outlined, as originally suggested by Price (1977). *C. mariae* (Carter and Hart 1977) also probably arose from *C. intermedia* during the middle Cenomanian.

Introduction

The following chapter represents a detailed study of the Ataxophragmiidae, particularly the arenobuliminids, in Albian and Cenomanian sediments throughout northwestern Europe. Figure 2.1 broadly outlines the geographical positions of both surface and subsurface sections used in this study from southern England, northern and central France, the Netherlands and northwestern Germany. The paper is a synthesis of, with subsequent amendments to, the works of Barnard and Banner (1981), Carter and Hart (1977), Frieg (1979), Gawor-Biedowa (1969) and Price (1977). It is also based largely on the works of Voloshina (1965, 1972), utilising Balahakmatova's (1972) classification of the Ataxophragmiidae. Voloshina has illustrated and described the very diverse arenobuliminids of the USSR. Her subgeneric classification is of great value in interpretation of their evolutionary development. However, the various generations of arenobuliminid species as described by Gawor-Biedowa (1969) from Poland, and to a lesser extent by Barnard and Banner (1981) from England, are not considered.

It would appear that the arenobuliminids reach their acme and maximum diversification in the European temperate realm during the latest Albian and early Cenomanian. Extremely diverse faunas occur in Poland and the western USSR. Although Haig (1979) correctly noted that records of this genus are to date confined to the northern hemisphere, he erroneously stated that it is limited to Europe alone. *Arenobulimina* has been found in the northwestern Atlantic (Hart 1976), in the western interior plains and the western coast of Canada (McNeil & Caldwell 1981), in the Gulf Coast of the USA (Cushman 1946) and in Alaska (Tappan 1962). Specimens similar to those of Tappan (1962) have also been found in the Mackenzie Delta and Beaufort Sea, Northwest Territories, Canada. Therefore, although the genus occurs in truly boreal environments, its maximum diversity is found in the temperate realm probably within mid-Cretaceous palaeolatitudes of 35–45°N. A marked decrease in species diversity is seen into southern Europe and the Tethyan province, judging by specimens illustrated by Moullade (1966), Magniez-Jannin (1975) and Robaszynski and Magniez-

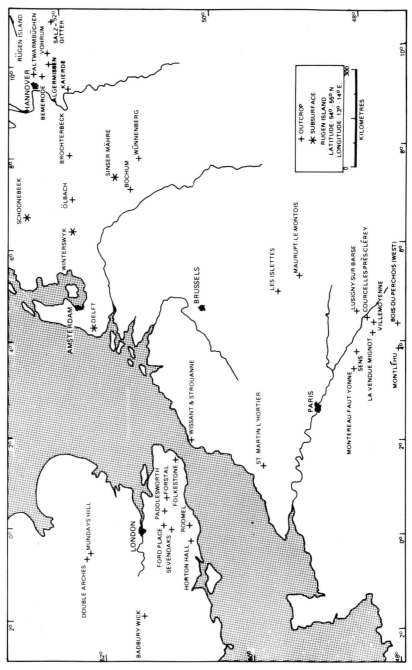

Figure 2.1 Albian and Cenomanian sampled localities within northwestern Europe.

Jannin (1980) from France. Macrofaunal zones and subzones for northwestern Europe referred to in the text are based on Owen (1975) for the Albian and Kauffman (1979) for the Cenomanian.

Systematics

The following classification, adopted from Balahakmatova (1972), is followed here:

Family: Ataxophragmiidae Schwager 1877, emended Balahakmatova 1972
Subfamilies: Verneuilininae Cushman 1911
Dorothiinae Balahakmatova 1972
Ataxophragmiinae Schwager 1877

Only species from the subfamilies Verneuilininae and Ataxophragmiinae are considered in this discussion. A full list of synonyms for each species is given, so as to clarify the taxa which the authors consider synonymous or otherwise with those described and illustrated.

Family: Ataxophragmiidae

Description: Test free, coiling plano-, trocho- or streptospiral in proximal portion with four or more chambers per whorl in the Ataxophragmiinae and Dorothiinae; triserial in the Verneuilininae. In all three subfamilies the distal stage can become tri-, bi- or uniserial: aperture in morphologically less complex genera interiomarginal, sometimes arcuate or extending with an areal portion. In more complex genera, the aperture becomes circular or loop-shaped, terminal and more complicated. The wall is composed of predominantly quartz or calcite grains, mostly with calcareous cement. Chamber interiors range from simple to infoldings of walls and development of radiate internal partitions, buttresses and chamberlets. The reader is also referred to Barnard and Banner (1981) for a detailed discussion of this family.

Subfamily: Verneuilininae

This subfamily is defined by both Loeblich and Tappan (1964) and Balahakmatova (1972) as possessing an initial triserial stage, sometimes becoming distally biserial or uniserial, with the number of chambers per whorl tending to decrease with growth. The aperture is simple.

The only species described in this subfamily is *Vialovella praefrankei nov. sp.* Its evolutionary relationship with *Crenaverneuilina intermedia* (ten Dam), *C. frankei* (Cushman) and *C. mariae* (Carter and Hart) is discussed below.

Subfamily: Dorothiinae

The new subfamily proposed by Balahakmatova (1972) includes some genera placed in the subfamilies Globotextulariinae, Valvulininae and Ataxophragmiinae by Loeblich and Tappan (1964). The reader is referred to Balakhmatova (1972) and Barnard and Banner (1981) for lists of genera incorporated into this subfamily.

Subfamily: Ataxophragmiinae

Balahakmatova (1972) noted the following developments within the Ataxophragmiinae: (a) reduction of the number of chambers per whorl from the morphologically simple to more complex forms, or from geologically older to younger species; (b) development from interiomarginal apertures to terminal and more complicated ones; (c) increasing complication of the chamber interior from simple to radiate pillar or partition structure through time. A fourth criterion should be added, namely the development of a more complex coiling pattern from trochospiral in *Arenobulimina* to streptospiral in *Ataxophragmium*. The latter combines streptospiral coiling with internal chamber division; this appears to be an evolutionary progression in the group.

The above four progressive developments within the arenobuliminids led to their subgeneric divisions and evolutionary interpretation. In the case of (c), the sudden development of internal partitions is considered here of generic importance, thus giving rise to *Hagenowina* Loeblich and Tappan 1961 and *Arenobulimina (Voloshinoides)* Barnard and Banner 1981. Barnard and Banner (1981) recognise a twofold subgeneric division of *Arenobulimina* into internally non-partitioned (simple) forms termed *Arenobulimina (Arenobulimina)* and simple to complex partitioned forms termed *Arenobulimina (Voloshinoides)*. In addition, they retain the subgenus *Hagenowina* Loeblich and Tappan 1961, differentiating it from *Voloshinoides* on the basis that the former has its planes of intercameral sutures virtually or wholly parallel to the axis of coiling. Similarly, they distinguish the genera *Hagenowella* Cushman 1933 and *Arenobulimina* Cushman 1927. We think that this feature is variable and is therefore not of generic importance.

The progression from stages one to four in the evolutionary development of the group suggests that the more natural classification of Balahakmatova is correct in that the development of radiate internal pillars does not warrant a different subfamily, but merely the progression from one genus to another, namely *Arenobulimina* to *Hagenowina*.

Subfamily: Verneuilininae Cushman 1911

Genus: Vialovella *Voloshina 1972*

Type-species: Vialovella oblonga *(Reuss 1865)*

Description. Test triserial and elongate in longitudinal section, outline subtriangulate with rounded edges in cross section; five to seven whorls in adult, chambers slightly inflated with distal portion uniserial and ultimate chamber often occupying one-third test length. Aperture variable, loop-shaped, crescentic or rounded, perpendicular to last suture, sometimes projecting. Wall surface rough, coarsely agglutinated, composed of quartz grains with calcareous cement; interior simple.

Discusssion. This genus is differentiated from *Arenobulimina sensu lato* Cushman 1927 by its clearly triserial and not trochospiral mode of growth. It differs from *Verneuilinoides* Loeblich and Tappan 1949 by its variable aperture and subtriangulate cross section, from *Verneuilina* d'Orbigny (in de la Sagra 1839) by the latter feature together with the uniserial last chamber in adults. The genus *Gaudryinella* Plummer 1931 also exhibits a triserial test in juvenile stages, but the adult form is clearly distinguished by its biserial or uniserial last chambers. In rare specimens of *Vialovella* the absence of the distinct subtriangulate cross-sectional development may make it superficially similar to *Plectina* Marsson 1878. This is particularly the case in the later Cretaceous, although it does occur in the Cenomanian, as can be seen by comparing *'A.' frankei* and *P. cenomana* as illustrated by Carter and Hart (1977, p. 18, Pl. 2, Figs 5 & 9, respectively).

Vialovella praefrankei *n.sp.*
See Plates 2.1a–c.

1972 *Vialovella frankei* (Cushman); Voloshina, p. 87, Pl. 8, Figs 5a&b
1977 *Arenobulimina frankei* Cushman; Price, p. 508, Pl. 59, Figs 5&9
1977 *Arenobulimina frankei* Cushman; Carter and Hart, p. 15, Pl. 1, Fig. 1, Pl. 2, Fig. 5

Description. Dimensions: maximum length 0.6 mm, maximum width 0.2 mm.

Test triserial and elongate in longitudinal section, gradually tapering proximally; subtriangulate in cross section with rounded edges. Chambers arranged triserially in three columns and sometimes slightly twisted in the vertical plane; in adult specimens each of the three vertical columns consisting of six chambers, uniserial last chamber of adult often enveloping underlying triserial portion and occupying one-quarter to one-third test length. Apertural face strongly convex; aperture loop-shaped, areal and only rarely connected to the last suture. Wall rough, coarsely agglutinated.

Identification. The characteristic subtriangulate cross-sectional outline of this species is its most distinctive feature. The only other species recorded to date belonging to this genus is *V. oblonga* (Reuss) (Voloshina 1972). Price (1977) has suggested that early forms now referable to *V. praefrankei* are rounded in cross section, only evolving into the characteristic subtriangulate forms during the latest Albian. Carter and Hart (1977) also illustrate a typical specimen.

The size of specimens studied confirms Carter and Hart's (1977) observation that they are only half the size of Cushman's (1936) type *Arenobulimina frankei (Crenaverneuilina frankei)*. Also, the aperture is not connected in all cases to the last suture.

Price (1977) suggested that *Crenaverneuilina intermedia* (ten Dam) arose from the species now named *V. praefrankei*, as opposed to the former arising from *Arenobulimina (Sabulina) sabulosa* (Chapman) as suggested by Carter and Hart (1977) and by Barnard and Banner (1981). Under the classification used herein, it would appear that the more logical morphological evolution is that of Price (1977) as both *Vialovella* and *Crenaverneuilina* should be assigned to the Verneuilininae.

The specimen of '*A*' *frankei* illustrated after Brotzen by Loeblich and Tappan (1964) is not related to the species described herein. This specimen is morphologically very different. The specimens of Gawor-Biedowa (1969) and Barnard and Banner (1981) have all the external morphological features of *frankei*, but in addition possess internal subdivisions of the chambers, a characteristic not only of *Crenaverneuilina* but also of *Hagenowina*. It is possible that a polyphyletic evolution occurs within some genera of the three subfamilies of the Ataxophragmiidae during the early Cenomanian, thus demonstrating their interrelationship to one another. None of the publications by Carter and Hart (1977), Frieg (1979), Price (1977) and Voloshina (1972) describe internal partitions within this species. However, the undisputable presence of internal partitions, as illustrated by Gawor-Biedowa (1969) and Barnard and Banner (1981), signifies another similar evolutionary development from simple *Vialovella* to partitioned *Crenaverneuilina*, paralleling that of *Arenobulimina* to *Hagenowina*.

The form figured by Cuvillier and Szakall (1949, Pl. 13, Fig. 10) from Llandes, France, is *Plectina ruthenica* (Reuss) 1846.

Stratigraphical range. This extends from the Late Albian (*Callihoplites auritus* subzone) to the earliest Cenomanian. It is of note that within southern England, Carter and Hart (1977) show the disappearance of *frankei* at the Albian/Cenomanian contact, whereas Barnard and Banner (1981) record it in the early Cenomanian. Frieg (1979) records *frankei* from the Cenomanian of Bochum, northwestern Germany, extending its range into younger sediments and thus confirming Cushman's (1936) observation of its age at Hildesheim, now Mierczany, Poland. However, it is now probable that Cenomanian forms are all referable to the internally partitioned *Crenaverneuilina frankei* and the Late Albian forms to the ancestral, simple *Vialovella praefrankei*.

Areal distribution. The species is widespread throughout northern Europe (Price 1977), being found in southern England (Carter & Hart 1977) and the western USSR (Voloshina 1972). It is of note that it does not occur in the Paris Basin, France (Magniez-Jannin 1975), possibly suggesting that the more Tethyan influence further south had a limiting effect on its distribution.

Subfamily: Ataxophragmiinae Schwager 1877, emended Balahakmatova 1972

Description. Test trochospiral to streptospiral, some genera low spired and involute; four to five chambers in last whorl sometimes decreasing to two or three in adults. Aperture variable from loop-shaped or slit-like to often V-shaped in evolute genera, sometimes with a tooth developed. Presence of radiate internal partitions dividing chambers in some genera either in the form of marginal infoldings from the walls or as pillars extending from the base to the top of each chamber.

Discussion. The definition of the subfamily by Balahakmatova (1972) allows a more natural classification of the arenobuliminids *sensu lato* than that of Loeblich and Tappan (1964), as it includes within this subfamily the evolving lineages of *Arenobulimina*, *Hagenowina* and *Ataxophragmium* that respectively evolve from one to another during the Albian and Cenomanian, as discussed later and illustrated in Figure 2.2.

Genus: Arenobulimina *Cushman 1927*

Type-species: Bulimina presli *Reuss 1846*

pars 1840 *Bulimina* d'Orbigny; d'Orbigny, p. 39
pars 1846 *Bulimina* d'Orbigny; Reuss, p 37
    1927 *Arenobulimina* Cushman, p. 80
    1941 *Ataxogyroidina* Marie, p. 54
    1948 *Ataxophragmoides* Brotzen, p. 35
    1953 *Ataxogyroidina* Marie; Barnard and Banner, pp. 177, 206
non 1971 *Arenobulimina* Cushman; Fuchs, p. 13
    1981 *Arenobulimina (Arenobulimina)* Cushman; Barnard and Banner, pp. 389, 398

Description. Test trochospiral, three to six whorls; initially with four, then with three to five chambers per whorl, number of chambers per whorl constant in most cases but may sometimes decrease distally, all visible. Interior simple, wall agglutinated and composed of quartz and/or calcite grains with calcareous cement. Aperture slit-like, loop or V-shaped with a small lateral tooth sometimes developed – not a valvuline tooth of distal origin.

Discussion. Voloshina (1965) subdivided *Arenobulimina sensu lato* into the following subgenera: *Arenobulimina (Arenobulimina)*, *Arenobulimina*

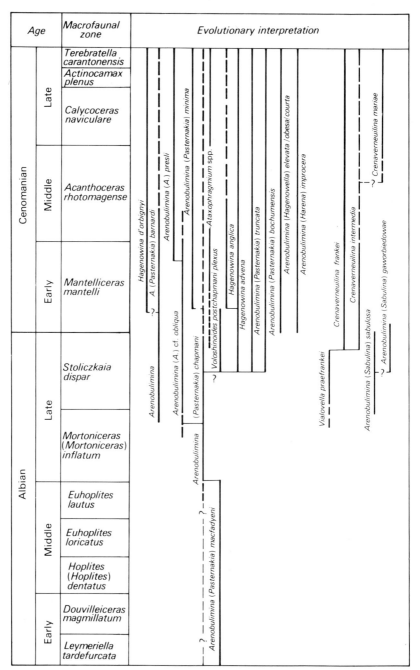

Figure 2.2 Evolutionary interpretation of Arenobuliminid subgenera, and their relationship to the genera *Vialovella*, *Crenaverneuilina*, *Hagenowina* and *Ataxophragmium* in the Albian and Cenomanian of northwestern Europe.

(Harena), *Arenobulimina (Novatrix)* and *Arenobulimina (Pasternakia)*. The additional subgenus *Arenobulimina (Sabulina)* is proposed herein. Voloshina also proposed the subgenus *Arenobulimina (Columnella)* for internally divided forms; this homonym was renamed *Arenobulimina (Voloshinoides)* by Barnard and Banner (1981) and it is herein put in synonymy with *Hagenowina* Loeblich and Tappan 1961. All the arenobuliminid subgenera are differentiated by their manner of trochospiral coiling, apertural development and chamber arrangement. In addition, the genus *Hagenowina* has a radiate internal partitioned or pillar structure within the chambers and is clearly an evolutionary development from the subgenus *Arenobulimina (Pasternakia)*. In the classification of Loeblich and Tappan (1964) *Hagenowina* is placed in the Ataxophragmiinae whereas *Arenobulimina* is assigned to the Globotextulariinae. In the classification of Balahakmatova (1972), followed herein, both are placed in the former subfamily. However, contrary to Voloshina (1965) and Barnard and Banner (1981), the presence of internal divisions and their stratigraphical importance is considered sufficient enough to merit generic status, as was originally proposed for *Hagenowina* by Loeblich and Tappan (1961).

Subgenus: Arenobulimina (Arenobulimina) *Cushman 1927*

    1927 *Arenobulimina* Cushman, p. 80
    1933 *Hagenowella* Cushman, p. 21
    1933 *Arenobulimina* Cushman; Galloway, p. 214
pars 1936 *Arenobulimina* Cushman; Cushman, p. 34
    1937a *Hagenowella* Cushman; Cushman, p. 172
pars 1941 *Arenobulimina* Cushman; Marie, p. 47
pars 1964 *Arenobulimina* Cushman; Loeblich and Tappan, p. 273
    1981 *Arenobulimina (Arenobulimina)* Cushman; Barnard and Banner, pp. 398–9

Description. Test trochospiral with three to four whorls, proximal portion slender and rapidly tapering, ultimate whorl inflated and convex; four to five chambers per whorl of rhomboidal outline in initial stages, later becoming elongated; sutures indistinct, septal sutures deeper than vertical ones. Aperture small, slit-like and narrow, parallel to the last suture with an areal process forming a V-shaped tooth. Wall of microcrystalline calcite often with small quartz grains; overall surface texture smooth.

Discussion. The type-species is *Arenobulimina (Arenobulimina) presli* (Reuss) 1846. This species exhibits the typical features of the subgenus, namely the agglutination of silt-sized quartz grains, smooth wall, very broad last whorl and V-shaped lateral apertural tooth. The precursor of this subgenus is *Arenobulimina (Pasternakia) chapmani* Cushman 1936, from which the oldest species of *Arenobulimina (Arenobulimina)*, namely *A. (A.) cf. obliqua* Price 1977, appears during the Late Albian.

Arenobulimina (Arenobulimina) presli *(Reuss 1846)*
See Plates 2.1d–g.

	1846	*Bulimina presli* Reuss, p. 38, Pl. 13, Fig. 72
non	1862	*Ataxophragmium preslii* (Reuss); Reuss, p. 31
non	1892	*Bulimina presli* Reuss; Chapman, p. 755, Pl. 12, Fig. 4
non	1897	*Bulimina preslii* Reuss; Perner, p. 17, Pl. 7, Fig. 15
non	1899	*Bulimina preslii* Reuss; Egger, p. 52, Pl. 15, Fig. 56
	1928	*Bulimina preslii* Reuss; Franke, p. 156, Pl. 14, Fig. 15
	1934	*Arenobulimina preslii* (Reuss); Dain, p. 17, Pl. 1, Fig. 8
	1935	*Arenobulimina preslii* (Reuss); Keller, p. 544, Pl. 1, Figs 5&6
	1936	*Arenobulimina preslii* (Reuss); Grékoff, p. 498, Pl. 1, Fig. 9
	1937b	*Arenobulimina presli* (Reuss); Cushman, Pl. 4, Figs 7&8
non	1949	*Arenobulimina preslii* (Reuss); Cuvillier and Szakall, p. 27, Pl. 12, Fig. 4
?	1953	*Arenobulimina preslii* (Reuss); Hagn, p. 22, Pl. 2, Fig. 4
	1957	*Arenobulimina preslii* (Reuss); Mikhailova-Jovtheva, p. 104, Pl. 1, Figs. 7, 8
	1957	*Ataxophragmium puschi* (Reuss); Hofker, p. 44, Fig. 31k
non	1962	*Arenobulimina preslii* (Reuss); Bartenstein and Bettenstaedt, p. 290, Pl. 41, Fig. 5
non	1964	*Arenobulimina preslii* (Reuss); Groiss, p. 10, Pl. 3, Figs 1&3
	1964	*Arenobulimina preslii* (Reuss); Durovskaja *et al.*, p. 204, Pl. 33, Figs 1–3
non	1967	*Arenobulimina preslii* (Reuss); Fuchs, p. 272, Pl. 4, Fig. 3
	1972	*Arenobulimina (Arenobulimina) preslii* (Reuss); Voloshina, pp. 59–60, Pl. 1, Figs 2&3
non	1974	*Arenobulimina preslii* (Reuss); Hercogova, p. 78, Pl. 3, Fig. 2
	1981	*Arenobulimina (Arenobulimina) presli* (Reuss); Barnard and Banner, Pl. 8, Figs 1&3

Description. Dimensions: maximum length 0.4 mm, maximum width 0.3 mm.

Test small, comprising three to four slowly tapering whorls, ultimate whorl forming two-thirds of test length; four to five chambers per whorl with four and a half to five in the ultimate whorl; chambers slender and elongated in outline, sutures narrow, indistinct; spiral sutures at an angle of 10–20°, septal sutures 35–50°. Aperture hook-shaped, interiomarginal–areal. Wall finely agglutinated, often from calcareous particles with much cement.

Identification. The specimens illustrated by Chapman (1892), Reuss (1862), Bartenstein and Bettenstaedt (1962) and Fuchs (1967) are all referable to *Arenobulimina (Pasternakia) chapmani*, discussed later. It is probable that Chapman (1892), following Reuss (1862), did not differentiate species (or subgenera) on apertural type and therefore placed all short, cuneiform arenobuliminids in the *presli* group. However, Cushman (1936) erected the new species *A. chapmani* for some Albian forms; in his description he does not mention the V-shaped aperture characteristic of *A. (A.) presli*, and the latter remains restricted to the post-Albian. In the specimen illustrated by Reuss (1846), the aperture appears semicircular, as in the lectotype chosen by Barnard and Banner (1981) from among the Reuss syntypes redrawn by Cushman (1937b). However, *A. (A.) presli* shows an evolutionary development of the aperture throughout the Late Cretaceous. The earliest specimens from the late early Cenomanian, *Mantelliceras dixoni* subzone, exhibit an interiomarginal process at the base of the areal, slit-like apertural opening (see Cushman 1937a, Pl. 4, Fig. 8a). This process increases in size laterally through Cenomanian time, leading to V-shaped apertures from the late Cenomanian onwards. Specimens from Campanian strata show a slight curvature in the areal part of the aperture, and at this level appear similar to *A. (A.) obliqua* (d'Orbigny 1840). However, they can be easily distinguished from this species by their more pointed conical form and large last whorl (see also Barnard & Banner 1981, Pl. 8, Figs 1–3).

Stratigraphical range. Early Cenomanian (*Mantelliceras dixoni* subzone) to Maastrichtian.

Areal distribution. The species is ubiquitous throughout northern Europe, including the western USSR. Initially recorded by Reuss (1846) from the Turonian Pläner marls of Czechoslovakia, it has subsequently been found throughout northern Europe, as seen by reference to those authors cited in the synonyms. Carter and Hart (1977) also record, but do not illustrate, the species from southern England.

Arenobulimina (Arenobulimina) cf. obliqua *Price 1977*
See Plate 2.1h.

1977 *Arenobulimina* cf. *obliqua* (d'Orbigny); Price, p. 510, Pl. 59, Fig. 10

Description. Dimensions: maximum length 0.4 mm, maximum width 0.3 mm.

Test small, low trochospire with three gradually tapering whorls, ultimate whorl comprises up to two-thirds test surface; four and a half to five chambers per whorl, all chambers of greater height than width; sutures narrow, indistinct, mostly interiomarginal; younger specimens with small lateral hook. Wall finely agglutinated with calcite grains and much calcareous cement.

Identification. This, the earliest known species of the *Arenobulimina (Arenobulimina)* subgenus, first appears with very rare, atypical specimens in the Late Albian (*Callihoplites auritus* subzone). These forms have a rounded outline with a strongly overlapping last chamber. During its phylogenetic development the test becomes very high-spired and the aperture develops an areal, lateral hook. Its differentiation from *A. (A.) presli* is often difficult, particularly in the late early Cenomanian, *Mantelliceras dixoni* subzone, where the forms are intergradational. However, the latter is confined to the Cenomanian and younger strata, as discussed previously. It is possible that *A. (A.) cf. obliqua* is the predecessor of *A. (A.) presli* and probably is itself derived from *A. (P.) chapmani*, with which it initially appears to be intergradational. It is not given new species status herein as it may represent a transitional form between *A. (P.) chapmani* and *A. (A.) presli*. It resembles *A. (A.) pseudalbiana* Barnard and Banner 1981, from which it is distinguishable by its more slowly tapering initial end, convex apertural face and less obliquely set chambers.

Stratigraphical range. Late Albian (*Callihoplites auritus* subzone) to early Cenomanian (*Mantelliceras dixoni* subzone); ?late Cenomanian and/or Turonian.

Areal distribution. To date the form has been recognised throughout the Albian of northwestern Europe (Price 1977) and in the early Cenomanian at Bochum, northwestern Germany (Frieg 1979).

Subgenus: Arenobulimina (Hagenowella) *(Cushman 1933)*

pars 1840 *Globigerina* d'Orbigny; d'Orbigny, p. 34
pars 1842 *Valvulina* d'Orbigny; von Hagenow, p. 570
pars 1959 *Trochammina* Parker and Jones; Maslakova, p. 92
    1965 *A. (Novatrix)* Voloshina, p. 149

Description. Test trochospiral, spherical to elliptical, of two or three strongly involute whorls; ultimate whorl comprises up to three-quarters test surface. Four chambers per whorl, chambers strongly inflated; sutures depressed. Wall finely agglutinated, smooth; aperture areal to interiomarginal.

Discussion. Cushman (1933) erected the genus *Hagenowella*. As a type-species he chose *Valvulina gibbosa* d'Orbigny 1840. Marie (1941) stated that the holotype of *V. gibbosa* did not have the radiating internal partitions that had been supposed to characterise this genus. Since the International Code of Zoological Nomenclature (ICZN) does not allow the changing of a type-species, Loeblich and Tappan (1961) erected the new genus *Hagenowina* for species with internal partitions, placing *Hagenowella* in junior synonymy of *Arenobulimina* Cushman 1927. As their type-species they chose *Valvulina quadribullata* von Hagenow 1842 from the chalk of Rügen Island, north-west Germany.

As the genus *Hagenowella* Cushman 1933 (with its type-species *V. gibbosa*) is a junior synonym at generic level of *Arenobulimina* Cushman 1927, and as this genus corresponds in every detail of significance to the subgenus *A. (Novatrix)* Voloshina 1965, it should be regarded as of subgeneric rank and as the senior name for *A. (Novatrix)* Voloshina 1965.

This subgenus appears suddenly in the earliest Cenomanian. The oldest species are from the *Hypoturrilites carcitanensis* subzone (Bed 8 of Amedro *et al.* 1978) of Cap Blanc Nez, France. To date the subgenus has not been found in Albian strata.

Arenobulimina (Hagenowella) courta (*Marie 1941*)
See Plate 2.1k.

     1941 *Hagenowella courta* Marie; p. 43, Pl. 7, Fig. 68
     1953 *Hagenowella courta* Marie: Barnard and Banner, p. 202, Text Figs 60o–j
pars 1965 *Arenobulimina (Novatrix) courta* (Marie); Voloshina, p. 80
non 1975 *Arenobulimina (Novatrix) courta* (Marie); Voloshina, p. 398, Pl. 5, Fig. 4

Description. Dimensions: maximum length 0.3 mm, maximum width 0.35 mm.

Test small, depressed trochospiral; two to three whorls of four strongly embracing chambers, sutures depressed. Aperture crescentic-shaped, interiomarginal to areal.

Identification. Marie's (1941) original description of this small species illustrates specimens with a highly cemented wall. However, Voloshina (1972) includes in her definition forms with tests up to 0.6 mm in length with coarsely agglutinated walls, and whose apertures are partly Y- or loop-shaped. Only the smaller specimens with short, stout tests should probably be assigned to this species.

Stratigraphical range. Cenomanian to Maastrichtian.

Areal distribution. The species is ubiquitous throughout northern Europe.

Arenobulimina (Hagenowella) elevata *(d'Orbigny 1840)*
See Plates 2.1l and 2.2a&b

     1840 *Globigerina elevata* d'Orbigny, p. 34, Pl. 3, Figs. 15&16
     1842 *Valvulina quadribullata* von Hagenow, p. 570.
     1891 *Haplophragmium inflatum* Beissel and Holzapfel; p. 19, Pl. 4, Figs. 41–5
?1892 *Bulimina brevicona* Perner; p. 54, Pl. 3, Figs 1a&b
     1928 *Bulimina subsphaerica* Reuss; Franke, p. 161, Pl. 14, Fig. 24
     1935 *Trochammina borealis* Keller; p. 546, Pl. 2, Figs 1&2
     1937a *Arenobulimina subsphaerica* (Reuss); Cushman, p. 41, Pl. 4, Fig. 17
     1941 *Hagenowella elevata* (d'Orbigny); Marie, p. 42, Pl. 7, Figs 66&67
?1953 *Hagenowella elevata* (d'Orbigny); Barnard and Banner, p. 202, Pl. 8, Fig. 11
     1953 *Arenobulimina subsphaerica* (Reuss); Hagn, p. 22, Pl. 2, Fig. 11
pars 1957 *Ataxophragmium subsphaerica* (Reuss); Hofker, p. 45, Text Fig. 32
     1959 *Trochammina borealis* Keller; Maslakova, p. 92, Pl. 1, Fig. 3
     1961 *Hagenowina quadribullata* (Hagenow) Loeblich and Tappan, p. 241
     1972 *Arenobulimina (Novatrix) elevata* (d'Orbigny); Voloshina, p. 78, Pl. 15, Fig. 3; Pl. 6, Fig. 1; Pl. 21, Fig. 1

Description. Dimensions: maximum length 0.8 mm, maximum width 0.4 mm.

Test a high trochospire of two to four whorls, ultimate whorl comprises more than one-third test length; sutures depressed; chambers inflated, four chambers in ultimate whorl. Wall smooth, interior simple. Aperture interiomarginal with lateral process.

Identification. The aperture of this species is described as semicircular by Voloshina (1972), and interiomarginal to areal by Marie (1941). Interiomarginal apertures with areal, lateral processes are found in Westphalia, northwestern Germany, and in specimens from the type locality near Sens, France. The description (in a modern sense incomplete) of *Valvulina quadribullata* by von Hagenow (1842) matches wholly that of *A. (H.) elevata* and *A. (H.) obesa* (Reuss). As both latter species occur at the same level as *Valvulina quadribullata* von Hagenow (1842), that species may well be a junior synonym of *A. (H.) elevata* (d'Orbigny 1840). (For full discussion see description of the genus *Hagenowina*.)

It is of note that the areal part of the apertures of both *A. (H.) elevata* and *A. (H.) obesa* increases in size from the Cenomanian to Campanian.

Stratigraphical range. Cenomanian to Maastrichtian.

Areal distribution. The species is ubiquitous throughout northern Europe, being recorded by those authors as listed in synonymy.

Arenobulimina (Hagenowella) obesa (Reuss 1851)
See Plates 2.2c&d and 2.3i.

```
1851   Bulimina obesa Reuss, p. 40, Pl. 4, Fig. 12; Pl. 5, Fig. 1
1899   Bulimina obesa Reuss, Egger, p. 53, Pl. 24, Fig. 4
1928   Bulimina obesa Reuss; Franke, p. 161, Pl. 14, Fig. 25
1937a  Arenobulimina obesa (Reuss); Cushman, p. 43, Pl. 4, Figs 26&27
1953   Arenobulimina obesa (Reuss); Hagn, p. 20, Pl. 2, Fig. 10
1957   Ataxophragmium puschi (Reuss); Hofker, p. 44, Text Figs 31g,h&i
1972   Ataxophragmium ?obesum (Reuss): Voloshina, p. 111, Pl. 13, Fig. 2
```

Description. Dimensions: maximum length 1.2 mm, maximum width 1.5 mm.

Test a low trochospire of two whorls; ultimate whorl comprises nine-tenths of test surface, initial whorl small, four inflated chambers per whorl. Aperture areal, slightly curved with a short interiomarginal process. Wall smooth, finely agglutinated, with abundant calcareous cement.

Identification. *A. (H.) obesa* is distinguished from *A. (H.) courta* by being larger and more spherical, while the latter is more elongate in outline. In assemblages studied, *A. (H.) obesa* is the dominant species. It is probable that only the larger forms, with a diameter of 0.4 mm or greater, belong to this species. It is also possible that some of those specimens described by von Hagenow (1842) as *Valvulina quadribullata* belong here; however, due to his lack of illustrations or specimen measurements, it is not possible to resolve this problem.

Stratigraphical range. Cenomanian to Maastrichtian.

Areal distribution. The species is ubiquitous throughout the Cenomanian to Maastrichtian in northern Europe, apparently reaching its acme during the Campanian. In addition to those authors listed above, d'Orbigny (1840) recorded the species from the Paris Basin, France; von Hagenow (1842) noted its presence in the Campanian and Maastrichtian of Rügen Island, northwestern Germany, while Kalinin (1937) and Dain (1934) recorded it from the Russian Platform.

Subgenus: Arenobulimina (Harena) *Voloshina 1965.*

pars 1851 *Bulimina* Reuss, pp. 17–52
    1965 *Harena* Voloshina, p. 149

Description. Test small, of five or six slowly tapering whorls with approximately four chambers per whorl; septa oblique. Wall agglutinated of calcite and calcareous cement, surface rough. Aperture areal with semicircular to loop-shaped outline.

Discussion. This relatively rare subgenus is differentiated from *Arenobulimina (Pasternakia)* by its slowly tapering whorls that give a somewhat cylindrical outline to the test. Also in *A. (Harena)*, the chambers are a little more inflated and exhibit better ordination in rows than *A. (Pasternakia)*. The chamber inflation in *A. (Harena)* is never so strong as that of *A. (Hagenowella)*.

Arenobulimina (Harena) improcera *Voloshina 1972*
See Plate 2.1j.

?1935 *Arenobulimina* aff. *d'orbignyi* (Reuss); Keller, p. 528, Pl. 3, Figs 10&11
  1972 *Arenobulimina (Harena) improcera* Voloshina, p. 71, Pl. 4, Fig. 2

Description. Dimensions: maximum length 0.4 mm, maximum width 0.2 mm.
  Small trochospiral test with almost cylindrical-shaped proximal portion; test composed of four to five low whorls, consisting of four chambers per whorl, exhibiting rhomboidal or subquadrate shape. Wall very finely agglutinated of calcite grains and calcareous cement; septa slightly depressed. Aperture a semicircular opening at the inner margin of the last chamber.

Identification. The species is distinguished by its relatively compressed and less inflated form in comparison to species of *A. (Hagenowella)*, but with stouter outline than species of *A. (Pasternakia)*.

Stratigraphical range. Cenomanian. The species is commonly found in the early middle Cenomanian, *Turrilites costatus* subzone, of northwestern Germany.

Areal distribution. The species has been found in northwestern Germany (Frieg 1979) and the western USSR (Voloshina 1972).

Subgenus: Arenobulimina (Pasternakia) *Voloshina 1965.*

1965 *Arenobulimina (Pasternakia)* Voloshina, p. 149

Description. Test a high trochospire, mostly of four to five whorls, with three to five chambers per whorl; sutures distinct. Wall of microcrystalline calcite with some quartz. Aperture semi-oval to slit-like and parallel to the last suture.

Arenobulimina (Pasternakia) barnardi *nov. sp.*
See Plate 2.2e.

1977 *Arenobulimina truncata* (Reuss); Price, p. 510, Pl. 59, Fig. 12
1977 *Arenobulimina* cf. truncata (Reuss); Carter and Hart, p. 11, Text Fig. 3

Description. Dimensions: maximum length 1.0 mm, maximum width 0.4 mm.

Test very high trochospire of three to four chambers per whorl, ultimate whorl always of three chambers; test gradually tapering proximally. Wall coarsely agglutinated. Aperture loop-shaped, apertural face horizontally truncated.

The patronym honours Professor Tom Barnard for his work on the Ataxophragmiidae.

Holotype. 'Arenobulimina truncata (Reuss)', Price 1977. Collections of Geology Department (Postgraduate Unit of Micropalaeontology), University College, London, England. UCL no. 152.

Paratypes. Two duplicate specimens of above in same collection, not bearing UCL numbers.

Identification. The three major differences between *A. (P.) barnardi* and *A. (P.) bochumensis* Frieg, described below, are the former's horizontally (not obliquely) truncated apertural face, a coarsely agglutinated rather than smooth wall, and the gradually tapering and distally wider test, as compared to the elongate, narrow outline of the latter species. Internal partitions are absent in *A. (P.) barnardi*; however, its remarkably similar morphological outline to *Hagenowina d'orbignyi* (Reuss 1846), particularly in their truncated apertural faces, leads to the suggestion, as discussed below, that the former gave rise to the latter.

Stratigraphical range. Late Albian to Turonian. It is fairly common from the Late Albian to early middle Cenomanian, after which its occurrence is more sporadic.

Areal distribution. The species is ubiquitous throughout the Late Albian and earlier Cenomanian of northwestern Europe.

Arenobulimina (Pasternakia) bochumensis *Frieg 1980*
See Plates 2.2f,g.

> ?1977 *Arenobulimina frankei* Cushman; Price, p. 508, Pl. 59, Fig. 6
> 1980 *Arenobulimina (Pasternakia) bochumensis* Frieg, Pl. 2, Figs 1–3
> 1981 *Arenobulimina (Arenobulimina) macfadyeni elongata* Barnard and Banner, pp. 403, 404, Pl. 2, Fig. 7; Pl. 6, Figs 2–4

Description. Dimensions: maximum length 0.9 mm, maximum breadth 0.3 mm.

Test a very high trochospire, normally of four whorls with four chambers per whorl, except in ultimate whorl where number of chambers may be reduced to three; apertural face obliquely truncated with an areal loop-shaped aperture.

Identification. This species is characterised by a very high trochospiral test such that most specimens are elongate and relatively narrow in longitudinal outline. Coiling is regular with four chambers per whorl, with three chambers often developed in the ultimate whorl. The test wall is predominantly finely agglutinated and smooth, often with a 'polished' appearance. The loop-shaped aperture is areal and in specimens with three chambers there is a tendency for it to become terminal. The species first appears within the Late Albian, *Mortoniceras (Mortoniceras) rostratum* subzone, probably arising from *A. (P.) chapmani*, discussed below.

This species closely resembles, and is probably synonymous with *A. (A.) macfadyeni elongata* Barnard and Banner (1981). However, as *A. (P.) macfadyeni* s.s. may not range above the Middle Albian (see later discussion), its evolutionary link is tenuous.

Stratigraphical range. Late Albian (*Mortoniceras (Mortoniceras) rostratum* subzone) to the late Cenomanian (?Turonian).

Areal distribution. The species to date has only been found in northwestern Germany, northern France and possibly southern England.

Arenobulimina (Pasternakia) chapmani *Cushman 1936*
See Plates 2.2h,i,j,m.

> 1892 *Bulimina preslii* Reuss; Chapman, p. 755, Pl. 12, Fig. 4
> 1936 *Arenobulimina chapmani* Cushman, p. 26, Pl. 4, Fig. 7
> 1937a *Arenobulimina chapmani* Cushman; Cushman, p. 36, Pl. 3, Figs 27&28
> 1947 *Arenobulimina chapmani* Cushman; Grékoff, p. 493, Pl. 1, Fig. 1
> ?1949 *Arenobulimina jurassica* Cushman and Glazewski, p. 8, Pl. 2, Figs 17&18

1950 *Arenobulimina chapmani* Cushman; ten Dam, p. 14
1955 *Arenobulimina preslii* (Reuss); Bettenstaedt and Wicher, p. 503, Pl. 4, Fig. 29
pars 1962 *Arenobulimina preslii* (Reuss); Bartenstein and Bettenstaedt, Pl. 4, Fig. 29
1965 *Arenobulimina chapmani* Cushman; Neagu, p. 10, Pl. 2, Fig. 9
1967 *Hagenowella chapmani* (Cushman); Kaptarenko-Chernousova, p. 25, Pl. 1, Fig. 5
1967 *Arenobulimina preslii* (Reuss); Fuchs, p. 272, Pl. 4, Fig. 3
?1969 *Arenobulimina chapmani* Cushman; Gawor-Biedowa, pp. 81–4, Pl. 5, Figs 1a,b, 2; Pl. 7, Figs 1a,b, 2; Text Figs 3&4
1975 *Arenobulimina chapmani* Cushman; Magniez-Jannin, p. 79, Pl. 7, Figs 9–18
1977 *Arenobulimina chapmani* Cushman; Price, p. 508, Pl. 59, Fig. 4
1977 *Arenobulimina chapmani* Cushman; Carter and Hart, p. 15, Pl. 1, Fig. 4
1981 *Arenobulimina (Arenobulimina) chapmani* Cushman; Barnard and Banner, p. 404, Pl. 4, Fig. 1; Pl. 7, Fig. 1

Description. Dimensions: maximum length 1.0 mm, maximum width 0.65 mm.

Trochospiral test, proximal end pointed, rapidly widening distally; longitudinal outline triangular, rounded in cross section. Test in the adult consists of five whorls with four chambers per whorl except the last which has five chambers, the final whorl comprises more than half length of test; sutures between chambers clearly visible and depressed. Aperture a loop-shaped opening in trough-like depression running perpendicularly from top of test to internal edge of ultimate chamber, surrounded by a narrow lip. Surface of the test rough, being quartz with large amounts of calcareous cement.

Identification. The most characteristic features of *A. (P.) chapmani* are its rapidly tapering test and large ultimate chamber. It is easily distinguished from its predecessor *A. (P.) macfadyeni* Cushman 1936 by being larger, more inflated and coarser walled.

According to Magniez-Jannin (1975), *A. (P.) chapmani* becomes shorter and broader during the Late Albian, finally developing radiate internal partitions or pillars during the latest Albian, late *Stoliczkaia dispar* zone. Walters (1958) also made this observation on specimens from the Folkestone, England, sequence. The description of *A. (P.) chapmani* by Gawor-Biedowa (1969) describes the presence of internal partitions in the later chambers. Barnard and Banner (1981) have erected the taxa *Arenobulimina (Voloshinoides) postchapmani* and *A. (V.) postchapmani praecursor* in an attempt to establish the evolutionary development from *A. (P.) chapmani*. The present authors have grouped these forms into their '*Hagenowina postchapmani*' plexus. Both Price (1977) and Carter and Hart (1977) also recognise the transition from simple interior *A. (P.) chapmani* to internal partitioned '*A.*' *advena* across the Albian/Cenomanian contact in southern England. They retained, however, the same generic name for both groups. This feature is considered herein to merit generic status, namely that of *Hagenowina* Loeblich and Tappan. The exact affinity of *A. (P.) chapmani* to *Hagenowina* is discussed below under *Hagenowina advena* (Cushman).

The specimen described by Cushman and Glazewski (1949) from the Late Jurassic is probably a contaminant from overlying Albian strata.

Stratigraphical range. ?Middle Albian (?*Anahoplites intermedius* subzone) to the Late Albian (?Cenomanian). Only very rare specimens have been reported by Price (1977) from Middle Albian sediments of southern England, which are possible contaminants from younger strata. The sudden appearance of this species in large numbers characterises the Late Albian (post-*Dipoloceras cristatum* subzone). It is found commonly throughout the *Mortoniceras (Mortoniceras) inflatum* and *Stoliczkaia dispar* zones. Its very rare recorded occurrences in the Cenomanian can now probably be related to Barnard and Banner's (1981) *Voloshinoides postchapmani* and *V. postchapmani praecursor*.

Areal distribution. The species is ubiquitous during the Late Albian and Cenomanian throughout northwestern Europe (Price 1977).

Arenobulimina (Pasternakia) macfadyeni Cushman 1936
See Plates 2.2k,l,n.

1892 *Bulimina orbignyi* Reuss; Chapman, p. 754, Pl. 12, Fig. 2
1936 *Arenobulimina macfadyeni* Cushman, p. 26, Pl. 4, Fig. 6
1937a *Arenobulimina macfadyeni* Cushman; Cushman, p. 35, Pl. 4, Figs 13&14
1947 *Arenobulimina macfadyeni* Cushman; Grékoff, p. 497, Pl. 1, Fig. 4
1965 *Arenobulimina macfadyeni* Cushman; Neagu, p. 10, Pl. 2, Figs 7&8
1975 *Arenobulimina macfadyeni* Cushman; Magniez-Jannin, p. 78, Pl. 7, Figs 1–8
1977 *Arenobulimina macfadyeni* Cushman; Price, p. 510, Pl. 59, Figs 7&8
1977 *Arenobulimina macfadyeni* Cushman; Carter and Hart, pp. 15–16, Pl. 2, Fig. 2
1981 *Arenobulimina (Arenobulimina) macfadyeni* Cushman; Barnard and Banner, pp. 402–3, Pl. 2, Figs 4–6; Pl. 6, Figs 1&5–7

Description. Dimensions: maximum length 0.6 mm, maximum width 0.35 mm.

Small trochospiral test, sometimes short but most commonly elongate; proximal end pointed, gradually widening distally; longitudinal outline triangular, rounded in cross section. Adult test consists of four to five whorls with three (rarely four) chambers per adult; ultimate whorl comprises approximately one-third of test length; sutures between chambers clearly visible and depressed. Aperture a loop-shaped, sometimes rounded, opening, running perpendicularly from top of test to internal edge of ultimate chamber, usually surrounded by a narrow lip (not always visible). Surface of test finely agglutinated with large amount of calcareous cement.

Identification. This species differs from its immediate successor *A. (P.) chapmani* in being more slender and much more finely agglutinated. It has the finest and most even textured agglutinated test of all the Albian arenobuliminids. Extremely small specimens occur in the Early Albian which do not exhibit the more typical elongate outline of the Middle Albian forms (see Pl. 2.2k, Early Albian; Pl. 2.2l&n, Middle Albian specimens). *A. (P.) macfadyeni* is predominantly of Early and Middle Albian age, being one of the oldest arenobuliminid species (with the exception of *Arenobulimina flandrini* Neagu 1972 (*non* Moullade 1966) found in the Hauterivian of Romania). Barnard, in Barnard and Banner (1981), records rare specimens of *A. macfadyeni* from the Aptian Farringdon Sponge Gravel (Berkshire, southern England).

Stratigraphical range. ?Aptian and the Early to Middle Albian (?early Late Albian) (*Otohoplites raulinianus* to *Anahoplites daviesi* subzones). Numerous authors including Magniez-Jannin (1975), Price (1977) and Carter and Hart (1977) recognise the stratigraphical importance of this Early to Middle Albian species. Although found sporadically in the early Late Albian, namely *Dipoloceras cristatum*, *Hysteroceras orbignyi* and *Hysteroceras varicosum* subzones, it is probably reworked at these horizons following the *D. cristatum* erosive phase. Often these Late Albian specimens are etched.

The larger forms termed *A. (A.) macfadyeni sensu lato* described by Barnard and Banner (1981) from the Late Albian and early Cenomanian of England have not been found by the present authors. Their Cenomanian subspecies *A. (A.) macfadyeni elongata* is placed here in synonymy of *A. (P.) bochumensis*.

Areal distribution. The species is ubiquitous in the Early and Middle Albian throughout northwestern Europe (Price 1977).

Arenobulimina (Pasternakia) minima *Vasilenko 1961*
See Plate 2.3a.

    1961 *Arenobulimina minima* Vasilenko, p. 19, Pl. 3, Fig. 2
    1972 *Arenobulimina (Pasternakia) minima* Vasilenko; Voloshina, p. 67, Pl. 3, Fig. 5

Description. Dimensions: maximum length 0.6 mm, maximum width 0.4 mm.

Test small, trochospiral, elongate, pointed proximally and conical; four whorls, ultimate one comprising one-third of test length; four nearly cubic-shaped chambers per whorl; sutures distinctly translucent, not depressed. Wall very finely agglutinated with much calcareous cement. Aperture an areal slit.

Identification. The short, stout outline of this species makes it fairly distinctive.

Stratigraphical range. Cenomanian to Campanian.

Areal distribution. The species, although probably ubiquitous throughout northwestern Europe, is fairly rare in any given assemblage. It has been recorded from northwestern Germany (Frieg 1979) and the western USSR (Vasilenko 1961, Voloshina 1972).

Arenobulimina (Pasternakia) truncata *(Reuss 1846)*
See Plates 2.3b,c.

    1846   *Bulimina truncata* Reuss; p. 37, Pl. 8, Fig. 73
    1928   *Bulimina truncata* Reuss; Franke, p. 158, Pl. 14, Fig. 17
    1937a *Arenobulimina truncata* (Reuss); Cushman, p. 40, Pl. 4, Figs 15&16
    1947   *Arenobulimina truncata* (Reuss); Grékoff, p. 499, Pl. 1, Fig. 10
non 1977   *Arenobulimina truncata* (Reuss); Price, p. 510, Pl. 59, Fig. 12

Description. Dimensions: maximum length 0.6 mm, maximum width 0.3 mm.

Test trochospiral, elongate, wedge-shaped or sometimes pointed to conical in outline; four whorls with three chambers per adult whorl, initially, and four per whorl distally; ultimate whorl comprises one-quarter test-length; chambers equi-dimensional, being as high as wide, with distinctly translucent sutures; apertural face even and truncated at an angle of approximately 45°. Aperture with an indistinct short tooth. Wall smooth, finely agglutinated with much calcareous cement.

Identification. The specimen illustrated by Price (1977) is not *A. (P.) truncata* (Reuss) 1846, as those of Reuss have an obliquely (not horizontally) truncated apertural face and smaller proximal angle. Price's (1977) figure (and also that referred to by Carter and Hart (1977) as *A.* cf. *truncata*) has been designated a new species herein, namely *A. (P.) barnardi*, as discussed previously.

Stratigraphical range. ?Late Albian to Turonian. This species is most characteristic of the silty marls and chalks of the Cenomanian.

Areal distribution. The species has been recorded to date only in France and Germany by the authors cited above.

Subgenus: Arenobulimina (Sabulina) *nov. subgen.*
Type-species: Arenobulimina (Sabulina) sabulosa *Cushman 1937*

pars 1892 *Bulimina* Chapman, p. 755
pars 1937a *Arenobulimina* Cushman, p. 80
pars 1976 and 1977 *Arenobulimina* Price, p. 510
pars 1977 *Arenobulimina* Carter and Hart, p. 14

Description. Test trochospiral, quadriserial, of four to five whorls; apertural face almost horizontally truncated; chambers exhibit a characteristic divergence away from the spiral axis of coiling near the spiral and septal sutures; interior simple. Wall very coarsely agglutinated with little calcareous cement. Aperture spatuliform, interiomarginal to areal.

Discussion. In the evolutionary interpretation following, it is suggested that the phylogenetic development of *Arenobulimina sensu lato* resulted in the evolution of the genus *Hagenowina* through *A. (Pasternakia)* during the Late Albian, latest *Stoliczkaia dispar* zone. However, the subgenus *A. (Sabulina)* shows little relationship to *A. (Pasternakia)*, a fact which led Price (1977) to question whether the former arose from the latter, as there appears to be no transition from the often smoother walled *A. (Pasternakia)* to the extremely coarse-walled *A. (Sabulina)*. Also, the latter shows no relationship to *Hagenowina*.

The agglutinated material (mostly quartz grains) in the wall is too coarse to assign it to any of the other arenobuliminid subgenera which all appear to develop from *A. (Pasternakia)* and have smooth to only fairly coarse walls with much calcareous cement. Also, *A. (Sabulina)* is quadriserial throughout and subquadrate in cross section, unlike the other arenobuliminids. Like all *Arenobulimina* subgenera, it is clearly distinguished from *Hagenowina*, by its lack of internal partitions.

Although the subgenus *Sabulina* has strong affinities with the other subgenera of *Arenobulimina*, it also has some similarities to both *Vialovella* and *Crenaverneuilina*, particularly its very coarse agglutination.

Arenobulimina (Sabulina) gaworbiedowae *nov. sp.*
See Plate 2.3d.

Holotype. '*Arenobulimina sabulosa* (Chapman)' Frieg 1979. Geological Collections, University of Münster, West Germany, only the specimen from Sinser Mähre, not those from Folkestone.

Paratypes. Personal collection of C. Frieg from Brochterbeck, northwestern Germany, Late Albian, Flammenmergel facies.

Description. Dimensions: maximum length 1.0 mm, maximum width 0.6 mm.

Test trochospiral, quadriserial; ultimate whorl triserial; rounded in cross section; later chambers slightly inflated, weakly projecting at sutures; ultimate whorl comprises half of test length. Apertural face strongly convex, aperture spatuliform, surrounded by a lip.

The patronym honours Dr Eugenia Gawor-Biedowa for her excellent contribution to the study of arenobuliminids in the Late Albian and Cenomanian of Poland.

Identification. *A. (S.) gaworbiedowae* is distinguished from *A. (S.) sabulosa* by its last-formed triserial whorl, as opposed to the large, single overlapping ultimate chamber of the latter. The ultimate whorl of *A. (S.) gaworbiedowae* has more inflated globose chambers than those developed in the distal part of *A. (S.) sabulosa*, together with a rounded, not subquadrate cross-sectional outline. The apertural face of the new species is also more convex and the aperture is spatuliform.

Stratigraphical range. Late Albian.

Areal distribution. The species has been recorded to date only from north-western Germany (Frieg 1979).

Arenobulimina (Sabulina) sabulosa *(Chapman 1892)*
See Plates 2.3e,f.

      1892   *Bulimina preslii* Reuss var. *sabulosa* Chapman, p. 755, Pl. 12, Fig. 5
     1937a   *Arenobulimina sabulosa* (Chapman); Cushman, p. 36, Pl. 3, Figs 29&30
non 1969   *Arenobulimina sabulosa* (Chapman); Gawor-Biedowa, pp. 77–80, Pl. 5, Fig. 3; Pl. 7, Figs 3a&b; Text Fig. 1
non 1972   *Arenobulimina (Columnella) sabulosa* (Chapman); Voloshina, pp. 75–6, Pl. 4, Fig. 6
   ?1972   *Arenobulimina convexocamerata* Voloshina, pp. 84–5, Pl. 7, Fig. 1a–c
   ?1972   *Arenobulimina vialovi* Voloshina, p. 85, Pl. 7, Fig. 2a–c
    1976   *Arenobulimina sabulosa* (Chapman); Price, p. 634, Pl. 1, Figs 5&6
    1977   *Arenobulimina sabulosa* (Chapman); Price, p. 510, Pl. 59, Fig. 11
    1977   *Arenobulimina sabulosa* (Chapman); Carter and Hart, p. 16, Pl. 1, Fig. 2
    1981   *Arenobulimina (Arenobulimina) sabulosa* (Chapman); Barnard and Banner, pp. 400–2, Pl. 1, Figs 1&2; Pl. 5, Fig 1

Description. Dimensions: maximum length 1.0 mm, maximum width 0.6 mm.

Short, broad trochospiral test, subtriangulate in longitudinal outline, gradually tapering distally with subparallel sides; subquadrate in cross section; test in adult consisting of four to five quadriserial whorls. Last chamber overlapping previous ones and extending completely across upper surface of test; sutures between chambers depressed but often obscured in very coarsely agglutinated specimens. Aperture loop-shaped, with lip sometimes visible. Wall very coarse with large quartz grains, little calcareous cement.

Identification. This species is the most coarsely agglutinated Albian arenobuliminid. In addition, its subquadrate cross-sectional outline make it easily identifiable.

Stratigraphical range. Late Albian (*Callihoplites auritus* subzone to latest *Stoliczkaia dispar* zone).

Areal distribution. Price (1976, 1977) concludes that this species is confined to southern England and the southern North Sea, as none of the publications of ten Dam (1950) or Fuchs (1967) in the Netherlands, Magniez-Jannin (1975) in the Paris Basin, France, or Neagu (1965) in Romania record the species. Its recorded occurrence in Poland by Gawor-Biedowa (1969) is re-interpreted herein as being *Crenaverneuilina intermedia* (ten Dam 1950); Dr Gawor-Biedowa kindly forwarded specimens of her '*A. sabulosa*' to the senior author (C.F.). Carter and Hart (1977) cite *A. sabulosa* occurring in the USSR; however, '*A. (Columnella) sabulosa*' as described by Voloshina (1972) from the western USSR has internal partitions typical of the genus *Voloshinoides*. Nevertheless, those specimens which Voloshina (1972) termed *A. convexocamerata* and *A. vialovi* are very similar to *A. (S.) sabulosa*. It is also of note that Hart (1976) does not record this species at Orphan Knoll, northwestern Atlantic.

Genus: Hagenowina *Loeblich and Tappan 1961*
Type-species: Valvulina quadribullata *von Hagenow 1842*

pars 1941 *Arenobulimina* Cushman; Marie, p. 43
pars 1941 *Hagenowella* Cushman; Marie, p. 41
non 1949 *Hagenowella* Cushman; Hofker, p. 431
     1953 *Arenobulimina* Cushman; Barnard, in Barnard and Banner, pp. 177, 206
     1961 *Hagenowella* Cushman; Vasilenko, p. 20
     1961 *Hagenowina* Loeblich and Tappan, p. 242
     1964 *Hagenowina* Lobelich and Tappan; Loeblich and Tappan, pp. C286–7
  ?1964 *Hagenowina* Loeblich and Tappan; Durovskaja *et al.*, p. 210
     1965 *Arenobulimina (Columnella)* Voloshina, p. 149 (*non* Levinson 1914)
pars 1969 *Arenobulimina* Cushman; Gawor-Biedowa, pp. 74–7
     1972 *Arenobulimina (Columnella)* Voloshina; Voloshina, p. 73
pars 1977 *Arenobulimina* Cushman; Price, pp. 510–14
pars 1977 *Arenobulimina* Cushman; Carter and Hart, pp. 14–17
     1981 *Arenobulimina (Voloshinoides)* Barnard and Banner, pp. 390, 398–400

Description. Test trochospiral, pointed proximally; cylindrical or ovoid in longitudinal outline; four to five chambers per whorl with three to five whorls. Walls of agglutinated quartz grains with varying amounts of calcareous cement, normally sparse; interior of adult chambers with secondary septae containing much quartz material, secondary septae form both vertical and horizontal chamber partitions or pillars appearing like radiating 'pigeon-holes' in test cross section. Aperture loop-shaped, semicircular, interiomarginal to areal.

Discussion. The genus was erected in 1961 by Loeblich and Tappan for internally subdivided arenobuliminids, as the type-species of *Hagenowella* Cushman 1933, namely *Valvulina gibbosa* d'Orbigny 1840, had been shown to have no internal partitions, placing *Hagenowella* in junior synonymy of *Arenobulimina* Cushman 1927. As type-species of the new genus, *Valvulina quadribullata* von Hagenow 1842 was chosen. As a representative of the type-species they took the specimen that had formerly been figured by Cushman (1933, Pl. 1, Figs 1–3) as *Hagenowella gibbosa*, and which was assumed by Cushman to have come from the Senonian of Sens, France. Loeblich and Tappan (1961) stated that Cushman's locality data were in error and that the specimen (Cushman collection 21213) had come from the Late Cretaceous (Senonian), Island of Rügen, Germany 'and should correctly be referred to *Valvulina quadribullata* von Hagenow'. Von Hagenow himself (1842) did not illustrate *V. quadribullata*, but merely described the species as trochospiral with thirteen chambers, four per whorl.

In two samples from the late Campanian to Maastrichtian of Rügen Island, kindly lent to the senior author (C.F.) by W. Koch, a few tests that match the description of *V. quadribullata* with thirteen chambers were found. Similarly, in three samples from the Campanian of Sens, France (type locality of '*Globigerina*' *elevata* d'Orbigny, type-species of *Novatrix* Voloshina) a few specimens were also found. These specimens compare in every detail with those from Rügen Island and von Hagenow's (1842) description of *V. quadribullata*. All the specimens are internally empty.

One now could suppose, as Voloshina (1972, p. 74) does, that *Hagenowina* Loeblich and Tappan 1961 must be invalidated for just the same reasons as *Hagenowella* Cushman 1933 (type-species being empty). But, *fidé* Voigt (1959) the Hagenow collection was burned in World War II in the Stettin Museum, so that the emptiness of the holotype of the type-species cannot be proved with the certainty that there is in the case of *Valvulina gibbosa*. Moreover, we must assume by article 70 of the ICZN rules that the original author of the genus was correct in his identification of the type-species of his new genus.

So, contrary to Volashina (1972), one has to conclude, if the Hagenow collection was burned (a fact that was not researched by the present authors), that the specimen Cushman Coll. 21213 should be the neotype of *Valvulina quadribullata* von Hagenow 1842. By this, the genus *Hagenowina* is valid and *Arenobulimina (Columnella)* Voloshina 1965, together with *Arenobulimina (Voloshinoides)* Barnard and Banner 1981, become for the authors of the present paper subjective synonyms of the former genus.

Barnard, in Barnard and Banner (1953), proposed that *Arenobulimina* be used to distinguish radiate internally partitioned forms, and *Ataxogyroidina* for simple interior forms. Their subsequent emendation of *Arenobulimina* (Barnard & Banner 1981) proposes the use of the subgenera *Arenobulimina (Voloshinoides)* and *Arenobulimina (Arenobulimina)* respectively for these different forms. Gawor-Biedowa (1969), Price (1977) and Carter and Hart (1977) did not recognise a generic difference between partitioned and simple (non-partitioned) forms, placing them all in *Arenobulimina*.

As stated earlier under discussion of the Ataxophragmiinae, the present authors considered the differentiation of both (genera) *Arenobulimina* from *Hagenowella* and (subgenera) *Hagenowina* from *Voloshinoides* as defined by Barnard and Banner (1981) by changes in inclination of the planes of intercameral sutures a variable feature, and not of generic or subgeneric importance. This major variance with the above work will, it is hoped, be resolved in the light of subsequent phylogenetic trends in the late Cretaceous by both Barnard and Banner in part 2 of their study on the Ataxophragmiidae (in preparation) together with Frieg and Jordan's studies in West Germany (in preparation).

Hagenowina advena *(Cushman 1936)*
See Plates 2.3g,h.

    1936 *Hagenowella advena* Cushman, p. 43, Pl. 6, Fig. 21
    1937a *Hagenowella advena* Cushman; Cushman, p. 174, Pl. 21, Figs 3&4
    1948 *Hagenowella advena* Cushman; Brotzen, p. 44, Pl. 1, Fig. 3
    1961 *Hagenowella chapmani* Cushman; Vasilenko, p. 22, Pl. 3, Fig. 3; Pl. 4, Figs 1–3
    1962 *Hagenowella advena* Cushman: Jeffries, Pl. 78, Fig. 13
    1964 *Hagenowella* aff. *advena* Cushman; Groiss, p. 12, Pl. 4, Figs 7&8
    1964 *Hagenowella pirum* Durovskaja *et al.*, p. 212, Pl. 24, Fig. 3
 ?1969 *Arenobulimina chapmani* Cushman; Gawor-Biedowa, pp. 81–4, Pl. 5, Figs 1a,b, 2; Pl. 7, Fig. 1a&b,l; Text Figs 3&4
    1969 *Arenobulimina advena* (Cushman): Gawor-Biedowa, pp. 86–90, Pl. 8, Figs 1–4; Text Fig. 78
    1975 *Arenobulimina* sp. Magniez-Jannin, p. 81, Pl. 7, Figs. 19–22
    1977 *Arenobulimina advena* (Cushman); Price, p. 508, Pl. 59, Fig. 3
    1977 *Arenobulimina advena* (Cushman); Carter and Hart, p. 14, Pl. 2, Fig. 4
    1981 *Arenobulimina (Voloshinoides) advena* (Cushman); Barnard and Banner, pp. 405–6, Pl. 4, Figs 6–8; Pl. 7, Figs 10–12

Description. Dimensions: maximum length 0.8 mm, maximum width 0.6 mm.

Test short, trochospiral, ovoid to elliptical in longitudinal outline, rounded in cross section; ultimate whorl has four chambers and comprises almost all test surface; apertural face obliquely truncated and flat. Sutures distinct, not depressed; radiate internal partitions or pillars developed within chambers. Wall coarsely agglutinated, coarsest near sutures. Aperture loop-shaped.

Identification. *H. advena* exhibits the typical internal structures associated with *Hagenowina*. It arose directly from *A. (P.) chapmani* during the late Late Albian. Barnard and Banner (1981) show three evolutionary lineages in which *Hagenowina* developed, namely *H. voloshinae praevoloshinae* and *H. voloshinae sensu stricto* developing from *A. (P.) macfadyeni*, while *H. postchapmani praecursor* and *H. postchapmani sensu stricto* together with *H. advena praeadvena* and *H. advena* developed from *A. (P.) chapmani*. Good illustrations of *H. advena* are to be found in Magniez-Jannin (1975) under '*Arenobulimina* sp.'. The major differences between *H. advena* and *A. (P.) chapmani* are the development of internal chamber partitions in the former and its more rounded to ovoid outline, as opposed to *A. (P.) chapmani's* distinctly pointed, rapidly tapering proximal portion and simple (non-partitioned) interior. Both Price (1977) and Carter and Hart (1977) recognise the transition from *A. (P.) chapmani* to *H. advena*, separating these species on outline and the presence or absence of internal divisions. *A. (P.) chapmani* is not supposed to possess internal divisions. However, Walters (1958) observed that specimens of a typically rapidly tapering *A. (P.) chapmani* in the Upper Gault Clay (Late Albian) at Folkestone, England possessed distinct internal divisions. Also, Gawor-Biedowa (1969) illustrates *A. (P.) chapmani* with well defined internal partitions, and these are referred by Barnard and Banner (1981) to *H. postchapmani praecursor*. She, like Price (1977), suggests that the internal structures are probably destroyed by recrystallisation of the test, although how agglutinated quartz grains can be recrystallised without trace is not clear. However, only those species possessing internal structure as described by Barnard and Banner (1981) and listed above should be assigned to the genus *Hagenowina*.

Barnard and Banner (1981) have shown the gradual development of internal divisions within the *A. (P.) chapmani* → *H. postchapmani* → *H. advena* evolutionary lineage in southern England. It would be of interest to know whether this same development occurs in the more diverse ataxophragmid faunas of Poland as described by Gawor-Biedowa (1969). Unfortunately, *A. (P.) macfadyeni* was not found here, as only Late Albian sediments were studied, although the *A. (P.) chapmani* lineage was well developed. In the western USSR, neither *A. (P.) chapmani* nor *A. (P.) macfadyeni* have been recorded by Voloshina (1972), although they would be expected to be present.

Stratigraphical range. Latest Albian (latest *Stoliczkaia dispar* zone) to the late Cenomanian.

Areal distribution. The species is ubiquitous throughout northern Europe, including the western USSR.

Hagenowina anglica *(Cushman 1936)*
See Plates 2.3i–m.

    1936 *Arenobulimina anglica* Cushman, p. 27, Pl. 4, Fig. 8
    1937a *Arenobulimina anglica* Cushman; Cushman, p. 37, Pl. 4, Figs 31–4
    1947 *Arenobulimina anglica* Cushman; Grékoff, p. 492, Pl. 1, Fig. 5
    1977 *Arenobulimina anglica* Cushman; Carter and Hart, p. 14, Pl. 2, Fig. 3
    1981 *Arenobulimina (Voloshinoides) anglica* Cushman; Barnard and Banner, pp. 407–8, Pl. 2, Fig 1–3; Pl. 6, Figs 12 & 13

Description. Dimensions: maximum length 1.4 mm, maximum width 1.0 mm.

Test trochospiral, elongate and conical, pointed proximal portion; ultimate whorl comprises about one-third test length; apertural face distinctly convex; four to three chambers per whorl. Wall coarsely agglutinated with coarser grains concentrated near the septa; radiate internal partitions distinct.

Identification. This species is distinguished from *H. advena* by its more pointed proximal portion, and from *H. d'orbignyi* (Reuss 1846) by a more stout, pointed, conical outline. *H. anglica* also has a tendency for the ultimate chamber to be uniserial. Carter and Hart (1977) note transitional forms between it and *A. (S.) sabulosa* in the latest Albian.

Stratigraphical range. Cenomanian to ?Maastrichtian.

Areal distribution. The species is ubiquitous throughout northern Europe.

Hagenowina d'orbignyi *(Reuss 1846)*
See Plates 2.3n,o.

     1846 *Bulimina d'orbignyi* Reuss; p. 38, Pl. 13, Fig. 74
     1862 *Ataxophragmium d'orbignyi* (Reuss); Reuss, p. 31
non 1892 *Bulimina d'orbignyi* Reuss; Chapman, p. 754, Pl. 12, Fig. 2
     1925 *Bulimina d'orbignyi* Reuss; Franke, p. 25, Pl. 2, Fig. 19
     1928 *Bulimina d'orbignyi* Reuss; Franke, p. 158, Pl. 14, Fig. 16
     1936 *Arenobulimina d'orbignyi* (Reuss); Brotzen, p. 42, Pl. 2, Fig. 9; Text Fig. 7
     1937a *Arenobulimina d'orbignyi* (Reuss); Cushman, p. 39, Pl. 4, Figs 9–12
?1941 *Arenobulimina pseudodorbignyi* Marie; p. 50, Pl. 4, Fig. 37
pars 1947 *Arenobulimina d'orbignyi* (Reuss); Grékoff, p. 497
     1953 *Arenobulimina d'orbignyi* (Reuss); Hagn, p. 21, Pl. 2, Fig. 7
     1957 *Arenobulimina d'orbignyi* (Reuss); Mikailova-Jovtheva, p. 105, Pl. 1, Figs 9–12
     1964 *Arenobulimina preslii* (Reuss); Groiss, p. 10, Pl. 3, Fig. 1; Text Fig. 3
     1964 *Arenobulimina d'orbignyi* (Reuss); Groiss, p. 11, Pl. 3, Fig. 2; Text Fig. 3
     1964 *Arenobulimina kempckei* Groiss; p. 12, Pl. 4, Fig. 12; Pl. 5, Fig. 15; Text Fig. 3
     1969 *Arenobulimina polonica* Gawor-Biedowa; pp. 90–3, Pl. 4, Fig. 3; Pl. 8, Figs 5–8; Text Fig. 9
non 1972 *Arenobulimina (Pasternakia) d'orbignyi* (Reuss); Voloshina, p. 65, Pl. 3, Figs 1&2
     1974 *Arenobulimina d'orbignyi* (Reuss); Hercogova, p. 408, Pl. 7, Fig. 1
?1981 *Arenobulimina (Voloshinoides) bulletta* Barnard and Banner, pp. 408–10, Pl. 3, Figs 1–6; Pl. 6, Figs 14–20

Description. Dimensions: maximum length 1.2 mm, maximum width 0.5 mm.

Large, elongate trochospiral test, parallel sided in upper portion, lowermost part conical; ultimate whorl comprises one-half of test; apertural face almost horizontal, slightly convex; three to four chambers per whorl. Wall coarse; radiate internal partitions indistinct.

Identification. According to Marie (1941), *A. pseudodorbignyi* is distinguished from *A. d'orbignyi* by its greater width, greater distal angle and less inclined sutures. Comparison with those specimens illustrated by Cushman (1937a) shows that the average width of the specimens comprises about 63% of the length. A similar figure of 60% is given by Marie (1941). With regard to the distal angle, this varies greatly in all arenobuliminids, and a difference in the inclination of the sutures is not considered herein of specific significance. Voloshina (1972) places her specimens in the subgenus *A. (Paster-*

nakia) but states that they are very small and smooth walled. This excludes them from the species as defined here.

H. d'orbignyi is morphologically very similar to *A. (V.) bulletta* Barnard and Banner (1981), although it has not been ascertained whether they are synonymous.

Stratigraphical range. Early Cenomanian to ?Campanian.

Areal distribution. The species is ubiquitous throughout northern Europe and the western USSR.

Evolutionary lineages

Price (1977) in the Albian, Carter and Hart (1977) in the Albian to early Turonian, and Barnard and Banner (1981) in the Albian and Cenomanian have suggested evolutionary lineages for the arenobuliminids. These interpretations, although generally similar, also have some significant differences. Outlined below is an interpretation resulting from a close study of the group and illustrated on Figure 2.2. The macrofaunal zonation referred to in Figure 2.2 is a composite of several authors' works as compiled by Owen (1975) and Kauffman (1979) for the Albian and Cenomanian respectively.

Although Neagu (1972) describes *A. flandrini* from the Hauterivian of Romania, no apparent relationship exists between it and the Aptian (*fide* Barnard, in Barnard & Banner 1981) to Middle Albian *A. (P.) macfadyeni*. The latter is considered the parent stock from which *A. (P.) chapmani* arose in the ?latest Middle or early Late Albian. Barnard and Banner (1981) also trace the development of *Hagenowina voloshinae praevoloshinae* and *H. voloshinae sensu stricto* from *A. (P.) macfadyeni*. These taxa were not recognised in this study. Their form *A. macfadyeni elongata* is considered synonymous with *A. (P.) bochumensis* whereas *H. bulletta* is possibly synonymous with *H. d'orbignyi*.

The sudden appearance of *A. (P.) chapmani* in large numbers during the early Late Albian heralds the rapid species diversification of this group in the late Late Albian and early Cenomanian. During the latest Albian, *A. (P.) barnardi, A. (P.) bochumensis, A. (P.) truncata, A. (S.) sabulosa, A. (A.)* cf. *obliqua, Vialovella praefrankei, Crenaverneuilina intermedia* and *Hagenowina advena* first appear.

H. advena arose from *A. (P.) chapmani*, probably through the subspecies *H. advena praeadvena* (Barnard & Banner 1981). Highest Albian *A. (P.) chapmani* has been observed to possess radiate partitions which would already place the species in the genus *Hagenowina*, with subsequent species in this plexus being termed *H. postchapmani praecursor* and *H. postchapmani* by Barnard and Banner (1981). The development of a divided interior therefore occurs in the latest Albian, and *A. (P.) chapmani*, which initially arose as a non-partitioned arenobuliminid, evolved into an internally divided form just prior to the appearance of its immediate descendant, *H. advena*. A similar development probably resulted in *A. (P.) barnardi* giving rise to *H. d'orbignyi* near the Albian/Cenomanian contact. *A. (P.) minima* also arose from *A. (P.) chapmani* at this level.

The first appearance of *A. (S.) sabulosa* in the Late Albian and its derivation are problematic. It does not appear to be directly related to *A. (P.) chapmani*, due to its completely different morphological features. Price (1976, 1977) believed that its occurrence is restricted to southern England and the southern North Sea. The specimen illustrated by Voloshina (1972) as *A. (C.) sabulosa* is not this species. However, her illustrations of *A. convexocamerata* and *A. vialovi* appear very similar to *A. (S.) gaworbiedowae* and *A. (S.) sabulosa* respectively, thus possibly extending their known geographical distribution. The relationship between *A. (S.) sabulosa* and *A. (S.) gaworbiedowae* is uncertain. The latter may be a geographical variant of the former, and, although they differ in certain morphologic aspects, the possibility of them being conspecific cannot be dismissed. However, their distribution is discontinuous between northwestern Germany and southern England, and indeed between northwestern Germany and the western USSR. Carter and Hart (1977) postulate that *Crenaverneuilina intermedia* arose from *A. (S.) sabulosa* in the latest Albian. However, Price (1977) suggests that *C. intermedia* arose through *Vialovella praefrankei*. The latter conclusion is again suggested herein, as *C. intermedia* and *V. praefrankei* are not Ataxophragmiinae but both belong to the Verneuilininae, as proposed by Balahakmatova (1972). Both species are triserial but not trochospirally coiled. *C. mariae* (Carter & Hart 1977) probably arose from *C. intermedia* in the middle Cenomanian. *A. (A.)* cf. *obliqua* arose directly from *A. (P.) chapmani* during the latest Albian. The former in turn gave rise to *A. (A.) presli* in the late early Cenomanian.

Within the early Cenomanian or possibly latest Albian another offshoot of the arenobuliminids arose, namely species of *Ataxophragmium* Reuss 1860. They probably arose from *Hagenowina*, as the genus possess radiate internal partitions, but are distinctly streptospiral, as opposed to the trochospiral *Arenobulimina* and *Hagenowina*. Also, from preliminary studies by the senior author (C.F.), it is possible that the next radiation of the Ataxophragmiinae occurred in the early to middle Campanian, when *Voloshinovella* Loeblich and Tappan and *Orbignya* von Hagenow appear, showing distinct relationships with previous genera.

Acknowledgements

It is with pleasure that both authors dedicate this chapter to the honour of Professor Tom Barnard. We hope our contribution, although at some variance with his latest study, augments his work on the Ataxophragmiidae.

The senior author (C.F.) thanks Professor Dr M. Kaever for supervision and constructive criticism of his doctoral thesis at the University of Münster, northwestern Germany (1979), upon which much of this paper is based. Funding and promotion of this work from the Deutsche Forschungsgemeinschaft is gratefully acknowledged. W. Koch (Bundesanstalt für Geowissenschaften und Rohstoffe, Hannover) and Dr E. Gawor-Biedowa (Geological Institute, Warsaw) kindly provided comparative material. Also, Dr A. M. Voloshina (Ukrainian Geological Science Investigation Institute) provided much useful information through personal communications. Dr W. Knauff (Geologische Landesamt Nordrhein-Westfalen) kindly provided the micrographs of *A. (S.) gaworbiedowae*.

The junior author (R.J.P.) thanks Professor T. Barnard for constructive criticism of his doctoral thesis at University College London, England (1975). Also the financial support of Robertson Research International Ltd is acknowledged, from whom a fellowship award was received, and without whose support the initial research would not have been undertaken. Thanks are extended to Dr H. G. Owen (Department of Palaeontology, British Museum (Natural History), London), Dr E. Kemper and Dr H. Bertram (Bundesanstalt für Geowissenschaften und Rohstoffe, Hannover), Dr P. Destombes (formerly of Institut Pasteur, Paris) and Dr M. B. Hart (School of Environmental Sciences, Plymouth Polytechnic, England) for their help and guidance during student tenureship. Also Shell Petroleum Company (N.A.M.) kindly made available borehole material from the Netherlands.

Amoco Canada Petroleum Company Ltd is acknowledged both for permission to publish this work and for providing drafting, photographic and financial support necessary for completion of the paper.

References

Akimetz, V. S. 1961. Stratigrafia i foraminiferi verkhnemelovich otlozhenii Belorussii. *Paleontologia i stratigrafia BSSR*. **3**, 1–245.

Amedro, F., R. Damotte, H. Manivit, F. Robaszynski and J. Sornay 1978. Echelles biostratigraphiques dans le Cénomanien du Boulonnais. *Géol. Médit.* **5**, 5–18.

Balahakmatova, V. T. 1972. K sistematike semeistva Ataxophragmiidae Schwager 1877. *Vopr. Mikropaleont.* **15**, 70–4.

Barnard, T. and F. T. Banner 1953. Arenaceous foraminifera from the Upper Cretaceous of England. *Q. J. Geol Soc. Lond.* **109**, 173–216.

Barnard, T. and F. T. Banner 1981. The Ataxophragmiidae of England: Part 1, Albian–Cenomanian *Arenobulimina* and *Crenaverneuilina*. *Rev. Esp. Micropalaeont.* **12**, 383–430.

Bartenstein, H. and F. Bettenstaedt 1962. Marine Unterkreide (Boreal und Tethys). In *Leitfossilien der Mikropaläontologie*, W. Simon & H. Bartenstein (eds), 224–97. Berlin: Gebrüder Borntraeger.

Beissel, I. and K. Holzapfel 1891. Foraminiferen der Aachener Kreide. *Abh. K. Preuss. Geol. Landesanst.* N.F., **3**.

Berthelin, G. 1880. Mémoire sur les foraminifères de l'étage Albien de Montcley (Doubs). *Mém. Soc. Géol. Fr. (3)* **1** (5), 1–84.

Bettenstaedt, F. and C. A. Wicher 1955. Stratigraphic correlation of Upper Cretaceous and Lower Cretaceous in the Tethys and Boreal by the aid of microfossils. In *4. Welterdölkongress* I/D, 493–516.

Brotzen, F. 1936. Foraminiferen aus dem schwedischen untersten Senon von Eriksdal in Schonen. *Sver. Geol. Unders.* C **396** (3), 1–206.

Brotzen, F. 1948. The Swedish Palaeocene and its foraminiferal fauna. *Sver. Geol. Unders.* **42** (2), 1–140.

Carter, D. J. and M. B. Hart 1977. Aspects of mid-Cretaceous stratigraphical micropalaeontology. *Bull. Br. Mus. Nat. Hist. (Geol)* **29** (1), 135 pp.

Chapman, F. 1892. The foraminifera of the Gault of Folkestone (Part 3), *J. R. Microsc. Soc. Lond.* 749–58.

Cushman, J. A. 1911. A monograph of the foraminifera of the North Pacific Ocean. (Part 2), Textulariidae. *Bull. US Natn. Mus. Wash.* **71** (2), 108 pp.

Cushman, J. A. 1927. An outline of a re-classification of the Foraminifera. *Contr. Cushman Lab. Foramin. Res.* **3** (1), 1–105.

Cushman, J. A. 1933. New American Cretaceous Foraminifera. *Contr. Cushman Lab. Foramin. Res.* **9** (3), 49–64.

Cushman, J. A. 1936. New genera and species of the families Verneuilinidae and Valvulinidae and of the subfamily Virgulininae. *Sp. Publs Cushman Lab.* **6**, 1–71.

Cushman, J. A. 1937a. A monograph of the foraminiferal family Verneuilinidae. *Sp. Publs Cushman Lab.* **7**, 1–157.

Cushman, J. A. 1937b. A monograph of the foraminiferal family Valvulinidae. *Sp. Publs Cushman Lab.* **8**, 1–210.

Cushman, J. A. 1946. *Upper Cretaceous foraminifera of the Gulf Coastal Region of the United States and adjacent areas.* US Geol Surv. Prof. Pap. 206.

Cushman, J. A. and K. Glazewski 1949. Upper Jurassic foraminifera from the Nizniow Limestone of Podole, Poland. *Contr. Cushman Lab. Foramin. Res.* **25** (1), 1–24.

Cuvillier, J. and V. Szakall 1949. *Foraminifères d'Aquitaine–Part 1: Reophacidae à Nonionidae.* Société Nationale des Pétroles d'Aquitaine.

Dain, L. G. 1934. Foraminiferi verchieyurskick i melovick otlozhenii mestopozhdenia Dshaksbai Temirskogo rayona. *Trudi Vses. Nauchn. Issled. Geol. Inst. A* **43**.

de la Sagra, R. 1839. Foraminifères. *Histoire Phys. Pol. Nat. de l'Îsle de Cuba* xlviii, 224 pp.

d'Orbigny, A. D. 1840. Mémoire sur les Foraminifères de la craie blanche du bassin de Paris. *Mém. Soc. Géol. Fr.* **4** (1), 1–51.

Durovskaja, N. F., V. V. Krivoborskii, V. I. Kuzina and N. N. Subbotina 1964. Otryad Ataxophragmiida. In *Foraminiferi melovikh i paleogenovikh otlozhenii zapadno-sibiriskoi nizmennosti*, N. N. Subbotina (ed.), *Trudi Vses. Nauchn. Issled. Geol. Inst.* **234** 193–304.

Egger, J. D. 1899. Foraminiferen und Ostrakoden aus den Kreide-mergeln der oberbayrischen Alpen. *Abh. Bayer. Akad. Wiss.* **21** (1), 1–230.

Franke, A. 1925. Die Foraminiferen der pommerschen Kreide. *Abh. Geol.-Pälaont. Inst. Griefswald* **6**, 1–96.

Franke, A. 1928. Die Foraminiferen der Oberen Kreide Nord-und Mitteldeutschlands. *Abh. Preuss. Geol. Landesanst* n.s. **111**, 1–207.

Frieg, C. 1979. *Systematische, biostratigraphische und polökologische Untersuchungen an agglutinierenden Foraminiferen des Cenomans in Bochum.* Dissertation, Naturwissenschaft Fakultät Westfallische Wilhelms Universität Münster.

Frieg, C. 1980. Neue Ergebnisse zur Systematik sandschaliger Foraminiferen im Cenoman des südwestlichen Münsterlandes. *Palaeont. Z.* **54** (3/4), 225–40.

Fuchs, W. 1967. Die Foraminiferenfauna eines Kernes des höheren Mittel-Alb der Tiefbohrung DELFT 2 – Niederlande. *Jb. Geol. Bundesanst. Wien* **110**, 255–341.

Fuchs, W. 1971. Eine alpine Foraminiferenfauna des tieferen Mittel-Barrême aus den Drusberg-Schichten vom Ranzenberg bei Hohenems in Vorarlberg. *Abh. Geol. Bundesanst. Wien* **27**, 49 pp.

Galloway, J. J. 1933. *A manual of foraminifera.* James Furman Kemp Memorial Series, Vol. 1, Bloomington, Ind.: Principia Press.

Gawor-Biedowa, E. 1969. The genus *Arenobulimina* Cushman from the Upper Albian and Cenomanian of the Polish Lowlands. *Rocz. Pol. Tow. Geol.* **39**, 73–104.

Grékoff, N. 1947. Répartition stratigraphique du genre *Arenobulimina* Cushman. *Rev. Inst. Fr. Pétrole.* **2** (10), 491–509.

Groiss, Th. 1964. Eine Mikrofauna aus der albüberdeckenden Kreide der südlichen Frankenalb. *Erlanger Geol. Abh.* **53**, 23 pp.

Hagn, H. 1953. Die Foraminiferen der Pinswanger Schichten (Unteres Obercampan). *Palaeontographica (A)* **104**, 1–119.

Haig, D. W. 1979. Global distribution patterns for mid-Cretaceous foraminiferids. *J. Foramin. Res.* **9** (1), 29–40.

Hart, M. B. 1976. The mid-Cretaceous succession of Orphan Knoll (north-west Atlantic): micropalaeontology and palaeo-oceanographic implications. *Can. J. Earth Sci.* **13**, 1411–21.

Hercogova, J. 1974. Foraminifera from the Cenomanian of the Bohemian Massif. *Sb. Geol. Ved.* **16**, 69–103.

Hofker, J. 1949. On *Hagenowella* and a new species. *J. Paleont.* **23**, 431–5.

Hofker, J. 1957. Foraminiferen der Oberkreide von Nordwestdeutschland und Holland. *Beih. Geol. Jb.* **27**, 464 pp.

Jefferies, R. P. S. 1962. The palaeoecology of the *Actinocamax plenus* Subzone (Lowest Turonian) in the Anglo-Paris Basin. *Palaeontology* **4**, 609–47.

Kalinin, N. A. 1937. Foraminiferi melovikh otlozhenii Baktygaryn (Aktjubinskaya oblast). *Etyudii po Mikropaleontologii* **1** (2), 7–60.

Kaptarenko-Chernousova, O. K. 1967. Foraminiferi nizhnoktei dovikh vikladiv Dniprovsko-Donetskoi zapadini. *Akad. Nauk URSR, Inst. Heologo Nauk* 125 pp.

Kauffman, E. G. Cretaceous; introduction – biogeography and biostratigraphy. In *Treatise on invertebrate paleontology*. Part A: *Introduction; fossilization (Taphonomy), biogeography and biostratigraphy*, R. A. Robinson & C. Teichert (eds), A418–87. Kansas City: Geological Society of America and University of Kansas Press.

Keller, B. M. 1935. Mikrofauna verchniego mela dnyeprovsko-donetzkoi badnii i nekotorikh drugikh sopredelnikh oblastei. *Byull. Mosk. Obsc. Ispit. Prir.* **23**, 522–88.

Loeblich, A. R., Jr and H. Tappan 1949. Foraminifera from the Walnut Formation (Lower Cretaceous) of northern Texas and southern Oklahoma. *J. Paleont.* **23**, 245–66.

Loeblich, A. R., Jr, and H. Tappan 1961. The status of *Hagenowella* Cushman 1933 and a new genus, *Hagenowina. Proc. Biol Soc. Wash.* **74**, 241–4.

Loeblich, A. R., Jr and H. Tappan 1964. Protista 2. Sarcodina chiefly 'Thecamoebian' and Foraminiferida. In *Treatise on invertebrate paleontology*, Part C, Vols 1 and 2, R. C. Moore (ed.), Kansas City: Geological Society of America and University of Kansas Press.

Loeblich, A. R., Jr and H. Tappan 1974. Recent advances in the classification of Foraminiferida. In *Foraminifera*, Vol. 1, R. H. Hedley & C. G. Adams (eds), 1–53. London: Academic Press.

McNeil, D. H. and W. G. E. Caldwell 1981. *Cretaceous rocks and their foraminifera in the Manitoba Escarpment.* Geol. Assoc. Can. Sp. Pap. no. 21.

Magniez-Jannin, F. 1975. *Les foraminifères de l'Albien de l'Aube. Paléontologie, stratigraphie, écologie.* Cah. Paléont., 351 pp.

Marie, P. 1941. Les foraminifères de la Craie à *Belemnitella mucronata* du Bassin de Paris. *Mém. Mus. Natn. Hist. Paris.*, n. sér. **12** (1), 1–296.

Marsson, T. 1878. Die foraminiferen der weissen Schreibkreide der Insel Rügen. *Mitt. Naturw. Ver. Neu-Vorpomm.* **10**, 115–96. I.

Maslakova, V. M. 1959. Foraminiferi. In *Atlas verkhnemelovoi fauni severno Kavkaza i Krima*, M. M. Moskvin (ed.), *Trudi Vses. Nauchn. Issled. Geol. Inst. Prir. Gasov* 87–129.

Mikjailova-Jovtheva, P. 1957. Sur la présence de quelques formes de Valvulinidae dans le Crétacé et le Tertiaire de la Bulgarie du Nord-Est. *God. Uprav. Geol. Min. Prouz.* **A7**, 97–134.

Moullade, M. 1966. *Etude stratigraphique et micropaléontologique du Crétacé inférieur de la 'fosse vocontienne'*. These Doct. Doc. Lab. Géol. Fac. Sci. Lyon no. 15.

Neagu, T. 1965. Albian foraminifera of the Rumanian Plain. *Micropalaeontology* **11**, 1–38.

Neagu, T. 1972. The Eo–Cretaceous foraminiferal fauna from the area between the Iacomitza and Prahova Valleys (Eastern Carpathians). *Rev. Esp. Micropalaeont.* **4**, 181–224.

Owen, H. G. 1975. The stratigraphy of the Gault and Upper Greensand of the Weald. *Proc. Geol Assoc.* **86**, 475–98.

Perner, J. 1892. Die Foraminiferen des böhmischen Cenoman. *Paleontographica Bohemiae* **1**, 1–65.

Perner, J. 1897. Die Foraminiferen der Weissenberger Schichten. *Paleontographica Bohemia* **4**, 1–73.

Plummer, H. J. 1931. Some Cretaceous foraminifera in Texas. *Publs Bur. Econ. Geol Univ. Tex.* **3101**, 109–203.

Price, R. J. 1976. Palaeoenvironmental interpretations in the Albian of western and southern Europe, as shown by the distribution of selected foraminifera. In *First international symposium on benthonic foraminifera of continental margins. Part B: Biostratigraphy and paleoecology*, C. T. Schafer & B. R. Pelletier (eds), 625–48. Halifax, Nova Scotia: *Maritime Sediments*.

Price, R. J. 1977. The evolutionary interpretation of the foraminiferida *Arenobulimina*, *Gavelinella* and *Hedbergella* in the Albian of north-west Europe. *Palaeontology* **20**, 503–27.

Reuss, A. E. 1846. *Die Versteinerungen der böhmischen Kreideformation*, Vol. 2. Stuttgart.

Reuss, A. E. 1851. Die Foraminiferen und Entomostraceen des Kreidemergels von Lemberg. *Naturw. Abh. Wien* **4** (1), 17–52.

Reuss, A. E. 1860. Die Foraminiferen der Westphälischen Kreideformation. *Sb. Akad. Wiss. Wien* **40**, 147–238.

Reuss, A. E. 1862. Die Foraminiferen des norddeutschen Hils und Gault. *Sb. Akad. Wiss. Wien* **46**, 5–100.

Robaszynski, F. and F. Magniez-Jannin 1980. Foraminifères. *Rev. Micropaléont.* **22**, 270–88.

Schwager, C. 1877. Quadro del proposto sistema de classificazione dei foraminiferi con guscio. *Boll. R. Com. Geol. Ital.* **7** (11–12), 475–85.

Tappan, H. 1962. Foraminifera from the Arctic Slope of Alaska. Pt. 3, *Cretaceous foraminifera*. US Geol Surv. Prof. Pap. 236C, 91–209.

ten Dam, A. 1950. Les foraminifères de l'Albien des Pays-Bas. *Mém. Soc. Géol. Fr.* **29** (63), 1–66.

Vasilenko, V. P. 1961. Foraminiferi verkhnego mela poluostrova Mangishlaka. *Trudi Vses. Nauchn. Issled Geol. Inst.* **171**, 487 pp.

Voigt, E. 1959. Revision der von F. v. Hagenow 1838–1850 aus der Schreibkreide von Rügen veröffentlichten Bryozoen. *Beih. Geol.* **25**, 1–80.

Voloshina, A. M. 1965. Sostoyanie izuchennosti nekotorikh rodov podsmeistva Ataxophragmiinae. *Vopr. Mikropaleont.* **9**, 147–56.

Voloshina, A. M. 1972. Ataxofragmiidae verkhnemelovikh otlozhenii volhinopodolski okrainii russkoi platformi. In *Materiali po paleontologii i stratigrafii neftegazonosnikh raionov zapadnikh oblastei USSR*, A. M. Voloshina & L. S. Pishvanova (eds), *Trudi Ukr. Nauch. Issled. Geol. Inst.* **64**, 55–130.

von Hagenow, F. 1842. Monographie der rügenschen Kreideversteinerungen, Abtheilung III: Mollusken. *N. Jb. Mineral. Geogn. Geol. Petrefaktenkd*, 528–75.

Walters, R. 1958 MS. *Investigations of the Albian foraminifera from south-east England*. PhD thesis, University of Wales, Aberystwyth.

Plates

Plate 2.1 (a)–(c) *Vialovella praefrankei nov. sp.*: Late Albian to Cenomanian. (a) Folkestone, southeastern England; GIM B1-396; ×45. (b), (c) Bemerode, northwestern Germany; UCL 129, 130; ×90, ×42 respectively.

(d)–(h) *Arenobulimina (Arenobulimina) presli* (Reuss): Cenomanian to Campanian. (d) Wünnenberg 2, northwestern Germany; GIM b1-403; ×36. (e) Rügen Island, northwestern Germany; GIM B1-406; ×57. (f), (h) Bochum, northwestern Germany; GIM B1-402; ×72; GIM B1-405; ×96 respectively. (g) Wissant, France; GIM B1-404; ×72.

(i) *Arenobulimina (Arenobulimina)* cf. *obliqua* (d'Orbigny): Late Albian to early Cenomanian; Schoonebeek, Netherlands; UCL 149; ×45.

(j) *Arenobulimina (Harena) improcera* Voloshina: Cenomanian; Bochum, northwestern Germany; GIM B1-429; ×84.

(k) *Arenobulimina (Hagenowella) courta* (Marie): Cenomanian to Maastrichtian; Bochum, northwestern Germany; GIM B1-422; ×78.

(l) *Arenobulimina (Hagenowella) elevata* (d'Orbigny): Cenomanian to Maastrichtian; Bochum, northwestern Germany; GIM B1-416; ×30.

Plate 2.2 (a),(b) *Arenobulimina (Hagenowella) elevata* (d'Orbigny): Cenomanian to Maastrichtian. (a) Rügen Island, northwestern Germany; GIM B1-418; ×24. (b) Montereau-faut-Yonne, France; GIM B1-417; ×30.

(c)–(e) *Arenobulimina (Hagenowella) obesa* (Reuss): Cenomanian to Maastrichtian. (c) Rügen Island, northwestern Germany; GIM B1-419; ×24. (d) Bochum, northwestern Germany; GIM B1-420; ×36. (e) Bochum, northwestern Germany; GIM B1-421; ×51.

(f) *Arenobulimina (Pasternakia) barnardi nov. sp.*: Late Albian to Turonian. Schoonebeek, Netherlands; UCL 152; ×33.

(g),(h) *Arenobulimina (Pasternakia) bochumensis* Frieg: Late Albian to late Cenomanian (early Turonian?). Bochum, northwestern Germany; GIM B1-432 A and B; ×30.

(i)–(k),(n) *Arenobulimina (Pasternakia) chapmani* Cushman: ?latest Middle Albian to ?Cenomanian, predominantly Late Albian. (i) Bochum, northwestern Germany; GIM B1-424; ×48. (j),(k),(n) Folkestone, southeastern England; GIM B1-425; ×42 from *Stoliczkaia dispar* zone exhibiting simple interior; GIM B1-423; ×39: UCL 143; ×30 respectively.

(l),(m),(o) *Arenobulimina (Pasternakia) macfadyeni* Cushman: (?Aptian) Early to Middle Albian. (l) Villemoyenne, France; UCL 147; ×57. (m), (o) Wissant, France; UCL 148; ×36: GIM B1-426; ×48 respectively.

Plate 2.3 (a) *Arenobulimina (Pasternakia) minima* Vasilenko: Cenomanian to Campanian. Bochum, northwestern Germany; GIM B1-431; ×72.

(b),(c) *Arenobulimina (Pasternakia) truncata* (Reuss): Late Albian? to Turonian. (b) Bochum, northwestern Germany; GIM B1-427; ×66. (c) Folkestone, southeastern England; GIM B1-428; ×54.

(d) *Arenobulimina (Sabulina) gaworbiedowae nov. sp.*: Late Albian. Brochterbeck, northwestern Germany; GIM B1-436A; ×48.

(e),(f) *Arenobulimina (Sabulina) sabulosa* (Chapman): Late Albian. (e) Folkestone, southeastern England; GIM B1-437; ×45. (f) Forstal, Aylesford, southeastern England; UCL 151; ×42.

(g),(h) *Hagenowina advena* (Cushman): latest Late Albian to late Cenomanian. (g) Schoonebeek, Netherlands; UCL 142; ×48. (h) Bochum, northwestern Germany; GIM B1-415; ×48.

(i)–(m) *Hagenowina anglica* (Cushman): Cenomanian to ?Maastrichtian. (i), (j), (m) Bochum, northwestern Germany; GIM B1-410; ×36; GIM B1-413; ×24; GIM B1-411; ×45 respectively. (k),(l) Strouanne, France; GIM B1-412; ×40; GIM B1-414; ×24 respectively.

(n),(o) *Hagenowina d'orbignyi* (Reuss): Cenomanian to ?Campanian. Bochum, northwestern Germany; GIM B1-409; ×24; GIM B1-407; ×24 respectively.

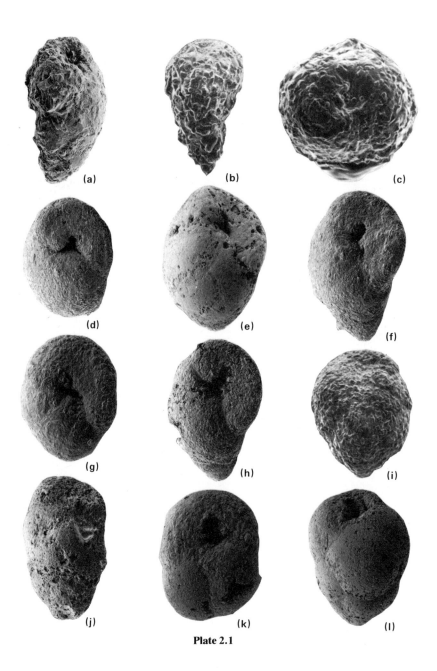

Plate 2.1

Plate 2.2

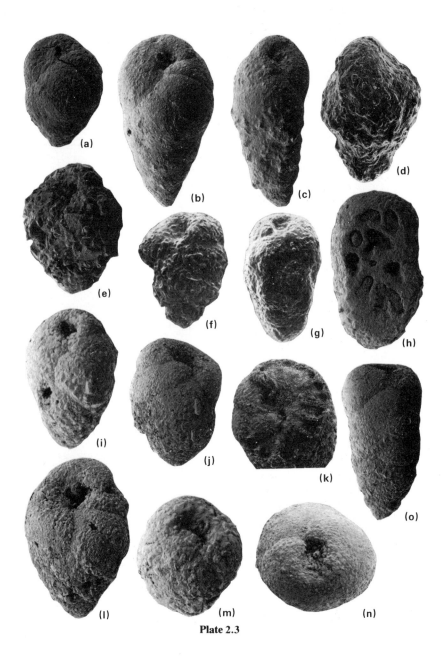

Plate 2.3

3 Middle Eocene assilinid foraminifera from Iran

Fathollah Mojab

An assilinid assemblage of ten species and four varieties including three new species, *Assilina burujenensis* n. sp., *A. hamzehi* n. sp. and *A. persica* n. sp., and one new variety, *A. orientalis* Douvillé var. *iranica* n. var., are described and illustrated from the Burujen area, Iran. The megalospheric (A-form) of *A. exponens* var. *tenuimarginata* Heim is also recorded for the first time. The Middle Eocene (Lutetian) age of the described assemblage is confirmed by associated foraminifera in the area and correlation of the fauna with reported assemblages from other parts of Iran and also abroad. Morphometric methodology applicable to *Assilina* is described and appled taxometrically.

Introduction

Though palaeontologists of the former Anglo-Iranian Oil Company did study the *Nummulites* and *Assilina* of the Bakhtiyari area in southwestern Iran (Taitt 1931), nothing was actually published by these pioneer palaeontologists. In recent decades, a few Iranian workers (Bozorgnia & Banafti 1964, Bozorgnia & Kalantari 1965, Rahaghi & Schaub 1976) published articles on these two genera mainly from central, northern and northeastern parts of the country. Sampò (1969) reported some *Assilina* and *Nummulites* species from the Jahrum Formation of the Zagros area located south and south-west of the present study area (Fig. 3.1). James and Wynd (1965) also reported a few species of these two genera from the type localities of Jahrum, Taleh Zang and Shahbazan Formations. The present work is a detailed account on assilinid fauna recorded from Middle Eocene (Lutetian) outcrops of Jahrum Formation at Hamzeh-Ali Mountain and adjacent areas west of Burujen, Iran (Fig. 3.1).

The Middle Eocene (Lutetian) age of this assemblage is confirmed by its correlation with the time-equivalent assemblages within Iran and also abroad. The Lutetian assilinids of northeastern Iran, described by Rahaghi and Schaub (1976, Fig. 7) and assilinid specimens illustrated by Bozorgnia and Banafti (1964, Pl. CXXIII, Fig. 2 from Sistan, Pl. CXXVI, Fig. 2 from

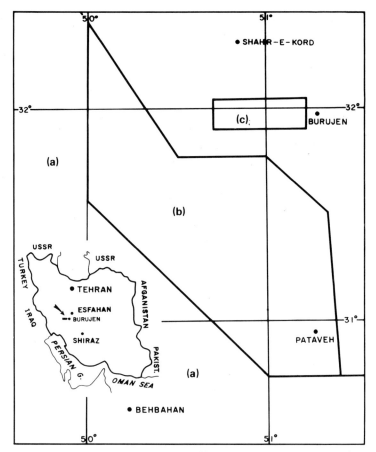

Figure 3.1 Diagrammatic sketch-map showing (a) part of the Iranian Oil Consortium Agreement Area, (b) the Agip Mineraria Area and (c) the present study area. Scale: 1:1 000 000. Source: *The geological map of south-west Iran*, Tehran: NIOC, 1969. Inset: Regional location map.

Jaz-Mourian, and Pl. CXXIX, Fig. 1 from Ashtian) could be correlated with the present fauna. Though the last of these specimens was believed by them to be Upper Eocene in age, it could be compared with *Assilina hamzehi* n. sp. (B-form) of this work and therefore be Lutetian in age. Generally assilinids are not recorded beyond Middle Eocene (Blondeau 1972, Fig. 25). The specimens in Plates CXXIII and CXXVI of Bozorgnia and Banafti (1964) are compared with *A. orientalis* Douvillé and *A. hamzehi* n. sp. (B-form) of this work respectively.

 A. exponens and *A. spira* are reported from the Middle Kirthar Series (Middle Eocene) of north-eastern parts of the Indian subcontinent (Nuttall 1926). In Turkey, *A. exponens* is recorded from Soleymaniyeh by Butterlin and Monod (1969), Haymana by Sirel and Gündüz (1976) and Katsamonu by Daci Dizer (1953), all of Lutetian age. *A. aspera* and *A. spira* are also reported from the last two localities, respectively, being associated with *A. exponens*.

The occurrence of *Truncorotaloides rohri* and *Nummulites aturicus* in association with these assilinids (Mojab 1978), also confirms the Lutetian age of the fauna.

The basal part of the section at Hamzeh-Ali Mountain of the study area is obscured (Fig. 3.2). The exposed 136 m of limestone, marly limestone and marls of Lutetian age of this locality corresponds with the upper two-thirds of the time-equivalent Taleh Zang Formation at its type locality. The root of the structure, however, is in contact with the Cretaceous Bangestan Group, through an oblique fault plane, due to the block-faulted nature of the area. The uppermost horizon of this sequence is overlain conformably by dolomitic

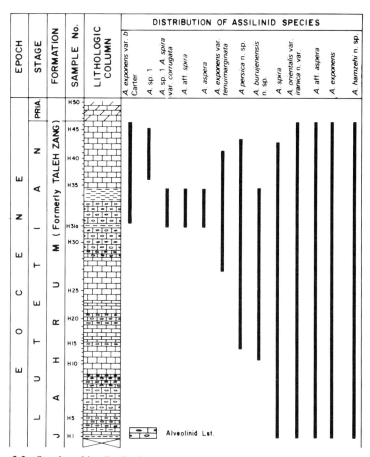

Figure 3.2 Stratigraphic distribution of assilinid fauna recorded from Middle Eocene (Lutetian) Jahrum Formation of Hamzeh-Ali Mountain, west of Burujen, Iran. Scale: 1:800.

limestone of Upper Eocene age (Priabonian) (Fig. 3.2). The assilinid forms are completely absent above this horizon.

This chapter seeks to exemplify quantitative studies and to show both how a simple application of morphometrics can usefully replace merely qualitative descriptions, and how the resulting taxometry can be used for biostratigraphic determination in assilinid studies. Some morphometry of some of the taxa is displayed graphically in Figure 3.3.

Descriptive micropalaeontology

Genus: Assilina d'Orbigny 1826

Morphometrics. In measuring specimens, Blondeau's method (Blondeau 1972) was generally used, except where more data was necessary in which case both the normal method of measurement and Blondeau's method were used (see *Assilina burujenensis* n. sp. A-form, *A. hamzehi* n. sp. A.-form and *A. persica* n. sp. A.-form). In Blondeau's method, the components of microspheric (B) and megalospheric (A) forms are separately measured and the ratio between some of these components are calculated. The following is an example for *Assilina aspera* Doncieux 1948:

Measurements

Form	D:T (mm)	W: R (mm)	Rate of spire opening	No. of chambers per quarter whorl	h:l	dm
B	8.5:1.6, 8.8: 2.1 (4.7)	11:4.9 (2.24)	1 (second whorl), 1.5, 2, 2, 3, 3, 3.5, 3.8, 3, 3	3 4 5 6 7 7, 8, 9, 9, 9, 8 9 10 11, 16, 15	>1	
A	7.8:2.0 (3.9)	7:3.5 (2)	1, 1.5, 1.2, 1.4, 2, 1.5, 1.75	1 2 3 4 5 6 3, 4, 6, 7.5, 11, 9	≥1	0.45

In this tabulation, B=microspheric, A=megalospheric, D=diameter of the test in millimetres, T=thickness of the test in millimetres, W=number of whorls, R=length of radius in millimetres. From these, average D/T and W/R values may be given (in parentheses). The rate of spire opening is equal to the proportion of the height of the whorls in relation to the height of the innermost measurable whorl. It is normally measured from the first whorl onward on A-form specimens and, if possible, from the second whorl onward in B-form specimens. If accurate measurement of the second whorl of a B-form specimen is not possible, it is made on the third or later whorls, and this is mentioned in brackets immediately after the relevant figure. Regardless of its actual length in millimetres, the first record is always

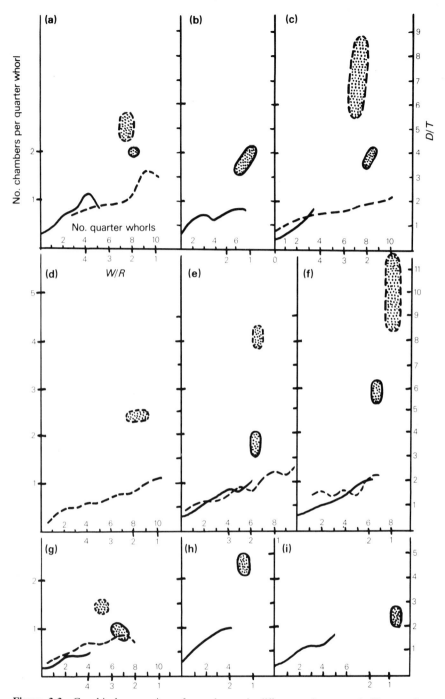

Figure 3.3 Graphical expression of morphometric differences between *Assilina* species. Continuous lines, A-forms; broken lines, B-forms. Graphs of chamber number per quarter whorl plotted for successive whorls, and envelopes (stippled) for $W/R:D/T$ ratios. (a) *A. aspera*; (b) *A. burujensis*; (c) *A. exponens* s.s.; (d) *A. exponens* b; (e) *A. exponens tenuimarginata*; (f) *A. spira* (A-form graph is for *A. spira corrugata*); (g) *A. hamzehi*; (h) *A. persica*; (i) *A. orientalis iranica*.

the base unit for later measurements and, therefore, it always appears as 1.

For example, if the height of the third whorl in a microspheric (B) form is 0.15 mm and the heights of the next two whorls (fourth and fifth whorls) are 0.45 mm and 0.90 mm respectively, then the rate of spire opening for those three whorls would be: 1 (third whorl), 3, 6. When taking the numbers of chambers in one-quarter of a whorl, the numbers of whorls are written in italic type in the top row and the relevant numbers of chambers in roman type in the bottom row: for example, consider the set *2 3 4 5 6*
                                                                                              4, 5, 5, 6, 7.
This means that there are four, five, five, six and seven chambers in one-quarter of the second, third, fourth, fifth and sixth whorls of the specimen.

h=height of an average adult chamber; l=length of an average adult chamber; dm=diameter of megalosphere proloculus, including chamber wall.

Taxonomy and taxometry. Types and most of the illustrated specimens are registered by the British Museum (Natural History), London, under BMNH P50554–P50603. A few specimens with the author's personal collection are numbered PCA 210 onward. Registration numbers are referred to in the explanations of the relevant plates.

Assilina aspera *Doncieux 1948*
See Plates 3.1b,e,f,j (B-form) and Plates 3.1a,c,d,g,h (A-form).

Assilina aspera Doncieux 1948, p. 25, Pl. 6, Figs 15–25 (A and B forms)
Assilina aff. aspera Doncieux; Sirel and Gündüz 1976, p. 37, Pl. 12, Figs 6–13 (A-form)

The Iranian specimens strongly resemble those described by Doncieux (1948) from the Middle Eocene of western Madagascar, although the average size of both A and B forms of Iranian specimens is slightly smaller. The specimens illustrated by Sirel and Gündüz (1976) from the Lutetian of the Haymana area, Turkey, under *A.* aff. *aspera* Doncieux (A-form only), resemble the A-forms illustrated here in Plate 3.1a,c,d,g,h.

Measurements

Form	D:T (mm)	W: R (mm)	Rate of spire opening	No. of chambers per quarter whorl	h:l	dm
B	8.5:1.6, 8.8: 2.1 (4.7)	11:4.9 (2.24)	1 (second whorl), 1.5, 2, 2, 3, 3, 3.5, 3.8, 3, 3	*3 4 5 6 7* 7, 8, 9, 9, 9, 8 9 10 11, 16, 15	>1	
A	7.8:2.0 (3.9)	7:3.5 (2)	1, 1.5, 1.2, 1.4, 2, 1.5, 1.75	*1 2 3 4 5 6* 3, 4, 6, 7.5, 11, 9	≥1	0.45

Material. A total of 324 specimens (A and B forms) was examined.

Assilina aff. aspera *Doncieux 1948*
See Plates 3.2b,h,i (A-form).

Associated with *A. aspera*, but this species is much smaller in size. Externally it resembles the species closely, except for a wider and less conspicuous mamelon which gradually decreases in thickness towards the short peripheral flange, in contrast to *A. aspera* A-form which has a smaller but more conspicuous mamelon and wider peripheral flange. Apart from the general axial profile of the test, which is more globose than that of *A. aspera* A-form, all other internal characteristics both in equatorial and axial sections are the same.

Measurements

Form	D:T (mm)	W: R (mm)	Rate of spire opening	No. of chambers per quarter whorl	h:l	dm
A	6.1:2.1, 6.3:1.6, 6.7:1.6 (3.7)	6:3, 6:3.2, 6:3.4 (1.88)	same as *A. aspera* A-form	same as *A. aspera* A-form	same as *A. aspera* A-form	0.32

Material. A total of 350 specimens was examined.

Assilina burujenensis *n. sp.* (A-form)
See Plates 3.2a,c,d.

Description. Test discoidal, large, diameter up to 14 mm, maximum thickness 3.6 mm; surface smooth; sides almost parallel, with slightly depressed central area; axial periphery rounded. In equatorial section, embryonic chambers bilocular, initial chamber circular, opening into a crescentic second chamber through a large aperture. Septa thin, radial, straight proximally, distally curved; spire somewhat irregular. Early chambers higher than long, later chambers vice versa. In transverse section, embryonic chambers bilocular, initial chamber circular, second chamber crescentic; chambers triangular, early ones slightly higher than later ones, basally curved; sides almost straight; marginal cord well developed, lateral laminae thick, composed of numerous thin laminae representing the shape of the test at different stages of growth. Early whorls involute, last two whorls semi-involute. Numerous pillars continuously extending from median layer to the surface (but not very well developed).

Measurements

Dimensions	Specimens Plate 3.2a	Plate 3.2c	Plate 3.2d	Not illustrated	
diameter (mm)	10.2	14.0	14.2	9.1	8.9
thickness (mm)	3.2	—	3.4	3.3	3.3
diameter of proloculus (mm)	0.5	0.7	—	—	—
no of chambers in first whorl	—	6	—	—	—
no of chambers in second whorl	—	15	—	—	—
no of chambers in third whorl	—	23	—	—	—

Discussion. The new species is associated with *Assilina orientalis* var. *iranica* n. var. but can be differentiated from it externally by its larger size, its being more compressed and its smooth unornamented surface (which does not show any indication of internal pillars). Internally, it can be contrasted with *A. orientalis* var. *iranica* n. var. by its irregularly opening spire, almost straight radial septa, lateral laminae which are not as thick and pillars which are only faintly represented. The specimens illustrated by Wagner (1964, p. 169, Figs 65 & 6) as typical *Assilina* are very similar in equatorial section to the present species; however, its axial section has a broad polar depression which is represented only faintly in a few specimens recorded in Iran.

The specimens in axial section illustrated in Figure 1 of Plate LXVIII of Cuvillier and Sacal (1951) as *Assilina leymerie* (d'Archiac and Haime), from the Lower Eocene of Aquitaine, exhibit almost the same profile as my specimen in Plate 3.2a but are smaller in size, have a less rounded axial periphery and relatively higher chambers.

Measurements

Form	D:T (mm)	W:R (mm)	Rate of spire opening	No. of chambers per quarter whorl	h:l	dm
A	8:2.4, 12:3 (3.67)	6:4, 7:7 (1.25)	1, 1.14, 1.14, 1.3, 1.42, 1.5, 1.64, 1.42	1 2 3 4 5 2.5, 6, 7, 6, 7.5, 6 7 8.5, 8	>1	0,7

Material. A total of forty specimens was examined.

Assilina exponens *(Sowerby 1840)*
See Plates 3.2e–g *(A-form)*, Plates 3.2j–k, Plates 3.3a–c,
Plates 3.7a–d *(B-form)*.

Nummularia exponens Sowerby, in Sykes 1840, p. 719, pl. 61, Figs 146c–d,f (B-form)
Nummulina mamillata (Sowerby), d'Archiac 1847, p. 1020 (A-form)

Assilina exponens (Sowerby), Carter 1861, pp. 373–6, Pl. XV, Figs 6a–d; Nuttall 1926, p. 142, Pl. 5, Figs 5&6, Pl. 6, Fig. 1 (B-form); Doncieux 1948, pp. 1–32, Pl. 6, Figs 1–5 (B-form); Schaub 1951, pp. 1–222, Pl. 211, Text Fig. 324 (A-form); Daci Dizer 1953, Pl. IX, Figs 3, 4, 7, 8 (B-form), Figs 5&6 (A-form)

The original description by Sowerby (1840) is incomplete. Nuttall's diagnosis for *A. exponens* (A and B forms) is more detailed (Nuttall 1926). The Iranian specimens fit well within the variations of the species. The B-form specimens (Pl. 3.2j–k and Pl. 3.3a–c) strongly resemble those of Nuttall (1926, Pl. 5, Fig. 5 and Pl. 6, Fig. 1) from western India. The A-form specimen (Pl. 3.2e) in axial section resembles those illustrated by Heim (1908, Pl. 7, Figs 24, 32 & 34) from Switzerland and those by Ziegler (1960, Pl. 4, Figs 7 & 8) from Germany. However, the very slow rate of spire opening of the first four whorls of the Iranian A-form specimens is remarkable.

It is believed that the slight variations of the external ornamentation, internal structure and the size of the specimens illustrated is due to the individual variations within the specific assemblage.

Measurements

Form	D:T (mm)	W:R (mm)	Rate of spire opening	No. of chambers per quarter whorl	h:l	dm
B	14.5:1.7, 15.9:2.8 (7.1)	14:5.8 (2.4)	1 (second whorl), 1.6, 3.6, 3, 5, 6, 8 9, 10, 8, 10, 7	2 3 4 5 6, 5, 6, 7, 7.5, 8, 7 8 9 10 11 8, 9, 9.5, 10, 12	>1	
A	4.8:1.2, 5:1.4 (3.8)	4:2.2, 5:2.5 (1.9)	1, 1.1, 1.1, 1.2	1 2 3 4 2, 3, 5, 7	≥1	0.3

Material. A total of 480 specimens (269 A-form, 211 B-form) was examined.

Assilina exponens *Sowerby aff. var.* b *Carter 1861 (B-form)*
See Plates 3.4a–c,e,f.

Assilina exponens Sowerby var. *b* Carter 1861. p. 58. Pl. 15. Figs 1 11a–c

The present form resembles those figured by Carter (1861, Pl. 15, Figs 1 & 1a) from the Eocene Nummulitic series of India in having a smooth surface with unraised indications of spiral and spetal sutures and a marked circular, solid, calcitic mass at the central part of the test. The Iranian form, however, is much smaller (almost half the size) and proportionally thicker. Also, it does not exhibit the undulating nature of the specimens described by Carter. Internally, however, it fits within the specifications given by Carter by being generally similar to the typical *A. exponens* but having larger chambers, thicker spiral laminae and two solid masses of calcitic material opposite each other on the polar parts of the test in axial section.

Measurements

Form	D:T (mm)	W:R (mm)	Rate of spire opening	No. of chambers per quarter whorl	h:l	dm
B	16:3.2, 16:3.3 (4.8)	13:8 14:8 (1.7)	1 (second whorl), 1.3, 1.6, 1.8, 2.3, 3 3, 3.6, 3.6 4, 3	2 3 4 5 6 7 3, 5, 5, 6, 6, 7, 8 9 *10 11* 8, 8, 10, 11	>1	

Material. A total of fifteen specimens, all B-form, was examined.

Assilina exponens *var.* tenuimarginata Heim 1908
See Plates 3.5b,c,e (A-form) and Plates 3.5a,d,f,g (B-form).

Assilina exponens var. *tenuimarginata* Heim 1908, p. 243, Pl. 7, Figs 19–22

As Heim (1908) stated, this variety is identical to the typical *Assilina exponens* (Sowerby) in equatorial section. Externally, however, it has an extremely narrow marginal zone, approximately half the diameter. The central inflated zone may be strongly granulated or smooth. In Heim's typical axial diagram (see Heim 1908, Fig. 22c) the polar region is slightly depressed.

Except for the thickness:diameter ratio (which is between 1:10 and 1:15 in Heim's specimen, and 1:8.5 in the Iranian form), the B-form specimens recorded here possess all other characteristics of the variety. The Iranian form is more stout and shorter in diameter. In accordance with Heim's collection, there are all kinds of transitional forms between the typical *A. exponens* B-form and the variety. The A-form companion of this variety is recorded here for the first time: Plate 3.5b,c,e. Externally, it strongly resembles the B-form, but is much smaller in size. It is also very similar to the B-form in transverse section, except that it is megalospheric and has a much stouter test. It is described as follows: embryonic apparatus bilocular, initial chamber circular, second chamber crescentic; chambers in transverse section triangular, higher than broad, with sides almost straight; lateral lamina of the central mamelon thick, composed of numerous layers of laminae which are almost absent in the thin marginal flange. It is traversed by numerous, continuous pillars extending from the median layer to the surface.

In equatorial section, spire loose, regular, opens very slowly; spiral laminae thin; number of whorls six or seven; chambers rectangular, higher than long, septa straight, distally slightly curved.

Measurements

Form	D:T (mm)	W:R (mm)	Rate of spire opening	No. of chambers per quarter whorl	h:l	dm
B	16.5:2.0, 14.4:1.8 (8.1)	15:8.2 (1.83)	1 (second whorl), 2.3, 3, 3.6, 4.3, 5, 5.3, 7.3, 9.3	2 3 4 5 6 5, 6, 6, 7, 9, 7 8 9 *10 11* 8, 11, 12, 11, 14	>1	
A	6.5:1.7, 7.0:2.2 (3.5)	7:3.8 (1.84)	1, 1.17, 1.4, 1.6, 1.7, 1.7, 1.6	*1* 2 3 4 5 3, 4, 6, 7, 8.5, 6 7 8, 10	≥1	0.53

Material. A total of twenty specimens (A and B forms) was examined.

Assilina hamzehi n. sp. (B-form)
See Plates 3.6a,b,h,m.

Description. Test lenticular, medium in size; surface smooth to weakly ornamented by pillars; septa and spire occasionally seen near the periphery; poles inflated, sloping gradually toward the subangular margin, or may be only slightly distinct from the rest of the test. In equatorial section, spire opens regularly except for the last whorl which is relatively much higher than the previous one. Chambers rectangular, higher than long; septa radial, almost straight, thin; spiral sheet well developed, with clear radial structure.

In transverse sections, chambers triangular, with both bottom and sides slightly curved, higher than wide; chamber walls and marginal cord thick, the latter with clear radial canals; lateral lamina quite thick with closely packed pillars extending from the median layer to the surface; parallel growth laminae visible.

Measurements

Form	D:T (mm)	W:R (mm)	Rate of spire opening	No. of chambers per quarter whorl	h:l	dm
B	7.2:2.6, 7.5:2.5 (2.9)	9:2.6 (3.46)	1 (second whorl), 3, 3.4, 4, 6, 7, 7.6 9.6, 13	2 3 4 5 3.5, 4.5, 5.5, 7, 6 7 8 9 6.5, 7.5, 8.8, 7	>1	

Material. A total of fourteen specimens was examined.

Discussion. The new species shows similarities to *Assilina laminosa* Gill 1953 (Lower Eocene) from the Kohat–Patwar Basin, northwestern Pakistan, in having a smooth unornamented surface (which is generally inflated in polar areas), an involute type of coiling, thick lateral laminae and straight septa with somewhat clubbed proximal ends. The new species, however, is much larger in size with fewer chambers in comparable whorls. It may be a larger descendant of *A. laminosa*.

Assilina hamzehi n. sp. (A-form)
See Plates 3.6c–g, i–l, n.

Description. The A-form companion is associated with *Assilina hamzehi* n. sp. B-form. It has a strongly inflated, lenticular test with slight indications of pillars, spiral and septal sutures on its smooth surface. Poles highly inflated, gradually sloping towards a very short peripheral flange. Internally, embryonic apparatus bilocular, initial chamber large and circular, second chamber small and crescentic spire somewhat irregular with chambers generally slightly higher than long. Septa slightly oblique, spiral sheet well developed. Chambers in transverse section triangular, higher than broad and almost arrow-shaped. Lateral lamina very thick except for the last whorl, transversed by numerous pillars extending from the median layer to the surface, growth laminae clearly visible.

Measurements

Dimensions	Plate 3.6e	Plate 3.6j	Plate 3.6c	Plate 3.6l	Plate 3.6g	Plate 3.6i	Plate 3.6d
diameter (mm)	2.90	2.90	3.70	4.25	3.50	3.80	4.60
thickness (mm)	1.80	1.90	1.80	2.40	1.80	—	—
diameter of proloculus (mm)	—	—	—	0.40	0.40	0.40	0.35
no. of whorls	—	—	—	5	5	5	5
no. of chambers in first whorl	—	—	—	—	—	6	6
no. of chambers in second whorl	—	—	—	—	—	12	12
no. of chambers in third whorl	—	—	—	—	—	22	16
no. of chambers in fourth whorl	—	—	—	—	—	24	19
no. of chambers in fifth whorl	—	—	—	—	—	—	20

Form	D:T (mm)	W:R (mm)	Rate of spire opening	No. of chambers per quarter whorl	h:l	dm
A	4.25:2.4, 3.5:1.8 (1.85)	5:2.1, 5:1.8 (2.58)	1, 1.3, 1.4, 2, 2.3	*1 2 3 4 5* 2, 2.3, 4.5, 4.5, 5	≥1	0.25

No published material was found to be closely similar to the present species.

Material. A total of 69 specimens was examined.

Assilina orientalis Douvillé var. iranica n. var. (A-form)
See Plates 3.7e–h.

Description. Test large, lenticular, margin subangular; diameter:thickness ratio between 3:1 and 2:1; surface smooth, with irregularly distributed round, white, unraised pillars. In equatorial section, spire regular, opens slowly, chambers longer than high. Septa slightly curved, inclined; embryonic apparatus bilocular, proloculus circular and larger than the second chamber; five to six whorls.

In transverse section, proloculus circular, second chamber subcircular; proloculus opens into the second chamber through a large aperture; shape of the chambers triangular, slightly higher than wide, sides and bottom slightly curved; chamber walls thick, marginal cord well developed with clear radial canals; lateral lamina very thick, consisting of numerous clear laminae and abundant, sharp, triangular pillars extending from the median layer to the surface continuously.

The variety differs from the typical *Assilina orientalis* by being larger in size, relatively compressed and having no granulations on the surface. It differs from the *Assilina orientalis* Douvillé var. *gargoensis* Samanta 1962, from Assam, by being much larger in size, having subangular margin, more inclined septa and no external granulations.

Measurements

Dimensions	Plate 3.7e	Plate 3.7g	Not illustrated	Plate 3.7f
diameter (mm)	10.1	10.3	7.5	9.5
thickness (mm)	3.6	4.2	3.2	—
D:T ratio	2.5:1	2.4:1	2.3:1	—
diameter of initial chamber (mm)	—	—	0.5	0.58
no. of whorls	—	—	4	6

Form	D:T (mm)	W:R (mm)	Rate of spire opening	No. of chambers per quarter whorl	h:l	dm
A	2.5:1, 2.3:1 (2.4)	5:4.6 (1.1)	1,1.1, 1.5, 1.8, 2.0	1 2 3 4 5 6 2, 3, 5, 6, 6, 8	≤1	0.58

Material. A total of forty-one specimens was examined.

Assilina persica n. sp. (A-form)
See Plates 3.8a,b,d,h.

Description. Test lenticular, surface smooth and uneven, with a wide crater-like depression which has a polar boss at the centre, periphery sharply angular. The evolute nature of the whorls is reflected on the surface by both spiral and septal sutures. Maximum thickness about two-thirds of the radius from the centre to the periphery. Maximum diameter 10.2 mm with an average of 8.4 mm, maximum thickness 2.2 mm with an average of 1.86 mm. In equatorial section, spire opens regularly; chambers numerous, rectangular, higher than long, septa radial, slightly oblique; embryonic apparatus bilocular, proloculus circular, larger than the second chamber; number of whorls, six.

In transverse section, polar area swollen with depressions on both sides; proloculus large, circular, second chamber smaller, almost crescentic, early chambers evolute, later party involute, chamber walls of the last chamber extends over three or four of the preceding whorls, chambers triangular with both bottom and sides curved; marginal cord well developed with radial canals; outer margin of chambers sharp; lateral lamina pillared in the centre, well developed at the maximum thickness.

Measurements

Dimensions	Plate 3.8d	Plate 3.8d	Plate 3.8h	Not illustrated			
diameter (mm)	8.7	9.0	9.1	8.5	7.2	9.9	10.2
thickness (mm)	1.7	1.9	—	1.9	2.0	1.9	2.2
diameter of proloculus (mm)	0.4	—	0.5	—	—	—	—
no. of chambers in first whorl	—	—	8	—	—	—	—
no. of chambers in second whorl	—	—	19	—	—	—	—
no. of chambers in third whorl	—	—	26	—	—	—	—

Material. A total of seven specimens was examined.

Discussion. The present species undoubtedly has some similarities in transverse section with *Assilina daviesi* de Cizancourt 1938, from the uppermost Laki Formation, Ypresian of Afghanistan, and with *Assilina daviesi* de Cizancourt var. *nammalensis* Gill 1953 (B-form), from the same horizon of north-west Pakistan. Being B-form, the latter cannot be the microspheric form of the present species because the average size is even smaller than the present form.

Assilina persica n. sp. (B-form)
See Plates 3.8e–g.

Description. No free microspheric specimen was recorded, therefore only axial sections are described and illustrated (Pls 3.8e–g). In axial section they closely resemble the megalospheric companion by having an inflated polar area bordered by a depression, but they are completely evolute with chambers increasing gradually in diameter (except for those of the last whorl which are slightly reduced in size).

Measurements

Form	D:T (mm)	W:R (mm)	Rate of spire opening	No. of chambers per quarter whorl	h:l	dm
B	18.5:2.6, 25:3.3 (7.35)	—	—	—	—	—
A	10.2:2.2, 8.4:1.9 (4.53)	6:4.4 (1.36)	1, 1.8, 2.3, 2.7, 2.8	1 2 3 4 5 3, 5.5, 7, 9, 10	>1	0.53

Material. A total of sixteen specimens was examined.

Assilina spira (de Roissy 1805)
See Plates 3.9a,c,d,h (B-form) and Plate 3.9e (A-form).

Nummulites spira de Roissy 1805, pp. 57–8 (B-form); d'Archiac and Haime 1853, p. 155, Pl. XI, Figs. 3, 3a, 4, 4a–b, 5a (A-form)
Assilina spira (de Roissy), Nuttall 1926, p. 143, Pl. 6, Figs 8–9 (B-form); Schaub 1951, p. 217, Fig. 335 (A-form); Sirel and Gündüz 1976, Pl. X, Figs 1–8 (B-form), Pl. XI, Figs 10–12 (A-form)

See de Roissy (1805, p. 155) for type description, and Nuttall (1926, p. 143) for amendments. Although several good axial sections of B-forms were recorded in thin sections of rock, only two free specimens were collected (Pls 3.9a&d). Plate 3.9c is a natural equatorial section found on the surface of the rock sample H-31. Several well preserved A-forms were also recorded (Pl. 3.9e). Externally, the Iranian B-form specimens strongly resemble the type-species in having a large, thin, flat, slightly undulated test with prominent, raised spiral laminae on the surface. Internally, they have thick, irregularly coiled spiral laminae, numerous high chambers and straight septa, distally slightly curved backward.

The megalospheric companion is much stouter and smaller in size with externally less prominent spiral laminae (Pl. 3.9e). Except for having a relatively large proloculus and regularly coiled spiral laminae, other characteristics are the same as the B-form. The present form seems to resemble closely *Assilina assamica* Samanta 1962, reported from the Siju Formation

(Kirthar) of Assam, except that the latter has granulations over the central part of the test.

Measurements

Form	D:T (mm)	W:R (mm)	Rate of spire opening	No. of chambers per quarter whorl	h:l	dm
B	25.3:2.9, 23.5:2.0 (7.74)	10:12, 12:14 (0.85)	1 (third whorl), 1.8, 2.3, 3, 4, 4.5, 6, 7, 9	3 4 5 6 7 8, 7, 8, 7, 11, 8 11	>1	
A	7.2:1.2 6.5:1.2 (5.71)	6:3.6 5:3.2 (1.62)				0.5

Material. A total of 54 specimens, both A- and B-forms, was examined.

Assilina spira *(de Roissy) var.* corrugata *de la Harpe (A-form)* Samanta 1962
See Plates 3.9b,f,g,i,j.

Assilina spira var. *corrugata* de la Harpe 1926, p. 93 (B-form)
Assilina subspira corrugata Samanta 1962, p. 10, Pl. 2, Figs 1–6, 9–11, 14–17 (A-form)

Both external and internal structures of the Iranian specimens are similar to those recorded by Samanta (1962) from the Eocene of Assam. Externally, both the spiral and septal sutures are raised except at the polar area. Internally, spire loose, regular, chambers almost as high as long, septa straight; embryonic apparatus bilocular, proloculus larger than the second chamber; spiral laminae well developed. Transverse section lenticular, chambers triangular with straight walls, much higher than broad; marginal cord well developed.

In order to avoid confirming a separate name given to the A-form of an already established species, the *Assilina subspira corrugata* of Samanta (1962) is replaced by *Assilina spira* (de Roissy) var. *corrugata* de la Harpe (A-form) Samanta.

Measurements

Form	D:T (mm)	W:R (mm)	Rate of spire opening	No. of chambers per quarter whorl	h:l	dm
A	6.9:0.9, (7.67)	6:3.5 7:3.5 (1.86)	1, 1.4, 1.4, 1.4, 1.6, 2.4, 2	1 2 3 4 5 3, 4, 5, 6, 7, 6 7 9, 10	≤1	0.31

Material. Five specimens were examined.

Assilina aff. spira *(de Roissy 1805)* A-form
See Plates 3.8c,i,j.

The present form resembles the *Assilina spira* (A-form) reported from Turkey by Sirel and Gündüz (1976, Pl. XII, Figs 1–5), although, internally, the spire is not as loose as those reported from Turkey. Chambers normally higher than long; septa straight with curvature at their distal end; nucleoconch bilocular with a circular proloculus which is larger than the second chamber.

Transverse section almost flat; chambers triangular with height almost twice the width; chamber walls thin and marginal cord weakly developed.

Measurements

Form	D:T (mm)	W:R (mm)	Rate of spire opening	No. of chambers per quarter whorl	h:l	dm
A	6.2:0.8, 6.5:0.9 (7.49)	6:3.1, 7:4 (1.85)	1, 1.5, 1.75, 2, 2, 2.75, 2.5	1 2 3 4 5 3, 5, 7, 8, 8, 6 7 8, 14	≥1	0.4

Material. A total of 204 specimens was examined.

Assilina *sp.1* (A-form)
See Plate 3.4d.

This is a relatively large megalospheric form. Only one specimen in perfect axial section was recorded. It is characterised in axial section by having a large circular proloculus and a crescentic second chamber; triangular chamber cavities with curved sides and bases; inflated central boss and thick chamber walls, both traversed by pillars; moderately well developed marginal cord.

The specimen is comparable with *Assilina persica* n. sp. A-form of this work, but is much larger in size and also has a larger central boss and a very weak depression around it. Its maximum thickness is at the centre, whilst it is near the periphery in *Assilina persica* n. sp. A-form. It may be considered as a mutant of the same species because it is associated with *A. persica* n. sp. B-form.

Measurements:
Length: nearly 10 mm
Central boss maximum thickness: 2.6 mm
Flange maximum thickness: 2.0 mm
Diameter of proloculus: 0.7 mm.

Material. Six specimens were examined.

Assilina sp. 2 (a-form)
See Plate 3.4g.

This species is an intermediate form between *Assilina spira* var. *corrugata* (A-form) and *Assilina* aff. *aspera* (A-form). Its internal characteristics and overall dimensions resemble *A. spira* var. *corrugata* (A-form) while its relatively distinct mamelon, broad peripheral flange and smooth unornamented surface makes it related to *A.* aff. *aspera* (A-form). Its stratigraphic range, however, corresponds with the former species.

Material. A total of twenty-six specimens was examined.

Acknowledgements

The author thanks the Managing Director of Exploration and Production of the National Iranian Oil Company Mr M. Parsi, and the Manager of Exploration Division of the NIOC, Mr J. Molanazadeh, for providing facilities for completion of this work. Several members of the Exploration Division of the NIOC were very helpful, especially Mr A. Kheradpir and Mr M. Bahri.

References

Blondeau, A. 1972. *Les nummulites*. Paris: Librairie Vuibert.
Bozorgnia, F. and S. Banafti 1964. *Microfacies and micro-organisms of Paleozoic through Tertiary sediments of some parts of Iran*. Tehran: National Iranian Oil Company.
Bozorgnia, F. and A. Kalantari 1965. *Nummulites of parts of Central and East Iran (a thin-section study)*. Teheran: National Iranian Oil Company.
Butterlin, J. and O. Monod 1969. Biostratigraphie (Paléocene à Eocéne Moyen) d'une coupe dans le Taurus de Beyshehir (Turquie). Étude de 'Nummulites cordeless' et révision de ce groupe. *Eclog. Géol. Helv.* **62**, 583–604.
Carter, H. J. 1861. Further observations on the structure of foraminifera and on the larger fossilised forms of Sind, etc., including a new genus and species. *Ann. Mag. Nat. Hist.* **3**, 309–33, 336–82, 446, 465.
Cuvillier, J. and V. Sacal 1951. *Correlations stratigraphiques par microfacies en Aquitaine Occidentale*. Leiden: E. J. Brill.
Daci Dizer, A. 1953. Contribution à l'étude paléontologique du Nummulitic de Kastamonu. *Istanbul Univ., Fen. Fak., Mecm. (Fac. Sci., Rev.)* **B/18**, 207–99.
d'Archiac, A. and J. Haime 1853. *Monographie de nummulites*. Paris: Gide et Baudry.
de Cizancourt, M. 1938. Nummulites et Assilines du Flysch de Gardez et du Khost, Afghanistan oriental. *Mém. Soc. Géol. Fr.* **17** (39), 1–44.
de la Harpe, P. 1926. Materiaux pour servir a une monographie de *Nummulites* et des *Assilines*. Rèdigé par P. Rozlozsnik, A. N. Kir Budapest. *Foldtani-Interet-Evkonyve* **27**, 1–102.
de Roissy, F. 1805. *Histoire naturelle, générale et particulière, des Mollusques (Baffon et Sonnini)*. Paris: Dufart.
Doncieux, L. 1948. Les foraminifères éocènes et oligocènes de l'ouest de Madagascar. *Madagascar Serv. Mines, Ann. Géol., Paris* **13**, 1–32.
Douvillé, H. 1912. Les foraminifères de l'ile de Nias. *Geol. Reichs-Mus. Leiden,ʼ Samml., Leiden* **1** (8/5), 263.

Gill, W. D. 1953. The genus *Assilina* in the Laki series (Lower Eocene) of the Kohat-Potwar Basin, north-west Pakistan. *Contr. Cushman Fdn Foramin. Res.* **4** (2), 1–86.

Heim, A. 1908. Die Nummuliten und Flyshbildungen der Schweizeralpen. *Abh. Schweiz. Palaeont. Ges. (Soc. Palaeont. Suisse Mem.)* **35**, 1–301.

James, G. A. and J. G. Wynd 1965. Stratigraphic nomenclature of the Iranian Oil Consortium agreement area. *Bull. Am. Assoc. Petrolm Geol.* **49**, 2182–245.

Mojab, F. 1978. *Tertiary foraminifera from Burujen Area, south-west Iran*. PhD thesis, University of London.

Nuttall, W. L. F. 1926. The zonal distribution and description of the larger foraminifera of the Middle and Lower Kirthar Series (Middle Eocene) of parts of western India. *Rec. Geol Surv. India* **59**, 115–24.

Rahaghi, A. and H. Schaub 1976. Nummulites et Assilines du N.E. de l'Iran. *Eclog. Géol. Helv.* **69**, 765–82.

Samanta, B. K. 1962. *Assilina* from the Eocene rocks of Garo Hills, Assam. *Bull. Geol Min. Metal. Soc. India* **27**, 1–20.

Sampó, M. 1969. *Microfacies and microfossils of the Zagros Area, south-western Iran (from Pre-Permian to Miocene)*, Leiden: Brill.

Schaub, H. 1951. Stratigraphie und Paläontologie de Schlierenflysches mit besonderer Berucksichtigung der Paleocaenen und untereocaenen Nummuliten und Assilien. *Schweiz Abh. Palaeont.* **68**, 1–222.

Sirel, E. and H. Gündüz 1976. Description and stratigraphical distribution of some species of the genera *Nummulites*, *Assilina* and *Alveolina* from the Ilerdian, Cuisian and Lutetian of Haymana region (S. Ankara). *Bull. Geol Soc. Turkey* **19**, 31–44.

Sykes, W. H. 1840. A notice respecting some fossils collected in Cutch, by Capt. Walter Seem, of Bombay Army. *Trans Geol Soc. Lond.* **2**, 715–19.

Taitt, A. H. 1931. *The Nummulites and Assilinas of the Bakhtiari country, S.W. Persia*. Anglo-Persian Oil Co. Report AHT8.

Wagner, C. W. 1964. *Manual of larger foraminifera*, The Hague: Bataafse International Petroleum Maatschappij NV.

Ziegler, J. H. 1960. Die Assilinen des Eozäns vom Kressenberg in oberbayern. *Geol. Bavarica* **44**, 209–31.

Plates

Plate 3.1 (a)–(i) *Assilina aspera* Doncieux, A- and B-form (late Lutetian). (a) BMNH P50558, equatorial section, A-form, ×7.65. (b) BMNH P50556, equatorial section, B-form, ×7.47. (c) BMNH P50540, surface, A-form, ×4.05. (d) BMNH P50559, axial section, A-form, ×11.7. (e) BMNH P50582, axial section, B-form, ×9.9. (f) BMNH P50555, axial section, B-form, ×8.28. (g) BMNH P50557, surface, A-form, ×5.85. (h) PC A210 oblique lateral view, A-form, ×6.3. (i) BMNH P50554 surface, B-form, ×4.23.

All specimens on this and on the following plates are from Hamzeh-Ali Mountain, west of Burujen, Iran.

Plate 3.2 (b),(h),(i) *Assilina* aff. *aspera* Doncieux, A-form (Lutetian). (b) PC A231 surface, ×4.68. (h) PC A232, axial section, ×10.08. (i) PC A233, axial section, ×13.32.

(a),(c),(d) *Assilina burujenensis* n. sp., A-form (mid-Lutetian). (a) BMNH P50562, axial section, paratype, ×9. (c) BMNH P50563, equatorial section, paratype, ×4.5. (d) BMNH P50561 surface, holotype, ×3.6.

(e),(f),(g),(j),(k) *Assilina exponens* Sowerby, A- and B-form (Lutetian). (e) PC A236 axial section, A-form, ×10.35. (f) PC A237 equatorial section, A-form, ×4.95. (g) PC A238 surface, A-form, ×5.13. (j) BMNH P50569, axial section, B-form, ×9. (k) PC A240, axial section, B-form, ×9.

Plate 3.3 (a)–(c) *Assilina exponens* Sowerby, B-form (Lutetian). (a) BMNH P50566, surface, ×5.4. (b) BMNH P50564, equatorial section, ×5.85. (c) BMNH P50565, axial section, ×7.83.

Plate 3.4 (a)–(c),(e),(f) *Assilina exponens* Sowerby aff. var. *b*. Carter, B-form (late Lutetian). (a) BMNH P50568, surface, ×4.32. (b) PC A241, axial section, ×6.3. (c) BMNH P50571, axial section, ×7.2. (e) PC A242, equatorial section, ×5.4. (f) BMNH P50570, axial section, ×5.22.

(d) *Assilina* sp. 1 A-form (late Lutetian). PC A246, axial section, ×6.3.

(g) *Assilina* sp. 2 A-form (late Lutetian). PC A249, axial section, ×9.45.

Plate 3.5 (a)–(g) *Assilina exponens* Sowerby var. *tenuimarginata* Heim, A- and B-form (mid to late Lutetian). (a) BMNH P50572, surface, B-form, ×3.78. (b) BMNH P50577, surface, A-form, ×5.76. (c) BMNH P50578, equatorial section, A-form, ×6.3. (d) BMNH P50573, equatorial section, B-form, ×3.78. (e) BMNH P50576, axial section, A-form, ×9.9. (f) BMNH P50574, axial section, B-form, ×6.66. (g) BMNH P50575, axial section, B-form, ×7.2.

Plate 3.6 (a)–(n) *Assilina hamzehi* n. sp. A- and B-form (Lutetian). (a) BMNH P50584, equatorial section, B-form, paratype, ×9. (b) BMNH P50580, surface, B-form, holotype, ×9. (c) BMNH P50586, surface, A-form, paratype, ×10.35. (d) BMNH P50588, equatorial section, A-form, paratype, ×10.35. (e) BMNH P50587, lateral view, A-form, holotype, ×9.18. (f) PC A250, axial section, A-form, ×23.4. (g) PC A251, axial section, A-form, ×11.25. (h) BMNH P50583, axial section, B-form, paratype, ×9. (i) BMNH P50590, equatorial section, A-form, paratype, ×9.9. (j) PC A256, surface, A-form, ×11.52. (k) PC A257, lateral view, A-form, ×8.1. (l) BMNH P50589, axial section, A-form, paratype, ×13.5. (m) BMNH P50581, sub-axial section, B-form, paratype, ×9.45. (n) BMNH P50585, oblique lateral view, A-form, paratype, ×12.15.

Plate 3.7 (a)–(d) *Assilina exponens* Sowerby B-form (Lutetian). (a) PC A260 surface, B-form, juvenile, ×4.32. (b) PC A261, surface, B-form, juvenile, ×4.32. (c) BMNH P50567, surface, B-form, juvenile, ×4.77. (d) PC A262, equatorial section, B-form, juvenile, ×5.85.

(e)–(h) *Assilina orientalis* Douville var. *iranica* n. var. A-form (Lutetian). (e) BMNH P50593, axial section, paratype, ×8.37. (f) BMNH P50592, equatorial section, paratype, ×5.85. (g) BMNH P50591, surface, holotype, ×4.14. (h) PC A265, axial section, ×23.4.

Plate 3.8 (a),(b),(d)–(h) *Assilina persica* n. sp. A- and B-forms (mid to late Lutetian). (a) PC A270, surface A-form, paratype, ×6.75. (b) BMNH P50594, surface, A-form, holotype, ×6.75. (d) BMNH P50595, axial section, A-form, paratype, ×9.45. (e) BMNH P50597, axial section, B-form, paratype, ×2.7. (f) BMNH P50598, axial section, B-form, holotype, ×4.05. (g) Perfect axial section of a B-form specimen photographed from the surface of an exposure of bedrock at the site of sample no. H28, ×2.7. (h) BMNH P50596, equatorial section, A-form, paratype, ×6.48.

(c),(i),(j) *Assilina* aff. *spira* (de Roissy) A-form (late Lutetian). (c) BMNH P50602, surface, ×3.87. (i) BMNH P50601, axial section, ×9.9. (j) PC A275, equatorial section, ×7.2.

Plate 3.9 (a),(c)–(e),(h) *Assilina spira* (de Roissy) A- and B-form (Lutetian). (a) BMNH P50599, surface, B-form, ×3.24. (c) PC A277, equatorial section, B-form, ×2.7. (d) BMNH P50599, surface, B-form, ×3.24. (e) BMNH P50601, surface, A-form, ×4.68. (h) BMNH P50599, axial section of the specimen in (a), ×2.79.

(b),(f),(g),(i),(j) *Assilina spira* (de Roissy) var. *corrugata* de la Harpe (A-form) Samanta (late Lutetian). (b) BMNH P50603, surface, ×5.85. (f) BMNH P50604, surface, ×5.64. (g) PC A280, surface, ×5.4. (i) PC A281, axial section, ×8.55. (j) PC A282, equatorial section, ×8.55.

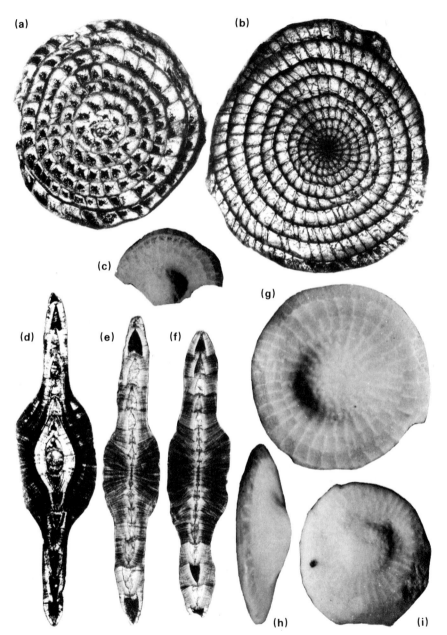

Plate 3.1

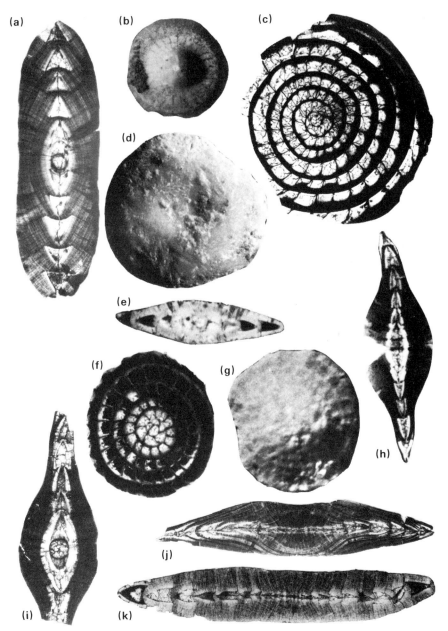

Plate 3.2

Plate 3.3

Plate 3.4

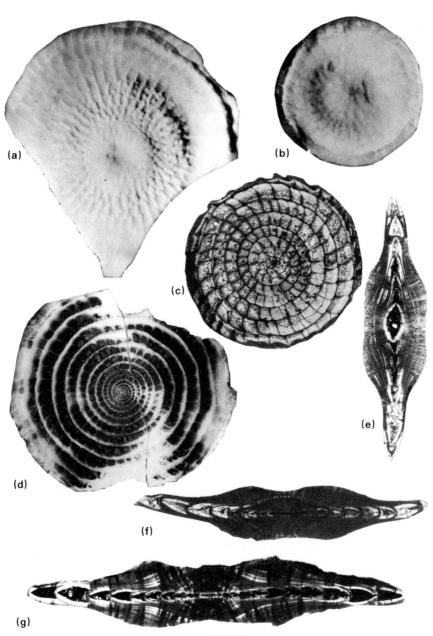

Plate 3.5

Plate 3.6

Plate 3.7

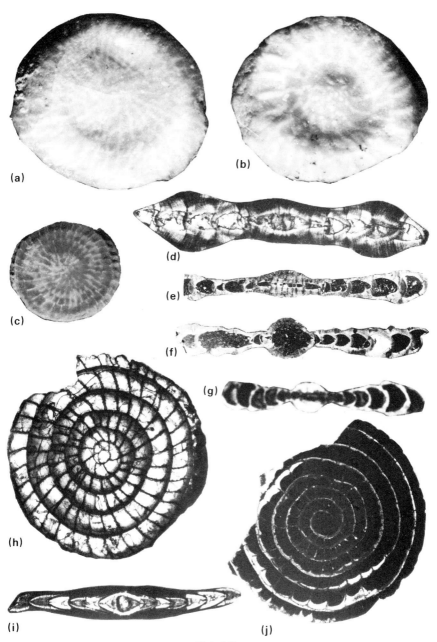

Plate 3.8

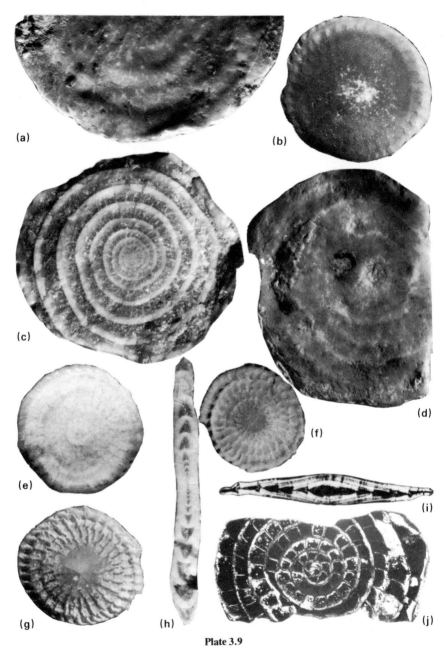

Plate 3.9

4 Ecology and distribution of selected foraminiferal species in the North Minch Channel, northwestern Scotland

P. G. Edwards

One hundred and seven species of Recent benthonic foraminifera are recorded from thirty sample stations sited in the North Minch Channel off the northwestern coast of Scotland. Of the physical parameters of depth, temperature, salinity, turbulence and substrate character, it is the nature of the substratum that appears to be of primary importance in limiting foraminiferal distribution. Although there undoubtedly exists a correlation between average grain size of the bottom sediment and the distribution of the foraminiferal faunas, the ecological factors of current action and the turbulence of the overlying watermass appear to be common denominators in producing this correlation. Several biofacies can be recognised from correlations between faunas and sediment type. *Cibicides* is the dominant genus in nearshore shallow water under generally high energy conditions. *Hyalinea balthica* dominates the offshore shelf areas in association with well sorted lithic sands and shell sands. The deep basins are typified by a proliferation of buliminid faunas. Heavy, thick-shelled forms, such as *Eponides* and miliolids, together with robust agglutinates such as *Gaudryina*, inhabit the high energy environments of the eastern Shiant gravel bank and the shallows to the north of Tiumpan Head.

Introduction

Sampling

The samples examined during the course of this study were mainly collected by the Continental Shelf and the Marine Geophysics Units of the Institute of Geological Sciences, Edinburgh, between August 1969 and July 1973. Additional information was obtained in December 1973 by a party from University College London, aboard the RRS *John Murray*. The results of the sedimentological studies carried out are presented by Bishop and Jones (1980).

Thirty samples were chosen for study to represent an even distribution over the North Minch Channel. All the samples contained empty foraminiferal tests. The majority of the tests examined displayed excellent preservation, especially those found in samples recovered from areas of low current activity and associated fine-grained sediments.

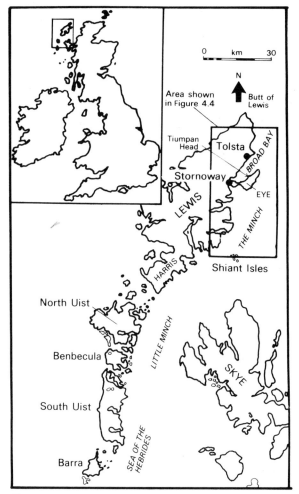

Figure 4.1 The Hebridean Islands (northern section) showing the location of the area of study. (Inset: Map of the British Isles.)

Limitations of the study

Since the samples were not treated with a protoplasmic stain, such as Rose Bengal at the time of their collection, a study of live:dead ratios cannot be carried out. The species distributions outlined in this account are necessarily arbitrary, due to the widespread location of the sample stations. A community of benthonic foraminifera may be regarded as a series of overlapping populations. Micro-environments are known to exist on the sea floor and these will inevitably be overlooked in such a widespread study.

Because of the essentially ecological nature of this study and in view of the extent of the relevant literature dealing with taxonomy, the author considers the inclusion of a species list, incorporated in the Appendix, to be an adequate reference to the nature of the foraminiferal fauna *in toto*.

Previous work

Ecological studies of foraminifera in British coastal waters have been, to date, mainly concerned with marginal marine environments (Murray 1965, 1968, 1970; Haynes & Dobson 1969; Haynes 1973). Early work carried out on Recent foraminifera by Williamson (1858), Brady (1887) and Heron-Allen and Earland (1909, 1913, 1916, 1930) involved collection and cataloguing specimens from around the British Isles. Such studies were of a taxonomic rather than an ecologic nature. Atkinson (1969, 1971) presented the first published account of a quantitative study in a British open marine area in his study of the ecology of Recent foraminifera in Cardigan Bay. No previous study of the ecology of foraminifera in the North Minch Channel has been carried out.

Geographic position and regional climate

The North Minch is that part of the Minch Channel lying between the Hebridean Isle of Lewis and the northwestern Scottish coast, and to the north-east of the Little Minch (Fig. 4.1). In the Hebrides, the mean summer temperatures range from 12.8 °C (55 °F) in the north, to 13.9 °C (57 °F) in the south. These temperatures are approximately 0.28–0.56 °C (0.5–1 °F) lower than those recorded for the eastern coast of Scotland. The winter temperatures lie between 5 °C (41 °F) and 5.6 °C (42 °F) for the whole length of the Hebrides, 1.7 °C (3 °F) greater than for the eastern coast.

Hydrographic conditions and their effect on the distribution of foraminifera

The hydrographic conditions in the North Minch Channel have been summarised by Craig (1951, 1958). Observations were made in this area during the autumn of 1951 in association with herring larvae surveys, and represent the most recent information available.

The North Minch is, on the whole, distinctly less oceanic in character than the northwestern North Sea. Such oceanic influence as exists takes the form of a powerful thrust into the South Minch. These thrusts, though dominating the hydrography of the region, are composed of oceanic water which has suffered appreciable dilution. Heavy dilution occurs in all the mainland estuaries and inlets; elsewhere the vertical gradients are practically nil.

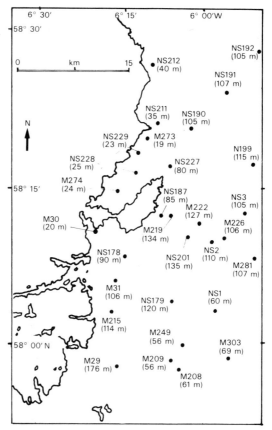

Figure 4.2 The geographical location of the sample stations, and the depth (in metres) at each station.

Temperature

It is the hydrography of the bottom water that is the significant factor affecting the benthonic populations. Temperature is no doubt an important factor in that it tends to limit the rate of metabolic activity in an organism. Information provided by the Institute of Oceanic Sciences data bank, which incorporates Craig's readings, shows that variations in temperature both regionally and vertically in the water column are very slight (0.2–0.5 °C) and may be discounted as a primary factor governing the distribution of foraminiferal faunas in the North Minch Channel. However, the general constitution of the faunas as a whole is doubtless controlled by the regional climate. The lack of variation in temperature on a regional scale is compensated for by a 6 °C seasonal variation. This variation is negligible in that the majority of the foraminiferal species encountered are tolerant of a wider range of temperature differences. Most of the species recorded are restricted to temperate latitudes.

Salinity

Salinity variations, both regionally and vertically in the water column, are comparatively small. Surface salinities average between 34 and 35‰, and any variations between these and bottom salinities are only in the order of 0.26–0.4‰. Such slight variations that do occur are due primarily to rainfall diluting the upper layers of the water column, combined with runoff effects from the surrounding land areas. The foraminifera encountered in the North Minch Channel are adapted to normal marine salinities of 34–35‰.

Oxygen and nitrogen

Data relating to concentrations of oxygen and nitrogen in the waters of the North Minch Channel were unavailable. Other workers have observed that foraminifera die within a very short period of time should they be buried in the sediment beyond a depth of a few centimetres and should the oxygen supply beneath the surface be insufficient to sustain life. This condition is made evident by the deposition of blue-black sulphides on the interior of the chamber walls of foraminiferal tests. This phenomenon was observed in certain tests collected from sample M31, which consists of moderately sorted mud. The presence of these sulphides was also noticeable, to a lesser degree, in the last-formed chambers of buliminids and thin-walled species of the genera *Melonis* and *Cancris* in other deep water samples. This feature is taken to be indicative of partially anaerobic conditions existing on the sea floor due to poor water circulation and negligible sediment resuspension.

pH

No direct data concerning pH were available. The effect of lowered pH on transported foraminiferal tests is to cause the solution of the newer chambers so that they become fragile (Murray 1967). Such effects were evident in specimens of *Hyalinea balthica* Schröter and *Planorbulina mediterranensis* d'Orbigny in which the last-formed chambers were frequently found in a fragmented state. At pH values below 7, etching and weakening of the tests takes place over a short period of time. Opaque, etched tests are common among individuals of *H. balthica*. In the absence of pH data it is impossible to distinguish between those tests which had undergone rapid etching, and those that had been subjected to a near neutral pH over a long period of time. However, pH is rarely a critical factor where the other ecological conditions are favourable, as they are in the North Minch Channel. Its importance lies in the fact that differential solution may take place, leading to the elimination of the thinner-walled tests.

Phosphates

Phosphates are among the trace compounds present in sea water that are

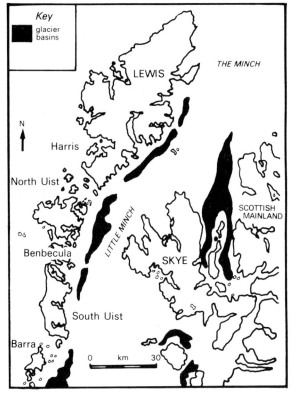

Figure 4.3 Glacier basins cut into the sea channels between the Hebridean Islands and the Scottish mainland.

taken up by organisms. Data relating to phosphate concentrations in the North Minch reveal increases away from the shoreline. This pattern roughly corresponds to a general increase in faunal diversity values which are noted to occur with increasing shore distance.

Bathymetry

The North Minch Channel is essentially a shelf sea, the continental platform being deeply incised in the east. Depths exceeding 120 m are quickly reached to the east of the Eye peninsula of Lewis (Fig. 4.4). The gradient is steepest to the north-west of the Shiant Isles where depths in excess of 200 m occur, marking the position of a north-northeasterly to south-southwesterly trending, glacially cut basin (Fig. 4.3). This basin follows the line of the Minch Fault–a geological structure formed during the Caledonian Orogeny. In the north and north-east the gradient of the platform is much shallower, falling away gradually to a maximum depth of 120 m. To the west and south-west the slope rises again gradually to a depth of 80 m, with isolated rises to 40 m below sea level.

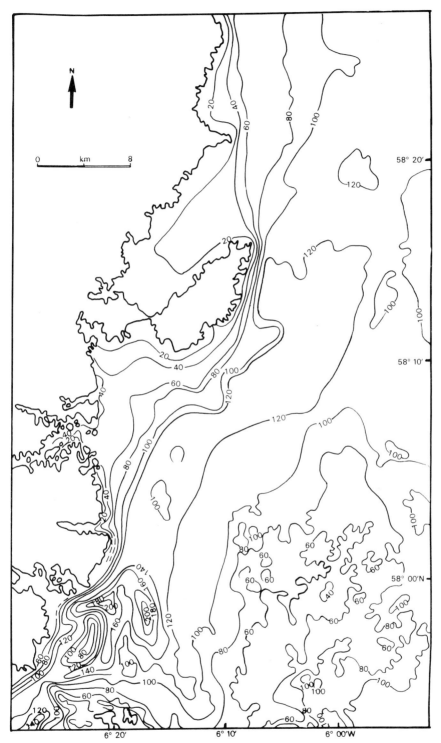

Figure 4.4 Bathymetry of the study area (depths in metres).

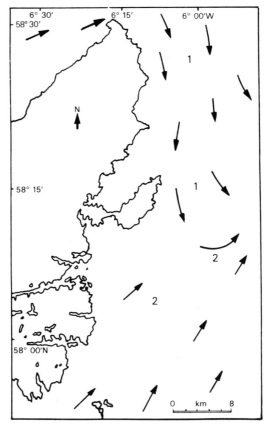

Figure 4.5 Estimated most usual net water movements near the bottom (after Craig 1958). Figures indicate approximate drift speeds (in km per day).

Depth is one of the more obvious variables of the marine environment, but does not appear to be solely the limiting factor in the distribution of foraminifera in the North Minch Channel. The major factor seems to be the interrelationship between depth and substrate type.

Currents

The North Minch Channel is characterised by vertically and horizontally homogeneous, diluted continental shelf water flowing north-east past Cape Wrath. Slightly more saline outer shelf water rounds the Butt of Lewis and enters the North Minch. Whereas there is a well defined northerly flow of deep water through the North Minch (Fig. 4.5), the surface waters appear to have a less clearly defined movement (Fig. 4.6).

In the Little and North Minches maximum tidal current speed is associated with high and low water. The strongest currents occur near the shallow,

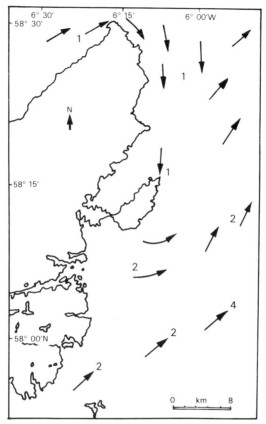

Figure 4.6 Estimated most usual net water movements at the surface (after Craig 1958). Figures indicate approximate drift speeds (in km per day).

shelving coasts from Tolston Head to the Butt of Lewis. Surface currents of 2 ms^{-1} occur during spring tides in the area of the eastern Shiant gravel bank and violent eddies and turbulence are noted around the Shiant Isles which act as obstructions to the paths of the currents.

Hydrodynamic factors of the sedimentation process indicate that areas of comparatively turbulent water also have quiescent water. Some foraminiferal species will be unable to inhabit the vigorous environment of turbulent water and strong current action but flourish in quiet water, e.g. *Bulimina marginata* with its delicate test and a lack of pronounced surface ornamentation. It is probable that some of the foraminifera that do inhabit the more turbulent waters may not be proportionally represented due to selective removal on death by transportation by currents or through fragmentation. This factor appears to be true of the well sorted lithic sands and shell sands of the open shelf area lying along the paths of converging regional currents, where the majority of the quinqueloculine fauna is in a fragmented state and

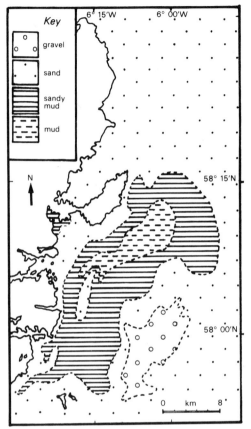

Figure 4.7 Distribution of sediment type in the North Minch (after Bishop & Jones 1980).

shows a high degree of abrasion, i.e. a death assemblage, having been transported from the coarse sands of the inner shelf areas.

Substrate

The sediment substrate is defined as an unconsolidated surface on or in which organisms live. Throughout the area of study (Fig. 4.7), the sediments exhibit a simple pattern of textural types comprising a range of gravels (over 2.00 mm diameter), sands (0.063–2.00 mm diameter) and silts (less than 0.063 mm diameter). The basis of the proposed classification of the textural groups is a triangular diagram (Folk 1968) on to which are plotted the proportions of sand, gravel and mud in the whole sediment. The gravel fraction, if present, is expressed as a percentage; the sand to silt–mud proportion is expressed as a ratio. Depending on the relative proportions of these three major constituents, fifteen textural groups are defined (Fig. 4.8).

The sediments of the North Minch, when plotted on the triangular diagram, occupy a field which encompasses the majority of the textural groups with the exception of mS, gM and mG.

Benthonic foraminifera may be regarded as constituents of the sediment and therefore behave as any other sedimentary particle in that they may be carried as sand grains by saltation through current action. This study takes into account that the present assemblages possibly do not represent a biocoenosis or life assemblage and might simply have been current drifted. In this respect, the degree of sorting of the sediments, as well as being used as a possible control on distribution, is used to determine the effects of currents in the area, i.e. a poorly sorted sediment is not subject to a great deal of current activity, hence the incorporated microfauna is taken to represent a biocoenosis. The sediments of the North Minch Channel exhibit a variety of degrees of sorting ranging from extremely poor to well sorted. In areas of high current velocities the bottom is one of shifting substrates and unstable conditions with a small proportion of mud and organic matter. Conversely, in areas of low current activity there is much mud, and hence a stable substrate.

The sediments of Broad Bay are poorly sorted. The degree of sorting

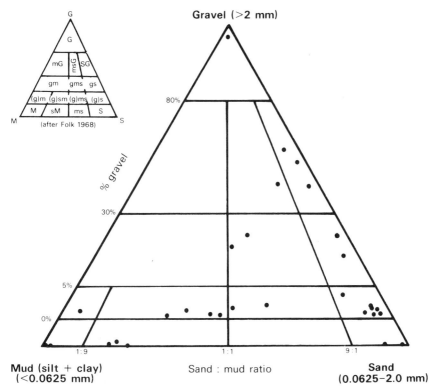

Figure 4.8 Triangular plot of the major textural groups of sediments in the North Minch Channel. G, g=gravel; S, s=sand; M, m=mud.

increases towards open shelf conditions. Thus in Broad Bay the faunal associations may be regarded as a life assemblage with little or no postmortem transportation. Over much of the continental shelf the sediments comprise well sorted lithic sands and shell sands. The high frequency of occurrence of certain foraminiferal species in the total assemblage at stations located in this area need not indicate true ecological dominance since the factors of post-mortem transportation and deposition may have led to a concentration of empty tests in areas other than the preferred habitats of the living organisms, i.e. these tests represent a thanatocoenosis or death assemblage. In the deep basins, floored with poorly to moderately sorted sediments, the faunal assemblages may be considered to represent a biocoenosis, with the tests accumulating on the sea floor under conditions of low current activity. The poorest sorting coefficients are attained in the sediments from the high energy environment of the eastern Shiant gravel bank, representing the other extreme of hydrodynamic–sediment–substrate conditions.

Grain size is the most obvious feature of a sediment and affects both directly and indirectly the organisms living in or upon it. Feeding methods, mode of attachment and movement of organisms are closely related to the grain size of a sediment. In the foraminiferal populations examined there appears to be a marked correlation between the median grain size of the sediment and the diameter of the tests.

Fine sand (0.18 mm average grain diameter) is more easily moved and better sorted than finer or coarser grades of sediment. Many small foraminiferal species inhabit the water-filled spaces between the sedimentary particles, i.e. an interstitial fauna. The large interstitial spaces present in lithic sands and shell sands provide living spaces for such a fauna. The size of the interstices is determined by the grain shape, packing and degree of sorting. The same sediment, particularly if it is poorly sorted, will have a range of different critical grain sizes for at least some of the interstitial species, e.g. *Rosalina globularis* d'Orbigny and *Cibicides lobatulus* (Walker & Jacob). The occurrence of foraminifera in the interstitial fauna has some bearing on their distribution patterns. An increase in the silt–clay content of the sediment effectively reduces the size of the interstices, and ultimately restricts the distribution of certain interstitial faunas to coarser-grained sediments. This is illustrated by the observable decrease in the frequency of attached forms with increasing depth and decreasing grain size of the sediments.

Distribution of sediment types in the North Minch Channel

Gravels

There is an extensive development of shell gravels and true gravels on the eastern Shiant bank over a depth range of 60–80 m. The proportion of gravel present is a function of the highest current velocity at the time of deposition

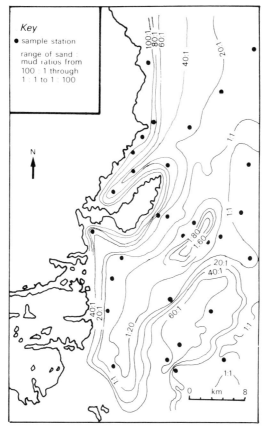

Figure 4.9 Distribution of sand:mud ratios in the sediments of the North Minch Channel.

in association with the maximum grain size of the detritus available. The sediments of the eastern Shiant bank comprise gravels with maximum grain diameters which reach cobble and pebble proportions (up to 256 mm diameter) on the Wentworth scale. The lithic constitutents of the gravels are large, well rounded pebbles and cobbles of Torridonian sandstone and Lewisian metamorphics derived from the mainland and the Hebridean islands, and a sandy matrix comprising angular to subangular quartz grains. The incorporated shell debris includes bivalved mollusc shell fragments, gastropod shell debris and worm tubes. The foraminiferal fauna inhabiting this turbulent environment comprises thick-shelled, robust forms with a predomination of attached species.

Material finer than 63µm diameter (Fig. 4.9)

The sand:mud ratio reflects the amount of winnowing that has taken place at the site of deposition. The significance of the amount of material finer than

63μm diameter present lies in its influence upon the density of the sediment. The greater the mud content, the lower will be the density of the unconsolidated sediment as a whole. The mean particle size of the shallow marine sediments decreases with increasing distance from the shoreline at which the sediment can be supposed to have originated. The highest ratio of mud to sand occurs in the centre of the glacially cut trough to the north-west of the Shiant Isles. There is a fall-off of sedimentary grain sizes towards the centre of the trough, with a passage from shell sands through sandy muds to true muds. The occurrence of fine-grained sediments in the trough may be the result of a combination of factors. First, there is a general decrease in grain size with increasing water depth, the greatest percentages by weight of fines occurring in the deepest areas on the shelf, e.g. 98.62% fines at Station M222 at a depth of 127 m. Secondly, the area of maximum concentration of fines is related to the general circulation, since the area to the north-west of the Shiant Isles is a point of convergence of the two trends of water circulation (Fig. 4.6) causing a decrease in current rates which results in the setting up of a current shadow where relatively quiet waters allow the fines to settle out (Bishop & Jones 1980). Finally, this is also an area of decreased maximum tidal current velocities, away from the maximum velocity areas near the Shiant Isles and the Butt of Lewis. The muddy sands, silts and true muds bear a characteristic thin-shelled fauna. Organically rich muddy sand was observed to provide a suitable medium for an abundant benthonic population.

Carbonate material

In the northern part of the North Minch Channel, the calcareous portion of the sediments is mainly composed of the remains of macrobenthos, comprising echinoderm tests and spine debris, and whole or comminuted shells of bivalved molluscs and of gastropod molluscs, together with bryozoans and worm tubes. These organisms contributed their skeletal remains to the sediment, forming shell sands and shell gravels. The total frequency of foraminiferal tests present in these sediments is less than 5% of the biogenic fraction. Hence the macrofaunas appear to dominate the benthonic population in the sublittoral zone of the marine environment, due to tidal currents and considerable vertical mixing of water providing the organisms with a maximum source of food.

The overall percentage component of the foraminifera in the bioclastic fraction of the sediments shows an increase towards open marine conditions. Under a low rate of sedimentation, as in the deep trough to the north-west of the Shiant Isles, in association with a lithology of muds, gravelly muds and sandy muds, foraminifera constitute the major calcareous constituent of the sediments.

The relationship between foraminiferal shell form and substrate type

Shell form is an adaptation of an organism to its environment and it is a direct reflection of that environment. The thickness of the test of foraminiferal species is observed to increase with increasing grain size of the sediments. From the observations made it is apparent that grain size is the predominating factor governing the character of the tests of benthonic foraminifera. The numbers of firmly attached species such as *Cibicides* spp., *Rosalina globularis* d'Orbigny and *Planorbulina mediterranensis* d'Orbigny increase with increasing coarseness of the sediments. These species are distinguished by their well marked, flattened attachment surfaces. Certain specimens are found still clinging to detrital material in the samples examined. The species *Acervulina inhaerens* Schultze is common in well sorted, coarse sands, where it is observed to encrust quartz grains and bioclastic material such as shell fragments and bryozoans. Such species are restricted to firm substrates that are suitable for attachment.

On a well sorted sandy substrate, biconvex tests offer very little resistance to current action. Vagrant benthonic forms such as miliolids and *Elphidium* species occur in association with attached species in high energy environments. Such free-living forms are generally thick-shelled to withstand vigorous current activity. The biconvex test of *Eponides repandus* (Fichtel & Moll) displays a secondary thickening of shell material on the exterior surface which serves to strengthen it against damage through current buffeting occurring in the high energy environment of the eastern Shiant gravel bank. Adaptations to survive or avoid turbulent transport are commonplace among faunas associated with coarse, well sorted sandy sediments. None of the tests of foraminifera occupying the current-swept environments bear ornament in the form of spines or other protrusions which may be broken during transport.

Under reduced current activity, and on a substrate of fine-grained sediments, there is a predominance of forms with a rounded to sub-rounded cross section, e.g. *Bulimina marginata* d'Orbigny, and of individuals with an elliptical to relatively flat cross section, e.g. *Cassidulina carinata* Silvestri. The intact preservation of thin-walled tests and of such delicate test features as the long slender necks of unilocular forms (such as *Lagena*) provides evidence of non-turbulent deposition. Hence, in such an environment, foraminifera with vastly differing test shapes are observed to occur in the same population.

Distribution of selected foraminiferal species

The dominance of a given species in a particular habitat can be correlated directly with the success of that species in adapting to its environment. The

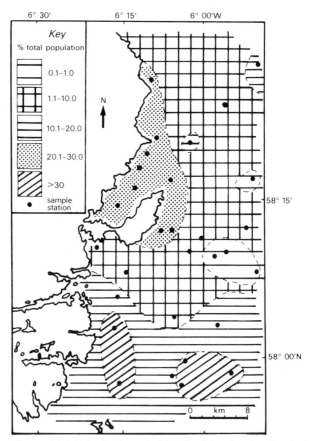

Figure 4.10 Distribution of *Cibicides lobatulus* (Walker and Jacob) in the North Minch Channel.

frequency variation of the foraminiferal species is gradual from one station to the next. Many of the species recorded occur so infrequently as to make their distribution meaningless. Seven commonly occurring shelf species are selected and their distributions analysed.

(a) *Cibicides lobatulus* (Walker & Jacob) (Fig. 4.10, Pls 4.1a–c). The observations made are in accordance with those of Nyholm (1961) who showed that this species exhibits a considerable range of morphological variation in the shape of its test caused primarily by the nature of the substratum. *C. lobatulus* is recorded in all types of sediment encountered in the North Minch Channel, but it is most abundant on a substrate of coarse sand. Occasional intergrown individuals are observed. This intergrowth is caused by an overlap of the territories occupied by the organisms in life. The area of maximum frequencies of occurrence extends from Broad Bay along the inner shelf zone off the northeastern coast

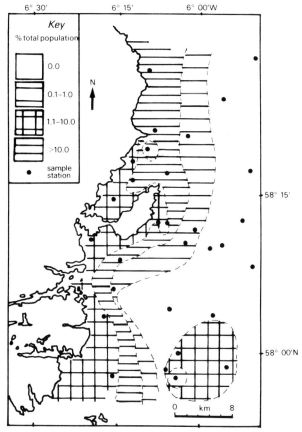

Figure 4.11 Distribution of *Rosalina globularis* d'Orbigny in the North Minch Channel.

of Lewis. The abundance of *C. lobatulus* populations decreases with increasing distance from the shore and with correspondingly greater depths. In the fine-grained lithologies flooring the deep basin this species has a patchy distribution; the individuals are small, more symmetrical, and do not show the degree of morphological variation witnessed in forms found on coarser substrates. The percentage occurrence of *C. lobatulus* increases once again towards the eastern Shiant gravel bank where it dominates the population in association with *Eponides repandus* (Fichtel & Moll) and *Gaudryina rudis* Wright. *C. lobatulus* also represents the dominant species of the microfauna on a lens of sand off the east coast of Lewis.

(b) *Rosalina globularis* d'Orbigny (Fig. 4.11, Pls 4.1d–f). The distribution of *R. globularis* is comparable to that of *C. lobatulus* to which it is subordinate in numbers. This is another substrate-conforming, attached species which is found clinging to sand grains, shell material, etc., in life.

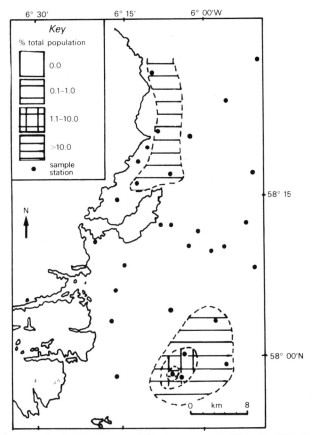

Figure 4.12 Distribution of *Eponides repandus* (Fichtel and Moll) in the North Minch Channel.

The maximum frequencies of occurrence of *R. globularis* are attained at stations in Broad Bay and off the northeastern coast of Lewis. This species is absent from the offshore shelf area of well sorted sands, and from the fine-grained sediments in the deepest areas of the North Minch Channel. Whereas *C. lobatulus* formed a major component of the microfauna on the eastern Shiant gravel bank, *R. globularis* displays an inverse distribution, with a fall-off in percentage frequencies towards the centre of the bank. This fall-off indicates that the relatively thin-walled test of this species is unable to withstand the turbulence caused by high current velocities across the bank.

(c) *Eponides repandus* (Fichtel & Moll) (Fig. 4.12, Pls 4.1g–i). *E. repandus* has a restricted distribution associated with coarse gravelly sands, well sorted sands and sandy gravels deposited under conditions of turbulence and high current velocities. The maximum frequency of occurrence of this species is attained on the very coarse sandy gravels of the eastern

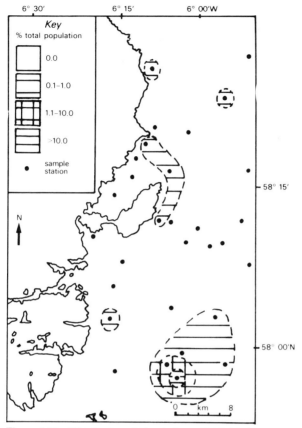

Figure 4.13 Distribution of *Gaudryina rudis* Wright in the North Minch Channel.

Shiant bank. There is a gradual fall-off in numbers of individuals away from the centre of the bank. The second area of occurrence lies to the north-east of the Isle of Lewis in a strip stretching southwards across the mouth of Broad Bay in the inner shelf zone. Its distribution overlaps on to the *C. lobatulus*-dominant zone. *E. repandus* is absent from the offshore, open shelf area, and from the deep basin.

(d) *Gaudryina rudis* Wright (Fig. 4.13, Pls 4.2d–f). The agglutinated test of *G. rudis* is conical and robust, with a wall composed of clear quartz grains. Murray (1971) suggests that this species may live by clinging to hard substrates. The overall distribution is similar to that of *E. repandus*. Maximum frequency of occurrence is recorded on the eastern Shiant gravel bank. The distribution of *G. rudis* in the northern part of the North Minch Channel is more patchy, being predominantly associated with a substrate of poor to well sorted gravel sands.

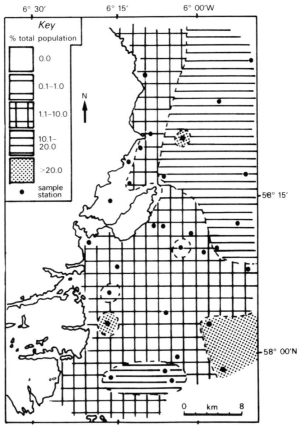

Figure 4.14 Distribution of *Textularia sagittula* Defrance in the North Minch Channel.

(e) *Textularia sagittula* Defrance (Fig. 4.14, Pls 4.2g,h). The *T. sagittula* species group is a complex of variable forms. The species is often confused with *Spiroplectammina wrightii* (Silvestri) (Pls 4.2i,j) which is planispiral in the initial portion of the test, whereas *T. sagittula* is biserial throughout. Both species have a comparable distribution. *T. sagittula* seems to have a wide range of adaptability, being found throughout the full spectrum of substrate types in the North Minch Channel. Low frequencies of occurrence are attained over much of the inner shelf zone, increasing gradually towards the open shelf zone offshore in association with well sorted sands and gravelly sands. Maximum frequencies are attained on a substrate of muddy sand at Stations M218 and M303. *T. sagittula* appears to display a preference in the size of quartz grains it agglutinates into its test wall, preferring grains of the very fine sand grade. The author suggests that this factor accounts for the low

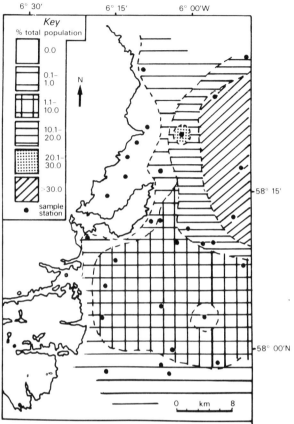

Figure 4.15 Distribution of *Hyalinea balthica* (Schröter) in the North Minch Channel.

frequencies of this species recorded over much of the North Minch Channel on substrates of coarse sands.

(f) *Hyalinea balthica* Schröter (Fig. 4.15, Pls 4.2a,b). The overall distribution of *H. balthica* shows a wide range of adaptability. The species is essentially absent from the innermost shelf areas, its numbers increasing towards the open shelf area. The maximum frequency of occurrence is reached at depths in excess of 100 m on the open shelf areas in the north and north-east of the North Minch Channel in association with substrates of well sorted sands. Moderate frequencies are attained to the south and south-west of the North Minch on substrates of finer-grained sediments.

(g) *Bulimina marginata* d'Orbigny (Fig. 4.16, Pl. 4.2c). A gradual increase in the total number of individuals of this species occurs with increasing water depth and corresponding decrease in sedimentary grain size.

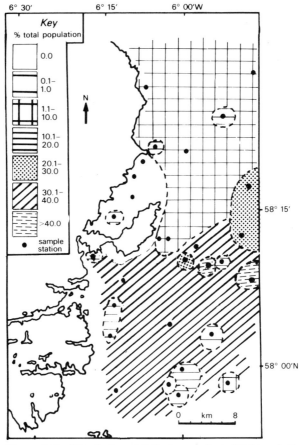

Figure 4.16 Distribution of *Bulimina marginata* d'Orbigny in the North Minch Channel.

Maximum frequencies are attained on substrates of poorly sorted sandy muds, silts and true muds deposited under quiescent conditions at depths in excess of 100 m.

Generic distributions and dominance

A biofacies may be defined as an assemblage of organisms, alive or dead, which are repeatedly found together and which are typical of certain kinds of environmental conditions. Six biofacies are recognised in the North Minch Channel. Figure 4.17 shows the distributions of areas in which the named genus constitutes the largest proportion of the microfauna on a numerical percentage basis. The six biofacies are as follows: *Cibicides*-dominated, *Eponides*-dominated, *Gaudryina*-dominated, *Textularia*-dominated, *Hyalinea*-dominated, *Bulimina*-dominated.

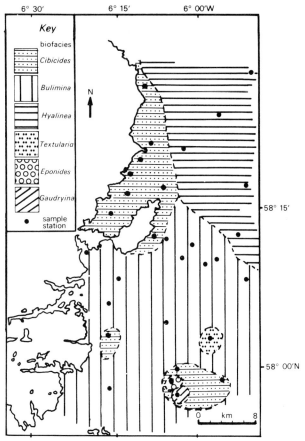

Figure 4.17 Distribution of generic dominance (areas in which the named genera constitute the largest portion of the fauna on a number percentage basis).

The *Cibicides* biofacies occupies the inner shelf area, associated with sand, gravelly sand and sandy gravel substrates. This genus dominates the microfauna at Station M215, on a lens of gravelly sand within the deep basin, which is essentially floored with fine-grained sediments. On substrates of gravelly sands and true sands showing various degrees of sorting, the microfauna is dominated by *Cibicides* spp. with subordinate numbers of *Rosalina* spp. The presence or absence of other genera is determined by the factor of current sorting of the sediment, i.e. a poorly sorted sediment contains a wider spread of subordinate genera. On the well sorted sandy substrates in the nearshore shelf areas, current sorting winnows away the tests of subordinate genera so that the fauna remaining comprises a dominance of substrate-conforming, attached forms which resist current transport.

The *Textularia* biofacies is related to an essentially mud–sand substrate.

Table 4.1 The relationship between foraminiferal and sedimentary parameters.

Station	Depth (m)	Dominant genus	Percentage of attached forms	Textural groups (Folk 1968)	Sorting	Median grain size (μm)	Percentage of material <63 μm
M273	19	*Cibicides*	86.28	(g)S	well sorted	177	0.01
M30	20	*Bulimina*	11.65	(g)sM	poorly sorted	31	78.07
NS229	23	*Cibicides*	90.58	(g)S	well sorted	210	1.15
M274	24	*Cibicides*	59.35	(g)S	poorly sorted	210	1.95
NS228	25	*Cibicides*	68.18	sG	poorly sorted	300	0.73
NS211	35	*Cibicides*	58.33	(g)S	well sorted	177	0.31
NS212	40	*Cibicides*	70.90	S	well sorted	300	0.0
M209	56	*Eponides*	16.28	sG	poorly sorted	420	1.14
M249	56	*Cibicides*	34.07	G	extremely poorly sorted	?	2.76
NS1	60	*Textularia*	22.47	msG	very poorly sorted	300	6.9
M208	61	*Gaudryina*	30.29	sG	poorly sorted	420	3.18
M303	69	*Cibicides*	27.42	gmS	very poorly sorted	350	17.12
NS227	80	*Cibicides*	71.33	(g)S	poorly sorted	420	2.64
NS187	85	*Cibicides*	71.63	(g)S	well sorted	177	2.57
NS178	90	*Bulimina*	8.63	(g)M	well sorted	88	93.89
NS192	105	*Hyalinea*	15.60	(g)S	well sorted	149	5.5
NS3	105	*Hyalinea*	1.33	(g)S	moderately sorted	210	7.27
NS190	105	*Hyalinea*	16.12	(g)S	moderately sorted	149	4.67
M31	106	*Bulimina*	15.52	M	moderately sorted	156	97.67
M226	106	*Bulimina*	0.94	(g)sM	poorly sorted	74	61.44
M281	107	*Bulimina*	0.32	(g)sM	poorly sorted	62.5	4.86
NS191	107	*Hyalinea*	7.44	S	well sorted	177	4.86
NS2	110	*Bulimina*	0.54	sM	well sorted	44	75.62
M215	114	*Cibicides*	23.44	(g)mS	poorly sorted	88	46.15
NS199	115	*Hyalinea*	0.65	(g)mS	poorly sorted	125	23.97
NS179	120	*Bulimina*	5.05	sM	moderately sorted	31	78.33
M222	127	*Bulimina*	2.23	M	well sorted	44	98.62
M219	134	*Bulimina*	30.33	(g)sM	well sorted	62.5	52.12
NS201	135	*Bulimina*	1.06	M	well sorted	37	96.6
M29	176	*Bulimina*	26.20	sM	poorly sorted	31	77.68

Hyalinea occurs in association with a well sorted gravel–sand substrate in the offshore, open shelf area to the north-east of the North Minch Channel. The *Eponides* and *Gaudryina* biofacies are associated with the coarse substrates of the eastern Shiant gravel bank, which are characterised by a fauna comprising a predominance of attached and thick-shelled genera. The *Bulimina* biofacies is the most extensive of the six biofacies outlined, occupying the central area and much of the southern portion of the North Minch Channel. The appearance of *Bulimina* and *Hyalinea* species in the fine-grained sediment fraction towards the margins of the gravel bank reflects a reduction in current activity.

Conclusions

In view of the information available, it appears that the bottom sediment type and the associated sedimentary parameters are the major limiting factors governing the distribution of Recent benthonic foraminifera in the North Minch Channel. The sediments of the continental shelf are patchy and variable, whereas the sediments of the deeper areas are more uniform throughout. The processes of transportation and deposition, as expected, are more variable on the shelf areas than in the bathymetrically deep areas. Under turbulent conditions, the foraminifera tend to be confined to those ecological niches that afford shelter, e.g. interstitial faunas, attached to detrital material in the inshore shelf areas.

The relative coarseness of the substrate is considered to be more important than water depth in determining the distributions of foraminiferal assemblages. Distributions are observed to correlate closely with the mean grain size of the sediment. Post-mortem transport of foraminiferal tests has been shown to vary according to test size and density. Comparison of the shell form of each separate species to the enclosing sediment generally revealed whether or not the specimens had been transported from another environment. Non-indigenous species are recognised by their poor state of preservation. On the other hand, a life assemblage or biocoenosis is recognised by the presence of a large number of species showing a range of test sizes and occurring in a poorly sorted sediment. Concentrations of tests in fine-grained substrates appear to be due to accumulation in a current shadow rather than to differential winnowing from coarser deposits.

The study reveals that the majority of foraminiferal species have a distinct preference for well sorted sediments which provide larger, inhabitable, interstitial spaces.

The present study is limited in scope, but the results and conclusions demonstrate that the methods of study outlined can be of use in the examination of fossil foraminiferal assemblages and thus contribute to a better understanding of problems in stratigraphy and palaeoecology where fossilised tests of foraminifera are the principle biotic constituents in geological marine sediments.

Acknowledgements

I would like to extend my thanks to the following people who have assisted during the course of this study: Professor T. Barnard (University College London) for his supervision and helpful criticism of the work; Dr P. Bishop (formerly of University College London) for his advice and useful communications throughout, and for making available certain maps and charts. I am also indebted to the Continental Shelf and Marine Geophysics Units of the Institute of Geological Sciences, Edinburgh, for availability of the samples examined during this study; to Mr M. Gay (University College London) for photography of specimens and to the Shell International Petroleum Company for financial assistance towards the cost of the scanning electron micrographs.

References

Atkinson, K. 1969. The association of living foraminifera with algae from the littoral zone of Cardigan Bay, Wales. *J. Nat. Hist.* **3**, 517–42.

Atkinson, K. 1971. The relationship of Recent foraminifera to sedimentary facies in the turbulent zone, Cardigan Bay, Wales. *J. Nat. Hist.* **5**, 385–439.

Bishop, P. and E. J. W. Jones 1980. Patterns of glacial and post-glacial sedimentation in the Minches, north-west Scotland. In *The North-west European shelf seas: the sea bed and the sea in motion*, Vol. 1, *Geology and sedimentology*. F. T. Banner, M. B. Collins & K. S. Massie (eds), 89–192. Amsterdam: Elsevier.

Brady, H. B. 1887. A synopsis of British Recent foraminifera. *J. R. Microsc. Soc.*, Ser. 2, **7**, 872–927.

Craig, R. E. 1951. Hydrographic conditions in the Minch, Sept./Oct. 1951. *Annls biol.* **10**, 20–1.

Craig, R. E. 1958. *Hydrography of Scottish coastal waters*. Scottish Home Dept Marine Res., Ser. 2.

Folk, R. L. 1968. *Petrology of sedimentary rocks*. Austin, Texas: Hemphill's Book Store.

Haynes, J. R. 1973. Cardigan Bay Recent foraminifera (cruises of the RV *Antur*, 1962–1964). *Bull. Br. Mus. Nat. Hist. (Zool.)* Suppl. 4, 1–245.

Haynes, J. R. and M. R. Dobson 1969. Physiography, foraminifera and sedimentation in the Dovey Estuary, Wales. *J. Geol.* **6**, 217–56.

Heron-Allen, E. and A. Earland 1909. On Recent and fossil foraminifera of the shoresands of Selsey Bill, Sussex. *J. R. Microsc. Soc.* **1909**, 306–66, 422–46.

Heron-Allen, E. and A. Earland 1913. On some foraminifera from the North Sea dredged by the Fisheries cruiser *Huxley*. *J. Quekett Microsc. Club*, Ser. *11* (73), 121–38.

Heron-Allen, E. and A. Earland 1916. The foraminifera of the west of Scotland collected by Prof. W. R. Herdman, FRS, on the cruise of the SY *Runa*, July–Sept. 1913. Being a contribution to *Spolia Runiana*. *Trans Linn. Soc. Lond.*, Ser. 2 (Zool.) **11**, 197–300.

Heron-Allen, E. and A. Earland 1930. The foraminifera of the Plymouth district. *J. R. Microsc. Soc.* **50**, 46–84.

Murray, J. W. 1965. Foraminifera of the Plymouth region. *J. Mar. Biol. Assoc. UK* **45**, 481–505.

Murray, J. W. 1967. Transparent and opaque foraminiferid tests. *J. Palaeont.* **41**, 791.

Murray, J. W. 1968. The living Foraminiferida of Christchurch Harbour, England. *Micropaleontology* **14**, 435–55.

Murray, J. W. 1970. Living foraminifers of the western approaches to the English Channel. *Micropaleontology* **16**, 471–85.

Murray, J. W. 1971. *An atlas of British Recent foraminiferids*. London: Heinemann.

Nyholm, K. G. 1961. Morphogenesis and biology of the foraminifer *Cibicides lobatulus*. *Zool. Bidr. Upps.* **33**, 159–96.

Williamson, W. C. 1858. *On the Recent foraminifera of Great Britain*. London: Ray Society.

Appendix: Total benthonic foraminiferal species recorded in the North Minch Channel

Acervulina inhaerens Schultze
Ammonia beccarii (Linné), *A.* sp.
Ammoscalaria sp.
Amphicoryna cf. *A. scalaris* (Batsch), *A.* sp.
Asterigerinata mamilla (Williamson)
Astacolus sp.
Brizalina pseudopunctata (Höglund), *B. spathulata* (Williamson), *B. variabilis* (Williamson)
Bulimina gibba/elongata Fornasini and d'Orbigny respectively; *B. marginata* d'Orbigny
Cancris auricula (Fichtel & Moll)
Cassidulina carinata Silvestri, *C. obtusa* Williamson
Cibicides fletcheri Galloway and Wissler, *C. lobatulus* (Walker & Jacob), *C. pseudoungerianus* (Cushman)
Cribrostomoides jeffreysii (Williamson)
Cyclammina cancellata Brady
Cyclogyra involvens (Reuss)
Dentalina subarcuata (Montagu), *D.* sp.
Discorbis sp.
Eggerella scabra (Williamson)
Elphidium articulatum (d'Orbigny), *E. crispum* (Linné), *E. excavatum* (Terquem), *E. gerthi* Van Voorthuysen, *E. williamsoni* Haynes, *E.* sp.
Epistominella vitrea Parker, *E.* sp.
Eponides repandus (Fichtel & Moll)
Fissurina lucida (Williamson), *F. marginata* (Montagu), *F. orbignyana* Sequenza, *F.* sp.
Fursenkoina fusiformis (Williamson)
Gaudryina rudis Wright
Gavelinopsis praeggeri Heron-Allen and Earland
Globocassidulina subglobosa (Brady)
Globulina gibba d'Orbigny, *G.* sp.
Guttulina lactea (Walker & Jacob)
Haplophragmoides bradyi (Robertson)
Haynesina germanica (Ehrenberg)
Hyalinea balthica (Schröter)
Lagena clavata (d'Orbigny), *L. interrupta* Williamson, *L. laevis* (Montagu), *L. perlucida* (Montagu), *L. semistriata* Williamson, *L. striata* (Walker), *L. substriata* Williamson, *L.* sp.
Lagenammina arenulata (Skinner)
Marsipella elongata Norman
Massilina secans d'Orbigny
Melonis pompilioides (Fichtel & Moll)
Miliammina fusca (Brady)
Miliolinella circularis (Bornemann) var. *elongata* Kruit, *M. subrotunda* (Montagu)
Nonionella turgida (Williamson)
Oolina hexagona (Williamson), *O. squamosa* (Montagu), *O. williamsoni* (Alcock), *O.* sp.
Pateoris hauerinoides (Rhumbler)
Planorbulina mediterranensis d'Orbigny
Pleurostomella sp.
Pseudopolymorphina cf. *novanglae* (Cushman), *P.* sp.
Pyrgo constricta Costa, *P. depressa* (d'Orbigny), *P. williamsoni* (Silvestri), *P.* sp.
Quinqueloculina aspera d'Orbigny, *Q. bicornis* (Walker & Jacob), *Q. bicornis* var. *angulata* (Williamson), *Q. dimidiata* Terquem, *Q. duthiersi* (Schlumberger), *Q. lata* Terquem, *Q. oblonga* (Montagu), *Q. rugosa* d'Orbigny, *Q. seminulum* (Linné).

Reophax fusiformis (Williamson), *R. scorpiurus* Montfort
Robulus sp.
Rosalina globularis d'Orbigny, *R.* sp.
Sigmorphina sp.
Spirillina vivipara Ehrenberg
Spiroloculina excavata d'Orbigny, *S.* cf. *S. rotunda* d'Orbigny
Spirophthalmidium acutimargo (Brady)
Spiroplectammina wrightii (Silvestri)
Stainforthia sp.
Textularia earlandi Parker, *T. sagittula* Defrance
Trifarina angulosa (Williamson), *T. bradyi* Cushman, *T.* sp.
Triloculina sp.
Trochammina rotaliformis Heron-Allen and Earland
Uvigerina peregrina Cushman

Plates

Plate 4.1 Scanning electron micrographs. (a)–(c) *Cibicides lobatulus* (Walker and Jacob). (a) Dorsal view; (b) side view; (c) ventral view. ×80. Station NS 211.

(d)–(f) *Rosalina globularis* d'Orbigny. (d) Ventral view; (e) side view; (f) dorsal view. ×80. Station M273.

(g)–(i) *Eponides repandus* (Fichtel and Moll). (g) Ventral view; (h) side view; (i) dorsal view. ×60. Station M249.

Plate 4.2 Scanning electron micrographs. (a), (b) *Hyalinea balthica* (Schröter). (a) General view; (b) side view. ×80. Station NS 192.

(c) *Bulimina marginata* d'Orbigny. General view. ×100. Station M219.

(d)–(f) *Gaudryina rudis* Wright. (d) General view; (e) view of interiomarginal aperture. ×60. Station M209. (f) General view. ×60. Station M208.

(g) (h) *Textularia sagittula* Defrance. (g) General view, showing biserial chamber arrangement; (h) apertural view. ×80. Station NS 190.

(i),(j) *Spiroplectammina wrightii* (Silvestri). (i) General view, showing planispiral initial portion; (j) apertural view. ×80. Station NS 1.

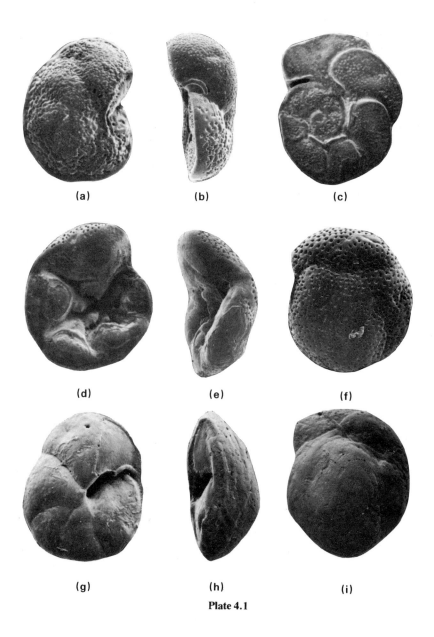

Plate 4.1

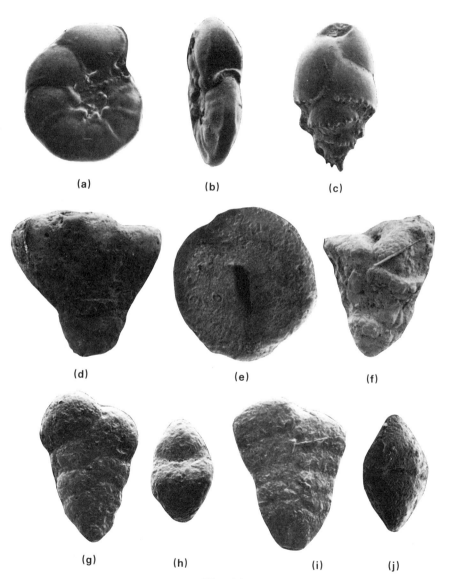

Plate 4.2

5 A classification and introduction to the Globigerinacea

F. T. Banner

> An index is given for all available genus-group and family-group names which are applicable to the Globigerinacea; the literary sources, the nature, source and depositories of the type specimens, and selected references to significant emendations, are given. The type-species of each genus-group name is illustrated, where possible with direct reference to type specimens. The taxa of the genus-group are reclassified in a discriminatory key, in which objective synonymy is noted and some subjective synonymy is suggested, and, from which, further subjective assessments are possible. The probable phylogenetic history of the Globigerinacea is shown graphically and is discussed with especial reference to the iterative evolution of gross homeomorphs in closely or more distantly related family groups.

Introduction

Modern attempts to systematise knowledge of the planktonic foraminifera really began following the success of the micropalaeontologists of the Trinidad Leaseholds Oil Company (now Texaco Trinidad Inc.), in the decade following World War II, in the biostratigraphic application of these microfossils in a geologically complicated area of highly faulted, thrust, slumped and reworked pelagic Cenozoic sediments (see, for example, Brönnimann 1950a,b, 1951b, 1952b,c; Brönnimann & Brown 1956). With few other stratigraphic tools to hand, it was shown that the species and genera of the Globigerinacea, if carefully discriminated, could provide surface-to-subsurface and well-to-well correlations to disentangle the complex structural and facies problems which beset exploration there. Many of the conclusions were published by Loeblich *et al.* (1957) and, in particular, the contribution by Bolli *et al.* (1957) set out the morphological bases for systematisation of the globigerines which had proved so useful. This, and subsequent papers (e.g., Bolli 1957a,b,c,d, 1958, 1959; Bolli & Bermúdez 1965; Blow 1956, 1959) marked the beginning of an explosion of research and publication on the systematics and stratigraphy of the Cenozoic globigerines, which paralleled that occurring, contemporaneously but independently, in Moscow and Leningrad, into the Mesozoic and Palaeogene Globigerinacea of the USSR (see, for example, Morozova 1948, 1957, 1958, 1959; Subbotina

1953; Berggren 1960). Since then, the expansion of micropalaeontological biostratigraphy in world-wide petroleum exploration and deep ocean drilling (see any of the volumes of the *Initial reports of the Deep Sea Drilling Project*, published by the US Government Printing Office, Washington, DC) and the research programmes which have been stimulated in universities, has led to the publication of vast numbers of assemblage descriptions, biostratigraphic reports, phylogenetic hypotheses and attempts to develop unambiguous and universally applicable systems of taxonomy and nomenclature, so that the descriptions, reports and hypotheses could be unequivocal, scientifically valid, and generally acceptable as bases upon which further work could proceed. The introduction of transmission and scanning electron microscopy, as well as improved optical photographic techniques, greatly enhanced both knowledge of the details of test structure and the quality of communication of the results between research workers; latterly, ecological (e.g. Reiss & Halicz 1976, Bé 1977), ontogenetic (e.g. Bé 1980), and biological studies (e.g. Anderson *et al.* 1979, Hemleben *et al.* 1977, 1979, Spindler *et al.* 1978, 1979) promise to yield knowledge of the purpose and function of the globigerinacean test itself. Even biochemistry has been used to test taxonomic relationships (King & Hare 1972).

When Cushman (1948) published the last edition of his textbook on the foraminifera, he recognised only twenty-two genera in three families (Globigerinidae, Hantkeninidae, Globorotaliidae) and, of these, three genera now must be excluded as benthic foraminifera. Four years later, Bermúdez's attempt (1952) to classify the rotaliform foraminifera recognised thirty-six genera in four families, but, as in Cushman's (1948) compilation, planktonic foraminifera were not systematically separated from benthonic ones and only thirty-four of the genera used by Bermúdez would nowadays be considered to be globigerinacean. In 1957, Bolli, Loeblich and Tappan produced their masterly paper, in which morphological terms were clearly defined, taxa were clearly described and furnished with a reliable nomenclature, and the true relationships between genera and family-groups began to emerge; they recognised thirty-two genera in four families, and both genera and families developed biostratigraphic value and meaning, with clear implications for phylogenetic hypotheses to support both the taxonomy and the stratigraphy. The following year, Sigal (1958), used six families. Banner and Blow (1959) attempted to build upon all of this work, using differentially diagnostic keys to the genera and subgenera of the Globigerinacea and graphic representation of the phylogenetic links between them; forty-five genera, grouped into three families, were used. Reiss (1963) reclassified all the Rotaliina to family and subfamily level, recognising forty-nine genera of the globigerines. It was from publications of this period that Loeblich and Tappan constructed the classification of the Globigerinacea which was used in the *Treatise* (Loeblich & Tappan 1964a), where seven non-heterohelicid families were used to incorporate forty-eight valid genera.

Since the first edition of the *Treatise*, major reviews of all known supraspecific groups of the Globigerinacea have been attempted by Postuma (1971), using Royal Dutch/Shell Group data, and by El-Naggar (1971a). Stainforth et al. (1975), using Exxon data, and Blow (1969, 1979), using BP data, published comprehensively upon Cenozoic genera and species, and other important surveys of Cenozoic globigerinacean genera included those by Lipps (1966) and Fleisher (1974). Palaeogene genera and species were the subject of reviews by, among others, Berggren (1966), McGowran (1968) and Steineck (1971). Bandy (1967) summarised the biostratigraphy and probable phylogeny of the Mesozoic planktonic foraminifera, and Longoria and Gamper (1975) revised their classification in the light of probable evolution. Very many species were described and discussed by Masters (1977), who also treated phylogeny, biogeography and nomenclature with great thoroughness. Many Late Cretaceous genera and species were carefully reviewed by Pessagno (1967, 1969), and Early and Middle Cretaceous taxa especially were also studied by Longoria (1974). Many thousands of other research papers have been published during this time, a growth of research activity indicated by the fact that Masters (1977) devoted forty-one pages to his cited references alone. There are now some 119 generic and subgeneric names now known to me to be available to accommodate the described species of the Globigerinacea, excluding typonyms, homonyms, names of dubious application, others which apply to species which are likely to have been benthonic, and all the Heterohelicacea.

In the last two decades, many zonal schemes have been produced in order to systematise the biostratigraphic value of globigerinacean species (e.g. Bolli, in Loeblich et al. 1957, Bolli 1966, 1969, Blow 1969, 1979, Bermúdez & Farias 1977, Berggren 1971b, 1978, Berggren & van Couvering 1974, Jenkins 1966, 1971, Pessagno 1967, 1969, Postuma 1971, Stainforth et al. 1975, and very many others). The purpose of this chapter is to review the applicability of the supraspecific taxa to nomenclature, taxonomy, classification and biostratigraphy. The morphological limits which can be given to any genus or subgenus must contain the morphological characters of the type-species, and that type-species, no matter how wide its morphological variation may be claimed to be, must be defined in such a way that all the characters of the type specimens of that species are included within its morphology. Just which morphocharacters of the type specimens are to be judged to be essential to the recognition of the species they represent, and just which common characters of a species group are to be used to define supraspecific categories, are decided, in fact, by the experience and judgement of each individual student, who may or may not follow those who have gone before. Certainly, in the study of the Globigerinacea, the choice of nomenclature and system of taxonomy widens with each and every new published contribution.

Therefore, I have endeavoured to compile a list of generic and subgeneric names available for the Globigerinacea, in each case noting the type-species, the type specimens, their origin and depositories, and giving reference to particularly significant comments, revisions and emendations which have been published about them. Figures 5.1–118 are drawings which are meant to represent the essential characters of each type-species of each available supraspecific name; each is drawn to about the same size for ready comparison, differences of scale (sometimes an order of magnitude) being indicated by scale bar-lines. Each drawing is based upon what is known of the type specimen or specimens of the type-species (as indicated in the keys to these figures), but, where the primary types are ill-preserved and no longer show the complete morphocharacters of the species (and of the genus), then this detail is added from better preserved specimens which I believe to be, beyond doubt, conspecific (wherever possible, using topotypes). The result should be drawings which should enable identification both of the type-species and also of the morphocharacters available for supraspecific classification–particularly for the recognition of species groups which may be named as subgenera, genera, subfamilies or families.

Although the scanning electron microscope has been used whenever possible for the elucidation of detailed structures, the practising micropalaeontologist is still necessarily committed to the optical microscope for his routine handling of samples and the identification of the species and genera they contain. The scientist at the well-head or on board ship will never have the facilities of a scanning electron microscope, nor the time to use them if he had. The micropalaeontologist has the great advantage, usually denied other palaeozoologists, of dealing with large assemblages of specimens, and it is impracticable, even to the favoured in research institutions, to electron-scan more than a few of them. Electron microscopy can explain the reasons for particular optical images and impressions, just as touch and further probing can confirm the evidence of our eyes that some everyday materials are wooden, while others are metallic or plastic. Once this is known, we do not need the analysis every time, once we know the ultrastructural reason for the visual impression we receive. Therefore, all supraspecific groups are defined here on characters recognisable, with a standard optical binocular microscope (with magnifications up to ×150 or ×200), in reasonably well preserved specimens. The result should be a practical one, although substantiated by fundamental ultrastructural research.

By no means all of the 118 available, usable names of the genus-group are useful in practice. Some, indeed, are very difficult to differentiate and result in monotypic genera with few or no features of phylogenetic, stratigraphic or taxonomic significance. Others represent groups of species which intergrade so frequently across the supposed supraspecific categories that differentiation between them as genera is often impractical; some workers may wish to regard them as being worthy only of subgeneric distinction, while others would consider them to be full synonyms. Such opinions are usually formulated according to the quality and quantity of the material available to the individual worker, to his needs and to the time he can afford to spend on the study. The species of the specialist tend to become the genera of the non-specialist, and the genera become the families, in each and every case. Consequently, almost every published study presents a view of supraspecific synonymy, differentiation and biostratigraphic application different from that of any other.

For this reason, a discriminatory key has been constructed, which attempts to distinguish between the available genus-group names as far as is possible, noting, of course, the objective synonyms (typonyms–generic names founded upon the same type-species) and reducing my own, subjective synonymy to a minimum (and then noted after the taxon has been discriminated). The user of the key can maintain this level of discrimination or reduce it, as he or she thinks is most nearly suited to the available evidence and to practical need. At least, when the taxa are subjectively united, that which is being put together is known; subjectively to synonymise during construction of the key would merely obscure possible distinctions and wilfully ignore the more or less informed opinions of others.

The key is cumulative in its diagnosis as well as discriminatory: for example, taxon 211 112 (*Biglobigerinella*) has the cumulative characters of 2.. ..., 21. ..., 211 ..., 211 1.., and 211 11., and is discriminated at the last order from 211 111 by the final digit. Thus, 2.. ... defines a major division of the superfamily, all members of the Planomalinidae have a code 21. ..., and those of the Planomalininae have characters discriminated within this by the prefix 211 ...; the numbers are grouped in threes merely for ease of reading. All available suprageneric names are noted in the key, those which I believe to represent useful families or subfamilies being written in capitals. The check list of nominally available genus-group names contains the available family-group names, each noted under its nominate genus: those with seniority are used in the key. Although by ICZN Article 23(d) (i) the suprafamilial 'Orbulinacea' (*nom. transl.* Orbulinida Schultze 1854) should have precedence over 'Globigerinacea' (*nom. transl.* Globigerinida Carpenter, Parker and Jones 1862), the well established usage of the latter in this case (e.g. Loeblich & Tappan 1964a) should determine its acceptance (ICZN Art. 23(d) (ii) and ICZN Op. 552).

In the key, the stratigraphic ranges of the taxa, *as discriminated*, are noted as they are understood by me. Combination of taxa would result, of course, in combination of the known stratigraphic ranges. For example, species and subspecies which would be *Hedbergella*, as defined by the type-species and by the diagnosis of the key, range from Aptian to Albian, but, if *Planogyrina* is to be considered to be its subjective junior synonym, then the range of the genus-group is as high as Maestrichtian. Some may prefer to keep them separate; others may wish to use *Planogyrina* as a subgenus of *Hedbergella* and enrich their stratigraphic records thereby; others may follow generally accepted current practice and consider them to be full synonyms. Here is, at least, a key with which to work, and one which is as objective, comprehensive and practical as I can make it.

The key was constructed while bearing morphology, stratigraphy and phylogeny almost equally in mind. The phylogenetic trees (Figs 5.122–126) are designed to show the affinities between generic and suprageneric groups and to indicate the biostratigraphy which can be based on these groups alone. It is clear that it is possible for the non-specialist to achieve a reasonable stratigraphy, to stage or to division of geological age, without identifying species at all. The suggested phylogenies must, necessarily, remain hypothetical, but are not unreasonable and, in many instances, are based upon the strongest evidence for progressive evolution that the fossil record can provide. Iterative evolution of species of genus A from species of genus B (for example, of *Globorotalia* spp. from *Turborotalia* spp., or of *Globigerina* from *Turborotalia*) does not create 'polyphyletic' genera; each species arose from one other species, each genus from one other genus (the curious case of *Globigerina* itself is discussed below), and each subfamily from one other subfamily.

Iterative evolution of this kind happened so frequently that it is quite impractical to provide the end stages of each phylogenetic reiteration with a separate generic name (cf. *Obandyella, Menardella, Fohsella*) for there would be many hundreds of them, each based on the *hypothesis* of separate ancestry from an ancestral genus, but none based on especial morphological characters objectively recognisable in a specimen. On the other hand, if the gradual changes found in the biostratigraphical record are to preclude the separation of earlier from later forms at generic level, then there would be no genera recognised, except for those species which are, at present, 'cryptogenic'. Where grades commonly occur (again, for example, from *Turborotalia* to *Globorotalia*) it may well be both convenient and scientifically honest to blur the genus-group distinction by the use of the subgenus category (*Globorotalia sensu stricto* and *G. (Turborotalia)* for example: intermediates would be *Globorotalia sensu lato*).

The genera and subgenera of the Globigerinacea

This list contains the generic and subgeneric names which have been proposed for the species of the Globigerinacea up to early 1981, excluding *nomina nuda* and those which are referable to the Heterohelicacea. The list is alphabetical, but with each name is the number which will locate it in the morphological discriminatory key. The species name is that of the type-species (O.D., original designation; S.D., subsequent designation; monotypy) in the form given it by its original author at its first valid publication. The strata, their age and the locality of provenance of the primary types of the type-species are noted in parentheses, followed by their depository, where it is known (AUGD, Alexandria University Geology Department; BMNH, British Museum (Natural History) London; DGUS, Department de Géologie, University of Strasbourg; EPRC, Exxon Production Research

Co., Houston; GOC, Gulf Oil Corporation, New York; GSENA, Geological Service, E.N. Adaro, Madrid; GSI, Geological Survey of Israel, Jerusalem; GSP, Geological Survey of Pakistan, Lahore; HCCU, Harris Collection, Cornell University, Ithaca, New York; HLSU, Howe Collection, Louisiana State University; IFP, Institut Français du Petrole, Paris; IGPS, Institute of Geology and Palaeontology, Tohoku University, Sendai; KFVNII, All-Union Oil–Gas Research Institute, Krasnodar Branch; KPUK, Katedra Paleontologie, University Karlovy, Prague; LPUB, Palaeontology Laboratory, University of Bucarest; LSUGM, Louisiana State University, Geological Museum; MCUT, Micropalaeontology Collection, University of Texas; MGIU, Mineralogical–Geological Institute, State University, Utrecht; MHN, Museum de l'Histoire Naturelle, Paris; MUPL, Moscow University, Palaeontological Laboratory; NHMB, Naturhistorisches Museum, Basel; ONGC, Oil and Natural Gas Commission, Dehra Dun; PCNMS, Paleozoological Collection, National Museum, Stockholm; PRII, Paleontological Research Institution, Ithaca, New York; UMML, University of Milan, Micropalaeontological Laboratory; USNM, United States National Museum, Washington; VNIGRI, Vsesoyuznogo neftyannogo naucho-issledovatel'skogo geologo-razvedochnogo instituta, Leningrad; WMUC, Walker Museum, University of Chicago).

Subjective interpretation of the stratigraphic range of the genus depends on its morphological limits and the species which may be included; the probable ranges are noted on the classificatory keys, rather than here. Senior synonyms of the type-species are noted, but subjective junior synonyms are not, except where they cross-refer, as subjective typonyms. Suprageneric groups which have been based on these generic names are noted. Finally, in parentheses, a highly selective list is given of additional references to redescriptions and reliable illustrations (where possible, produced by different methods, such as by drawings, photography, electron microscopy) of hypotypes of the type species, and references to significant notes on the characteristics of the genus or its types.

Nominally available names of the genus-group

Abathomphalus Bolli, Loeblich and Tappan 1957, p. 43; *Globotruncana mayaroensis* Bolli 1951, O.D., p. 198 (Guayaguayare formation, Maestrichtian; subsurface, south Trinidad), USNM. Abathomphalidae proposed by Pessagno (1967, p. 371); genus emended by Longoria and Gamper (1975, p. 93). (*Vide et* Corminboeuf 1961; Hofker 1963b, pp. 281–2; Postuma 1971, p. 50; Berggren 1971a, Pl. 3; Marks 1972, Pl. 5; Masters 1977, pp. 591–3.) Figure 5.117: Key 322.

Acarinina Subbotina 1953, p. 219; *A. acarinata* Subbotina 1953, p. 219, O.D. (early Eocene; near Nal'chik, northern Caucasus), VNIGRI. Subjective junior synonym of *Globigerina nitida* Martin 1943, p. 115 (Lodo formation, early Eocene; Fresno County, California), USNM, *teste* Stainforth *et al.* (1975, pp. 208–9). Acarininae proposed (as Acarininae) by Subbotina (1971). (*Vide et* McGowran 1968, p. 190; Fleisher 1974, pp. 1012–4; Blow 1979, pp. 900–3.) Figure 5.38: Key 113 221 11.

Anaticinella Eicher 1973 (1972), pp. 185–6; *Globorotalia? multiloculata* Morrow 1934, p. 200, O.D. (Hartland shale, Greenhorn formation, latest Cenomanian; Hodgeman County, Kansas) USNM, holotype refigured by Brönnimann and Brown (1956 (1955), p. 534, Pl. 22); topotypes illustrated by Loeblich and Tappan (1961a, Pl. 6, Fig. 13) and by Masters (1977, p. 512–3). Eicher's delayed publication (11 January 1973) still predated the junior typonym *Pseudoticinella* Longoria 1973 (q.v.). (*Vide et* Longoria 1974, Pl. 26; Longoria & Gamper 1975, Pl. 6.) Figure 5.103: Key 222 22.

Applinella Thalmann 1942, p. 812; *Hantkenina dumblei* Weinzierl and Applin 1929, p. 402, O.D. (Yegua formation, early Middle Eocene; subsurface, South Liberty Dome, Liberty County, Texas), USNM, lectotypified by Bolli *et al.* (1957, pp. 26–8) (*vide et* Blow 1979, pp. 1164–5, 1169–70). Figure 5.74: Key 131 221 21.

Aragonella Thalmann 1942, p. 811; *Hantkenina mexicana* Cushman var. *aragonensis* Nuttall 1930, p. 284, O.D. (Aragon formation, Early Eocene; La Antigua, Tampico, Mexico), USNM; lectotypified by Bolli *et al.* (1957, pp. 26–8) (*vide et* Blow 1979, p. 1164–8). Figure 5.75: Key 131 221 12.

Archaeoglobigerina Pessagno 1967, p. 315; *A. blowi* Pessagno 1967, p. 316, O.D. ('Lower Taylor Marl', Taylor formation, early Campanian, *teste* Pessagno (1969, p. 89); Tradinghouse Creek, McLennan County, Texas), USNM; topotypes figured by Robaszynski and Caron (1979b, Pl. 79) (*vide et* El-Naggar 1971b, Pl. 4). Figure 5.104: Key 311 11.

Asterohedbergella Hamoui 1965, p. 133; *Hedbergella (A.) asterospinosa* Hamoui 1965, p. 133, O.D. ('Subunit C5', late Cenomanian; Hamakhtesh Haqatan, Negev, southern Israel), GSI (*vide et* Saint-Marc 1970, p. 92). Figure 5.91: Key 221 112 2.

Astrorotalia Turnovsky 1958, p. 81; *Globorotalia (A.) stellaria* Turnovsky 1958, p. 81, O.D. (Middle Eocene; Saray by Mihallıççik, Vilâyet Ankara, Turkey) 'Collection Dr Wedding'. Subjective junior synonym of *Globorotalia palmerae* Cushman and Bermúdez 1937b, Pl. 2 (Eocene; Habana province, Cuba) (*vide et* Cushman & Bermúdez 1949, pp. 31–2, Pl. 6), USNM. *G. palmerae* has a poorly preserved holotype, *teste* Blow (1979, pp. 1388–90), who considered hypotypes from the Eocene of Jamaica to be referable to the benthonic rotaliacean genus *Pararotalia*; in contrast, Stainforth *et al.* (1975, p. 212) showed that near-topotypic specimens of *G. palmerae* (Early Eocene, Cuba) are globigerinacean. (M. Toumarkine (unpublished manuscript) believes *Globorotalia pseudoscitula* Glaessner to have been its ancestor: *vide et Planorotalites* Morozova and Hillebrandt 1976). Figure 5.37: Key 113 212 2.

Beella Banner and Blow 1960b, p. 26; *Globigerina digitata* Brady 1879, p. 16, O.D. (Recent, sediment: 'Challenger' 338, North Atlantic, 1990 fathoms = 3639 m), BMNH; lectotypified by Banner and Blow (1959, p. 13) (*vide et* Blow 1965, p. 365; Blow 1979, p. 847; Saito *et al.* 1976, p. 280; Bé 1977, Pl. 8). Figure 5.13: Key 111 122 212 11.

Berggrenia Parker 1976, p. 258; *Globanomalina praepumilio* Parker 1967, p. 148, O.D. (Pliocene, sediment: DODO core 117, 164–166 cm, 18°21′S, 62°04′E), USNM. Figure 5.69: Key 131 121.

Biglobigerinella Lalicker 1948, p. 624; *B. multispina* Lalicker 1948, p. 624, O.D. (Marlbrook Marl, 'Campanian' (early Maestrichtian, *teste* Smith and Pessagno (1973, p. 38)), 1½ miles north of Saratoga, Howard County, Arkansas), USNM; holotype refigured by Bolli *et al.* (1957, p. 25, Pl. 1) (topotypes illustrated by Masters (1977, Pl. 8)). Subjective junior synonym of *Globigerinella abberanta* Netskaya 1948, p. 220 (late Senonian; Gan'kino, western Siberia), VNIGRI, *teste* Masters (1977, p. 401–3) (*vide et* Hofker 1960a, p. 315–24; Smith & Pessagno 1973, p. 38) and of *Phanerostomum asperum* Ehrenberg 1854, p. 23, Plate 30, Figure 26b (late Campanian–early Maestrichtian, Rügen, DDR), lectotypified by Pessagno (1967, pp. 274–5); lectotype and strict topotypes refigured by Masters (1980, pp. 96–7, Pl. 1). Figure 5.79: Key 211 112.

Biorbulina Blow 1956, p. 69; *Globigerina bilobata* d'Orbigny 1846, p. 164, O.D. (Miocene, 'Tortonian'; near Nussdorf, Austria), MHN; lectotypified by Banner and Blow (1960, p. 2) (*vide et* Bolli *et al.* 1957, Pl. 7; Hofker 1968b, pp. 144–8, Pl. 23; Hofker 1969; Blow 1969, p. 334, Pl. 23). Figure 5.60: Key 123 121 221.

Biticinella Sigal 1956, p. 35; *Anomalina breggiensis* Gandolfi 1942, p. 120, O.D. (Scaglia bianca, late Albian; Breggia River section, southern Switzerland), NHMB; holotype refigured by Caron and Luterbacher (1969, p. 25, Pl. 7) (near topotypes figured by Sigal (1966a, Pl. 1) and by Masters (1977, Pl. 35)) (*vide et* Postuma 1971, pp. 14–15; Longoria 1974, pp. 93, 95; Longoria & Gamper 1975, p. 77, Pl. 1; Petters 1977, Pl. 3). Figure 5.81: Key 211 12.

Blowiella Kretzschmar and Gorbachik (in Gorbachik) 1971, p. 135; *Planomalina blowi* Bolli 1959, O.D. (upper Cuche formation, Aptian; Piparo River, Central Range, Trinidad), USNM (*vide et* Longoria 1974, pp. 76, 82; Masters 1977, p. 399). Figure 5.78: Key 211 111 2.

Globigerinidae (1)
Caucasellinae (111 11): scale bar-line = 0.1 mm (100 μm)

Figure 5.1 (a–c) *Conoglobigerina dagestanica* Morozova; based on holotype and topotypes (111 111 111).

Figure 5.2 (a–c) *Globuligerina oxfordiana* (Grigelis) Bignot and Guyader; based on holotype and topotypes (111 111 12).

Figure 5.3 (a–c) *Caucasella hoterivica* (Subbotina) Longoria; bulliform structure as holotype, umbilicus and aperture indicated as paratype (111 112 11).

Figure 5.4 (a–c) *Woletzina jurassica* (Hofman) Fuchs; based on primary types and Polish hypotypes (111 111 112).

Figure 5.5 (a–c) *Reticuloglobigerina washitensis* (Carsey) Reiss; based on neotype (of Plummer) and topotypes (111 112 12).

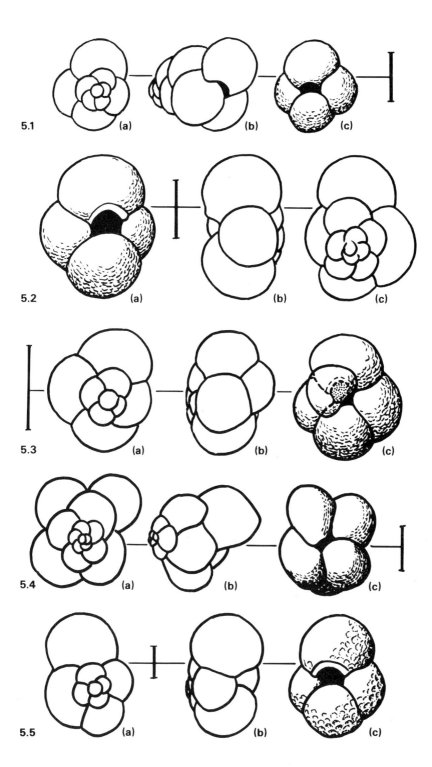

Bolliella Banner and Blow 1959, p. 12; *Hastigerina (Bolliella) adamsi* Banner and Blow 1959, pp. 12–14, O.D. (Recent, sediment; 'Challenger' 191A, off Kai (Ewab) Islands, 580 fathoms = 1061 m), BMNH (*vide et* Saito *et al.* 1976, p. 281, Pls 1, 6, 8; Bé 1977, Pl. 8). Figure 5.70: Key 131 113 2.

Bucherina Brönnimann and Brown 1956, p. 557; *B. sandidgei* Brönnimann and Brown 1956, p. 557, O.D. (Maestrichtian marl; construction pit for Grand Masonic Temple, Habana, Cuba), GOC. Figure 5.110: Key 312 122.

Candeina d'Orbigny 1839a, p. 107; *C. nitida* d'Orbigny 1839a, p. 107, monotypy (original type series from beach sands of Cuba; primary types are lost and original sample no longer contains topotypes, *teste* Le Calvez (1977, pp. 21–2); neotype *here proposed* is the specimen figured by Bolli *et al.* (1957, p. 35, Pl. 6, Fig. 10), USNM P.3924, from Recent sediment, 'Albatross' 2660, 28°40′N, 78°46′W, 504 fathoms = 922 m). Candeininae proposed by Cushman (1927a, p. 90). (*Vide et* Blow 1969, Pl. 23.) Figure 5.26: Key 111 122 231.

Candorbulina Jedlitschka 1934, p. 20; *C. universa* Jedlitschka 1934, p. 20, O.D. (syntypes (Jedlitschka 1934, p. 21) from Neogene sands and clays of 'Ost au-Karviner Kohlenrievers . . . am Rande der Olmützer Bucht'; lectotype, *here designated*, that specimen represented by Figure 19, p. 21 of Jedlitschka (1934)), subjectively a senior synonym of *Orbulina suturalis* Brönnimann 1951b, p. 135, and therefore a secondary subjective homonym of *O. universa* d'Orbigny, type-species of *Orbulina*, q.v. (*Vide et* Cushman & Dorsey 1940; Blow 1956, p. 66; Bolli *et al.* 1957, pp. 35–6; Subbotina *et al.* 1960, p. 63, Pl. 12.) Figure 5.61: Key 123 121 222.

Catapsydrax Bolli, Loeblich and Tappan 1957, p. 36; *Globigerina dissimilis* Cushman and Bermúdez 1937a, p. 25, O.D. (Eocene; 1 km north of Arroyo Arenas, Habana Province, Cuba), USNM; holotype refigured by Bolli *et al.* (1957, Pl. 7). Catapsydracinae proposed by Bolli *et al.* (1957, p. 36). (*Vide et* Beckmann 1954, p. 391, Pl. 25, Text Fig. 16; Blow & Banner 1962, p. 106, Pl. 16D; Cati & Borsetti 1968, p. 396; Blow 1969, Pl. 25; Blow 1979, pp. 538, 1328–32, Pl. 241; Fleisher 1974, p. 1015, Pl. 4.) Figure 5.15: key 111 122 212 211.

Caucasellinae (111 11) (cont.): scale bar-line = 0.1 mm (100 μm)

Figure 5.6 (a–c) *Globastica daubjergensis* (Brönnimann) Blow; based on holotype (111 112 13).

Globigerininae (111 12): scale bar-line = 0.1 mm (100 μm)

Figure 5.7 (a–c) *Subbotina triloculinoides* (Plummer) Brotzen and Pozaryska; based on holotype and topotypes (111 121 1).

Figure 5.8 (a–c) *Dentoglobigerina galavisi* (Bermudez) Blow; based on holotype (111 121 2).

Figure 5.9 (a–c) *Muricoglobigerina soldadoensis* (Brönnimann) Blow; based on holotype (111 122 11).

Figure 5.10 (a–c) *Pseudogloboquadrina primitiva* (Finlay) Jenkins; based on holotype (111 122 121).

Figure 5.11 (a–c) *Neoacarinina blowi* Thompson, based on holotype (111 122 122).

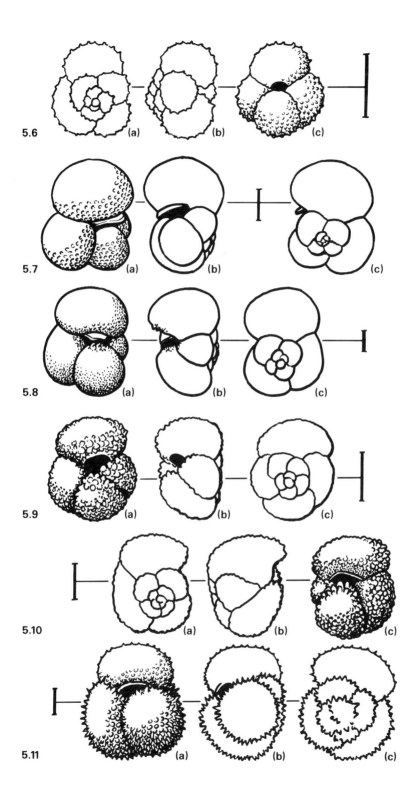

Caucasella Longoria 1974, p. 48; *Globigerina hoterivica* Subbotina 1953, p. 50, O.D. (*nom. corr.* Longoria 1974 *hauterivica*, not acceptable by ICZN *Règles*, Art.32(a)(ii)). (Hauterivian; River Pshish, northern Caucasus), VNIGRI. Caucasellidae proposed by Longoria (1974). (*Vide et* Longoria & Gamper 1975, p. 66; Grigelis & Gorbatchik 1980, Pl. 1, Fig. 7.) Figure 5.3: Key 111 112 11.

Clavatorella Blow 1965, pp. 367–8; *Hastigerinella bermudezi* Bolli 1957a, p. 112, O.D. (Miocene marl, Cipero formation, Retrench Member, type locality of the *Globorotalia fohsi barisanensis* zone of Bolli; Hermitage Quarry, southern Trinidad), USNM (*vide et* Saito *et al.* 1976, p. 288, Pl. 4; Blow 1969, p. 358, Pl. 41). Figure 5.44: Key 113 231 12.

Clavigerinella Bolli, Loeblich and Tappan 1957, p. 30; *C. akersi* Bolli, Loeblich and Tappan 1957, p. 30, O.D. (Navet formation, Eocene; Brasso–Tamana Road, Central Range, Trinidad), USNM (*vide et* Stainforth *et al.* 1975, p. 167, Fig. 32; Blow 1979, p. 1201, Pl. 157). Figure 5.71: Key 131 21.

Clavihedbergella Banner and Blow 1959, pp. 8, 18–19; *Hastigerinella subcretacea* Tappan 1943, p. 513, O.D. (Washita Group, Duck Creek formation, Albian; Horseshoe Bend, Red River, Love County, Oklahoma), USNM; refigured by Loeblich and Tappan (1964a, p. C658, Fig. 527.2) (topotype figured by Masters (1977, p. 446, Pl. 19)). Figure 5.90: Key 221 112 1.

Claviticinella El-Naggar 1971a, p. 436; *Ticinella raynaudi* var. *digitalis* Sigal 1966a, p. 202 (Late Albian: Diégo boring, north of Mont-Raynaud, Madagascar), IFP (*vide et* Caron 1971, Fig. 21) (trinomen *incorrectly* considered to be a junior subjective synonym of *Ticinella primula* Luterbacher (in Renz *et al.* 1963) by Masters (1977)). Figure 5.99: Key 222 112.

Conoglobigerina Morozova (in Morozova & Moskalenko) 1961, p. 24; *Globigerina (C.) dagestanica* Morozova (in Morozova & Mosalenko 1961, p. 24) (Early Bathonian; Chakh, Dagestan, USSR), VNIGRI. Topotypes illustrated by Grigelis and Gorbatchik (1980, Pl. 1, Figs 1, 2). Figure 5.1: Key 111 111 111.

Cribrohantkenina Thalmann 1942, pp. 812, 815; *Hantkenina (C.) bermudezi* Thalmann 1942, p. 815, O.D., 'new name' for *H. (Sporohantkenina)*

Globigerininae (111 12) (cont.): scale bar-line = 0.1 mm (100 μm)

Figure 5.12 (a–c) *Globoturborotalita rubescens* Hofker) Hofker; based on holotype (111 122 211).

Figure 5.13 (a–c) *Beella digitata* (Brady), Banner and Blow; based on lectotype (111 122 212 11).

Figure 5.14 (a–c) *Globigerina bulloides* d'Orbigny; based on lectotype and topotypes (111 122 212 12).

Figure 5.15 (a–c) *Catapsydrax dissimilis* (Cushman and Bermúdez) Bolli, Loeblich and Tappan; based on holotype, with umbilicus and aperture based on hypotypes (111 122 212 211).

Figure 5.16 (a–c) *Globigerinita naparimaensis* Brönnimann; based on holotype (b, c) and near-topotype (a); umbilicus and primary aperture indicated from near-topotypes (111 122 212 212 1).

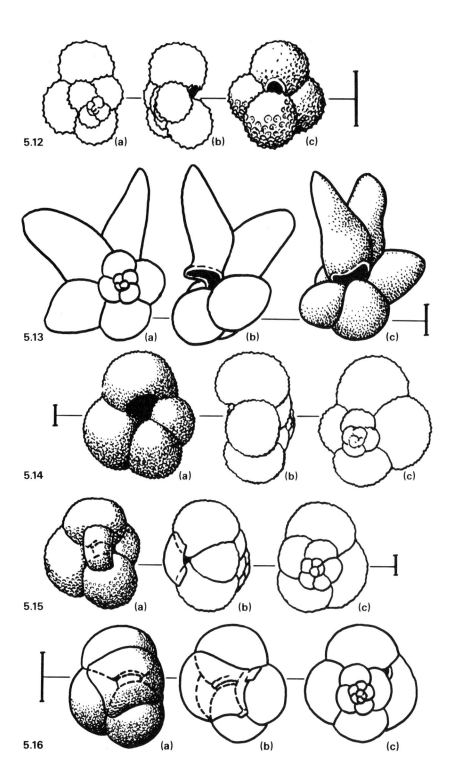

brevispina of Bermúdez (1937), *non H. brevispina* Cushman 1925a (Chapapote formation, Late Eocene; Romal Juan Criollo, Central Jatibonico, Camagüey Province, Cuba); Bermúdez collection; lectotypified by Brönnimann (1950a, p. 417); topotypes figured by Spraul (1963, Pl. 41) (PRII), and by Postuma (1971, pp. 134–5). Subjective junior synonym of *H. danvillensis* Howe and Wallace 1934, p. 37 (Jackson formation; Late Eocene; Danville Landing, Ouachita River, Catahoula Parish, Louisiana), LSUGM; holotype refigured by Spraul (1963, Pl. 41); also of *H. inflata* Howe 1928, p. 14 (Late Eocene, reworked into basal Vicksburgian, Oligocene; Old Fort Street, St Stephen's Bluff, Washington County, Alabama), LSUGM; holotype refigured by Spraul (1963, Pl. 41). (*Vide et* Barnard 1954; Bolli *et al.* 1957, pp. 28–9; Hofker 1962, pp. 127–8; Blow & Banner 1962, p. 145; Blow 1979, pp. 1170–4.) Figure 5.76: Key 131 222.

Dentoglobigerina Blow 1979, p. 1298–301; *Globigerina galavisi* Bermúdez 1961, p. 1183, O.D. (Upper Jackson formation, Late Eocene; Frost Bridge, Mississippi), USNM; holotype redrawn in Blow (1969, Pl. 5). Figure 5.8: Key 111 121 2.

Dicarinella Porthault (in Donze *et al.*) 1970, p. 70: *Globotruncana indica* Jacob and Sastry (December) 1950, p. 267, O.D. (Uttatur stage, 'Upper Albian–Cenomanian', *recte* Early or Middle Turonian, associated with *Helvetoglobotruncana helvetica*; Trichinopoly, southern India), ONGC. Topotypes, provided by V. Narayanan, ONGC, indicate that *G. indica* may be a subjective junior synonym of *G. imbricata* Mornod 1950 (1949), p. 581, as neotypified by Caron (1976, p. 332) (middle–late Turonian; Corvayes Brook, north of Cerniat, Montsalvens region, Switzerland), NHMB; alternatively, it may be senior synonym of the very closely related *Praeglobotruncana hagni* Scheibnerova 1962, pp. 219–21, 225–6 (middle Turonian; Horné Srnie, Slovakia), Dept of Palaeontology, J. A. Comenius University (*vide et* Schcibnerova 1962; Caron 1966, Pl. 6, Fig. 4; Postuma 1971, pp. 46–7; Sastry & Mamgain 1972; Longoria & Gamper 1975, pp. 86–9; Masters 1977, p. 533; Robaszynski & Caron 1979b, pp. 55, 85, 92). Figure 5.94: Key 221 121 21.

Eoglobigerina Morozova 1959, p. 1113; *Globigerina (E.) eobulloides* Morozova 1959, O.D. ('lower substage of the Danian stage'; Tarkhankut,

Globigerininae (111 12) (cont.): scale bar-line = 0.1 mm (100 μm)

Figure 5.17 (a–c) *Tinophodella ambitacrena* Loeblich and Tappan; based on holotype, with umbilicus and primary aperture indicated from hypotypes (111 122 212 212 2).

Figure 5.18 (a–c) *Polyperibola christiani* Liska; based on holotype (and paratypes) (111 122 212 22).

Figure 5.19 (a–c) *Guembelitrioides higginsi* (Bolli) El-Naggar; based on holotype (b) and hypotypes (a, c) (111 122 221).

Figure 5.20 (a–c) *Globicuniculus mitra* (Todd) Saito and Thompson; based on holotype (b) and hypotypes (a, c) (111 122 222).

Figure 5.21 (a–c) *Globigerinoides ruber* (d'Orbigny) Cushman; based on lectotype (a, c) and hypotype (b) (111 122 223 11).

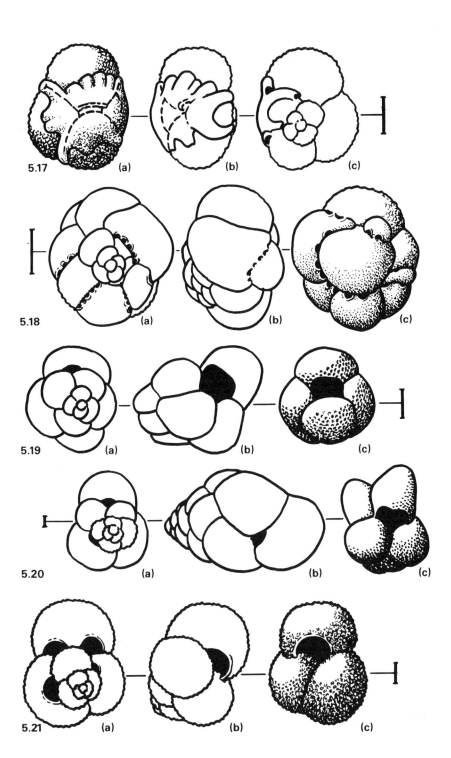

Crimea), VNIGRI (S.D. of *G. (E.) balakhmatovae* Morozova 1961 (in Morozova & Moskalenko 1961, pp. 23–4), invalid by Art. 68(a) of the ICZN *Règles*). Eoglobigerinidae proposed by Blow (1979, p. 1203), who emended the genus (Blow 1979, p. 1205). (*Vide et* Lipps 1964, p. 129; Hofker 1978, p. 56.) Figure 5.34: Key 113 211 1.

Eohastigerinella Morozova 1957, p. 1112; *Hastigerinella watersi* Cushman 1931, p. 86, O.D. (Austin Chalk, Senonian; near Howe, Grayson County, Texas), USNM (*vide et* Bolli *et al.* 1957, Pl. 1, topotype; Frerichs *et al.* 1977, Pl. 2). Figure 5.85: Key 212 32.

Falsotruncana Caron 1981, p. 66; *F. maslakovae* Caron 1981, p. 67, O.D. (basal Aleg formation, late Turonian; Pont du Fahs, Tunisia), NHMB. Figure 5.118: Key 221 121 22.

Favusella Michael 1973, pp. 212–16; *Globigerina washitensis* Carsey 1926, p. 44, O.D.; junior objective typonym of *Reticuloglobigerina* Reiss 1963, q.v., although Favusellidae was proposed by Longoria (1974, p. 74).

Fissoarchaeoglobigerina Abdel-Kireem 1978, p. 58; *F. aegyptiaca* Abdel-Kireem and Abdou (in Abdel-Kireem 1978, pp. 58–63) (late Maestrichtian; Gebel Thelmut, Eastern Desert, Egypt), AUGD. Abdel-Kireem (1978) designated as type-species *Rugoglobigerina aegyptiaca* Abdel-Kireem and Abdou, 1976, 'in press', *7th African Micropal. Colloq. Nigeria (Abstr.)* but the species name became valid only with the first publication of the species description, in 1978, by Art. 13(a) of the ICZN *Règles*, and *F. aegyptiaca* became the type-species by monotypy, see Art. 68(c). Figure 5.105: Key 311 12.

Fohsella Bandy 1972, p. 297; *Globorotalia (G.) praefohsi* Blow and Banner 1966, p. 295, O.D. (Husito Marly Clay member, Pozón formation, Middle Miocene, zone N.11; El Mene–Pozón road, eastern Falcón, Venezuela), BMNH (*vide et* Blow 1979, pp. 38–40; Stainforth *et al.* 1975, pp. 270–6). Figure 5.47: Key 113 231 212.

Globanomalina Haque 1956, p. 147; *G. ovalis* Haque 1956, p. 147, O.D. (Laki series, Palaeocene or Early Eocene; Nammal Gorge, Salt Range, Pakistan), holotype GSP, paratypes BMNH; subjective junior synonym of *Anomalina luxorensis* Nakkady 1950, p. 691 (highest Esna shale, Late Palaeocene or Early Eocene; Abu Durba, western Sinai), BMNH. *Not* synonymous with *Globorotalia chapmani* Parr 1938 (p. 87), see topotypes

        Globigerininae (111 12) (cont.): scale bar-line = 0.1 mm (100 μm)

Figure 5.22 (a–c) *Globigerinoidesella fistulosa* (Schubert) El-Naggar; based on hypotypes (111 122 223 12).

Figure 5.23 (a–c) *Globigerinanus sudri* Ouda; based on holotype (111 122 223 211).

Figure 5.24 (a–c) *Velapertina iorgulescui* Popescu; based on holotype; primary aperture indicated from paratype (111 122 223 212).

Figure 5.25 (a–c) *Globigerinoita morugaensis* Brönnimann; based on holotype (111 122 223 22).

        Candeininae (111 122 23): scale bar-line = 0.1 mm (100 μm)

Figure 5.26 (a–c) *Candeina nitida* d'Orbigny; based on neotype; position and shapes of septal aperture indicated from hypotypes (111 122 231).

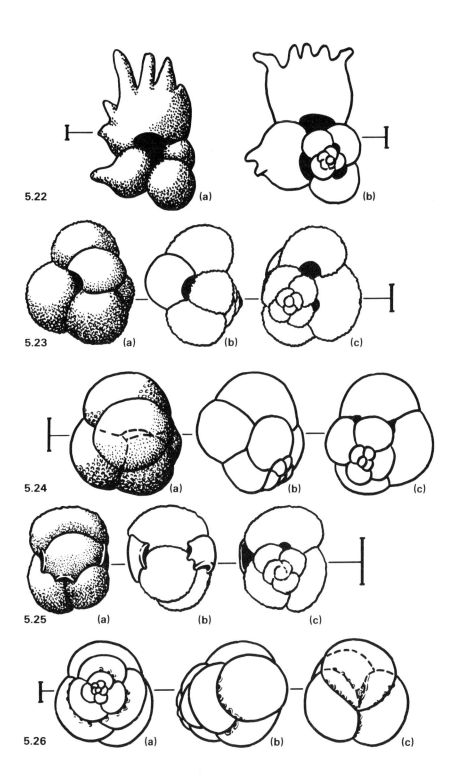

in McGowran (1964, Pl. 1), although '*G. (Turborotalia) chapmani*' sensu Berggren *et al.* (1967, text Figs 1, 3, 4) and Blow (1979, Pl. 116) is referable to *Globanomalina ovalis* and to *G. luxorensis* (Nakkady). (*Vide et* Nakkady 1959; Fleisher 1974, p. 1017.) Figure 5.35: Key 113 211 2.

Globastica Blow 1979, pp. 1232–5; *Globigerina daubjergensis* Brönnimann 1953, p. 340, O.D. (Danian; Daubjerg, Denmark), USNM (*vide et* Hofker 1960b, pp. 77–8; Hofker 1968a, pp. 84–5; Hofker 1976a; Postuma 1971, pp. 148–9; Moorkens 1971, pp. 858–9; Stainforth 1975, p. 181). Figure 5.6: Key 111 112 13.

Globicuniculus Saito and Thompson (in Saito *et al.*) 1976, p. 287; *Globigerinoides mitrus* Todd 1957, p. 302 (*nom. corr. ex mitra* Todd), O.D. (Donni sandstone member, Tagpochau limestone series, Miocene; northeastern central Saipan), USNM (*vide et* Bolli 1957a, p. 114, Pl. 26) (note that the specimens figured by Saito and Thompson (in Saito *et al.* 1976, Pl. 5) are unlikely to be conspecific with the type-species). Figure 5.20: Key 111 122 222.

Globigerapsis Bolli, Loeblich and Tappan, 1957, p. 33; *G. kugleri* Bolli, Loeblich and Tappan, 1957, p. 34 (Penitence Hill Marl, Navet formation, Eocene, reworked into Nariva formation, Point-à-Pierre, Trinidad), USNM; emended by Bolli (1972, p. 128) and by Blow (1979, pp. 1133, 1143). Globigerapsidae proposed by Blow (1979, p. 1117). (*Vide et* Cordey 1968, Text Fig. 1; Postuma 1971, pp. 138–9.) Figure 5.54: Key 122 2.

Globigerina d'Orbigny 1826, p. 277; *G. bulloides* d'Orbigny 1826, S.D. Parker *et al.* (1865, p. 36); lectotypified by Banner and Blow (1960a, pp. 3–4, Pl. 1) ('Subappenine', Recent sediment; Rimini, Adriatic Sea), MHN; topotypes figured by Le Calvez (1974, pp. 13–16, Pl. 2) and by Blow (1969, Pl. 14). Globigerinida proposed by Carpenter *et al.* (1862, p. 171), *nom. conserv.* by ICZN Op. 552. (*Vide et* Towe 1971; Fleisher 1974, p. 1018; Reiss *et al.* 1974, Pl. 1; Blow 1979, pp. 846–7, 1352.) Figure 5.14: Key 111 122 212 12.

Globigerinanus Ouda 1978, p. 358; *G. sudri* Ouda 1978, p. 361, O.D. (Ayun Musa formation, Miocene, 'U. Burdigalian'; subsurface, Sudr East Well No. 2, eastern Gulf of Suez), GSENA. Figure 5.23: Key 111 122 223 211.

Globigerinatella Cushman and Stainforth 1945, p. 68; *G. insueta* Cushman

Sphaeroidinellinae (111 2): scale bar-line = 0.1 mm (100 μm)

Figure 5.27 (a–c) *Prosphaeroidinella disjuncta* (Finlay) Ujiié; based on holotype (111 211).

Figure 5.28 (a–c) *Sphaeroidinellopsis subdehiscens* (Blow) Banner and Blow; based on holotype (111 212).

Figure 5.29 (a–c) *Sphaeroidinella dehiscens* (Parker and Jones) Cushman; based on lectotype (a, c) and hypotypes (b) (111 22).

Globorotaloidinae (112): scale bar-line = 0.1 mm (100 μm)

Figure 5.30 (a–d) *Globorotaloides variabilis* Bolli; based on holotype (a–c) with intraumbilical primary aperture indicated from paratype; stage with extraumbilical aperture (d) based on paratypes (112 1).

Figure 5.31 (a–c) *Globoquadrina dehiscens* (Chapman, Parr and Collins) Finlay; based on holotype and hypotypes (112 2).

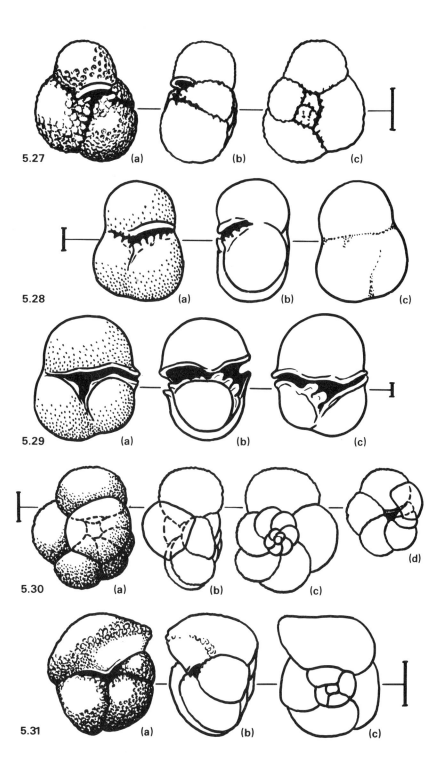

and Stainforth 1945, p. 68, O.D. (Cipero formation, Early Miocene; Cipero Coast section, Trinidad), USNM; paratypes refigured by Bolli *et al.* (1957, p. 38–9, Pl. 8); near-topotypes figured by Brönnimann (1950b, Pls 13, 15), by Stainforth *et al.* (1975, pp. 286–9) and by Blow (1969, Pl. 26). Figure 5.59: Key 123 121 21.

Globigerinatheka Brönnimann 1952, p. 27; *G. barri* Brönnimann 1952, p. 27, O.D. (Navet formation, Middle Eocene, reworked into Mount Moriah formation, Late Eocene, *teste* Proto Decima and Bolli (1970, p. 889); Harmony Hall well No. 2, Trinidad), USNM. Emended by Proto Decima and Bolli (1970, p. 888), by Bolli (1972) and by Blow (1979, pp. 865–8). (*Vide et* Cordey 1968, pp. 372, 374; Cati & Borsetti 1968, p. 391; Postuma 1971, pp. 164–5; Stainforth *et al.* 1975, pp. 170–1, Fig. 37.) Figure 5.56: Key 123 111 2.

Globigerinella Cushman 1927a, p. 87; *Globigerina aequilateralis* Brady 1879, p. 285, O.D. (Recent, sediment; 'Challenger' 224, northern Pacific, 1850 fathoms = 3383 m), BMNH, lectotypified by Banner and Blow (1960b, p. 22); subjective junior synonym of *Globigerina siphonifera* d'Orbigny 1839, p. 83 (Recent, sediment; Cuba), MHN, lectotypified by Banner and Blow (1960b, p. 22), lectotype and topotype photographed by Le Calvez (1977, p. 33–4, Pl. 6) (*vide et* Bé 1969 Pls 1, 2; Le Calvez 1974, pp. 19–20; Reiss *et al.* 1974, Pl. 4; Saito *et al.* 1976, pp. 281–2, Pls 3, 6). Figure 5.68: Key 131 113 1.

Globigerinelloides Cushman and ten Dam 1948, p. 42; *G. algeriana* Cushman and ten Dam 1948, p. 42, O.D. (*algerianum, nom. corr.*, this binomen) ('Upper Cretaceous', *recte* Aptian, blue marls; Djebel Menaouer, western Algeria); holotype may be lost, according to Longoria (1974, p. 76); paratypes USNM, figured by Bolli *et al.* (1957, pp. 20–3, Pl. 1) and by Longoria (1974, Pl. 6). Globigerinelloididae proposed by Longoria (1974, p. 76). (*Vide et* Glintzboeckel & Magné 1955; Ruggieri 1963; Gorbachik 1964.) Figure 5.77: Key 211 111 1.

Globigerinita Brönnimann 1951a, P. 16; *G. naparimaensis* Brönnimann 1951a, O.D. (Lengua formation, Middle Miocene; Naparima area, Trinidad), USNM; holotype refigured by Bolli *et al.* (1957); species

Tenuitellinae (113 1): scale bar-line = 0.1 mm (100 μm)

Figure 5.32 (a–c) *Parvularugoglobigerina eugubina* (Luterbacher and Silva) Hofker; based on holotype, detail amplified according to hypotypes (113 11).

Figure 5.33 (a–c) *Tenuitella gemma* (Jenkins) Fleisher = *T. postcretacea* (Myatliuk); based on holotype of *T. gemma* (113 12).

Eoglobigerininae (113 211): scale bar-line = 0.1 mm (100 μm)

Figure 5.34 (a–c) *Eoglobigerina eobulloides* Morozova; based on holotype and hypotypes (113 211 1).

Figure 5.35 (a–c) *Globanomalina ovalis* Haque = *G. luxorensis* (Nakkady); based on holotype of *G. ovalis* (a), paratype of *G. luxorensis* (c) and hypotype (b) (113 211 2).

Planorotalitinae (113 212): scale bar-line = 0.1 mm (100 μm)

Figure 5.36 (a–c) *Planorotalites pseudoscitula* (Glaessner) Morozova; based on holotype, near-topotype and hypotypes (113 212 1).

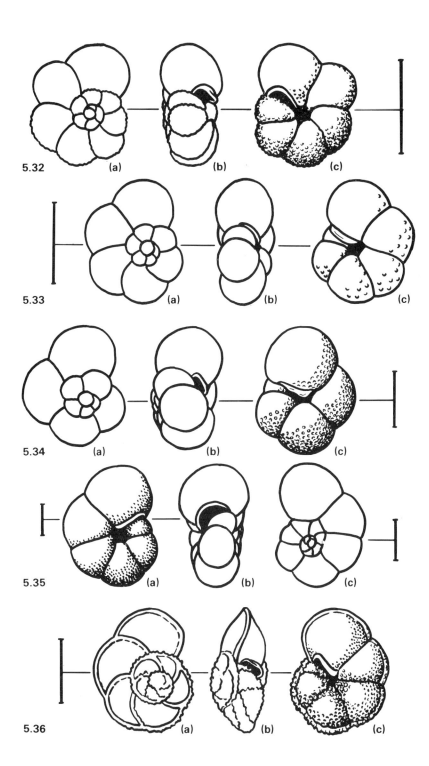

emended by Loeblich and Tappan (1957b) and by Blow and Banner (1962, p. 102) (*vide et* Stainforth *et al.* 1975, pp. 295–7, Fig. 129). Figure 5.16: Key 111 122 212 212 1.

Globigerinoides Cushman 1927a, p. 87; *Globigerina rubra* d'Orbigny 1839a, p. 82, O.D. (Recent, sediment; Cuba), MHN; lectotypified by Banner and Blow (1960a, pp. 19–21, Pl. 3) and lectotype photographed by Le Calvez (1977, Pl. 5, Fig. 5); topotype also by Le Calvez (1977, Pl. 5, Fig. 2) (*vide et* Hofker 1959, Figs 7–8; Cordey 1967; Orr 1969; Blow 1969, p. 326, Pl. 21; Reiss *et al.* 1974, Pl. 6; Blow 1979, p. 861, Pl. 262; Thompson *et al.* 1979; Bé *et al.* 1980). Figure 5.21: Key 111 122 223 11.

Globigerinoidesella El-Naggar 1971a, pp. 451, 476; *Globigerina fistulosa* Schubert 1910, O.D. (Neogene '*Globigerina* marls'; New Guinea); depository of type series unknown; specimens (syntypes?) refigured by Schubert (1911) (*vide et* Banner & Blow 1965, pp. 111–12; Parker 1967, Pl. 21, Fig. 6; Bolli 1970, Fig. 201; Jenkins & Orr 1972, Pl. 13, Figs 1, 3; Stainforth *et al.* 1975, pp. 406–8). Figure 5.22: Key 111 122 223 12.

Globigerinoita Brönnimann 1952a, p. 26; *G. morugaensis* Brönnimann 1952a, O.D. (Lengua formation, Middle Miocene; Moruga area, Trinidad), USNM, holotype redrawn by Bolli *et al.* (1957, Pl. 8, Fig. 3) (*vide et* Hofker 1959; Cati & Borsetti 1968, p. 395; Orr 1969). Figure 5.25: Key 111 121 223 212.

Globigerinopsis Bolli 1962, p. 281; *G. aguasayensis* Bolli 1962, p. 232, O.D. (Oficina formation, Middle Miocene; Aguasay well no. 1, Monagas, eastern Venezuela), NHMB (*vide et* Blow 1979, pp. 177–8). Figure 5.51: Key 121 1.

Globigerinopsoides Cita and Mazzola 1970, p. 470; *G. algeriana* Cita and Mazzola 1970, p. 472, O.D., *recte algerianus* in this binomen (Middle Miocene, 'Tortonian' marls; Melouza River section; Hodna, Algeria), UMML. Figure 5.63: Key 123 2.

Globoquadrina Finlay 1946, p. 290; *Globorotalia dehiscens* Chapman, Parr and Collins 1934, p. 569, O.D. (Balcombian, 'Oligocene', *recte* Early Miocene; Port Phillip area, Victoria, Australia), 'Chapman collection'; near-topotype figured by Bolli *et al.* (1957, Pl. 5); genus emended and Globoquadrinidae proposed by Blow (1979, pp. 1294, 1350) (*vide et* Bolli *et al.* 1957, p. 31; McGowran 1968, Pl. 3, Fig. 15; Jenkins 1971, p. 165, Pl.

Planorotalitinae (113 212) (cont.): scale bar-line = 0.1 mm (100 μm)

Figure 5.37 (a–c) *Astrorotalia stellaria* Turnovsky = *A. palmerae* (Cushman and Bermúdez); based on metatypes and hypotypes of *A. palmerae* (113 212 2).

Truncorotaloidinae (113 22): scale bar-line = 0.1 mm (100 μm)

Figure 5.38 (a–c) *Acarinina acarinata* Subbotina; based on holotype and hypotypes (= *A. nitida* (Martin)) (113 221 11).

Figure 5.39 (a–c) *Testacarinata inconspicua* (Howe) Jenkins; based on topotype (113 221 12).

Figure 5.40 (a–c) *Truncorotaloides rohri* Brönnimann and Bermúdez; based on holotype and hypotypes (113 221 2).

Figure 5.41 (a–c) *Morozovella velascoensis* (Cushman) McGowran; based on topotypes (113 222).

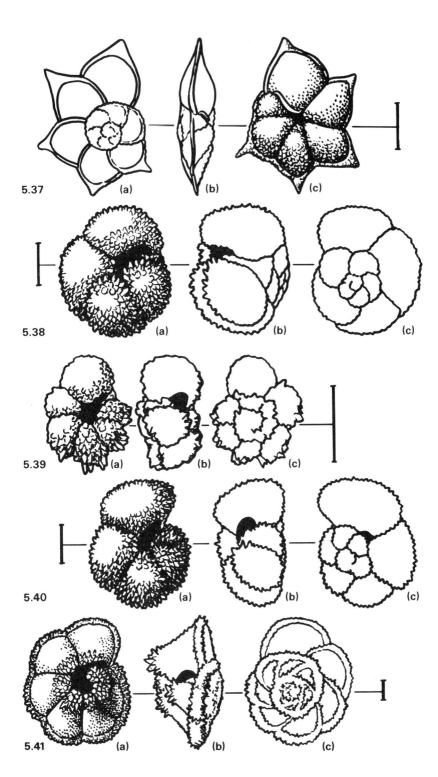

20; Stainforth *et al.* 1975, p. 266, Fig. 113; Scott 1976; Saito *et al.* 1976, p. 286, Pl. 4; Hornibrook 1978; Blow 1979, Figs Riv–v, Si–iv). Figure 5.31: Key 112 2.

Globorotalia Cushman 1927a, p. 91: *Pulvinulina menardii* (d'Orbigny) var. *tumida* Brady 1877, p. 534, O.D. (exotic block of Late Miocene or Pliocene; New Ireland), BMNH, lectotypified by Banner and Blow (1960a, p. 26, Pl. 5); topotype, effective paralectotype figured by Bolli *et al.* (1957, Pl. 10, Fig. 2). Globorotaliidae proposed by Cushman (1927a, p. 91); genus emended by Blow (1979, p. 879) (*vide et* Banner & Blow 1967, p. 160, Pl. 4; Blow 1969, Pl. 49; Bandy 1972, p. 357–8; Jenkins & Orr 1972, Pl. 34; Bé 1977, Pl. 12). Figure 5.46: Key 113 231 211 2.

Globorotaloides Bolli 1957a, p. 117; *G. variabilis* Bolli 1957a, p. 117, O.D. (Lengua formation, Middle Miocene, Concord area, Point-à-Pierre, Trinidad), USNM. Globorotaloidinae proposed by Banner and Blow (1959, p. 7). (*Vide et* Blow 1979, p. 1355–60.) Figure 5.30: Key 112 1.

Globotruncana Cushman 1927a, p. 91; *Pulvinulina arca* Cushman 1926, p. 23, O.D. (Mendez shale, Maestrichtian; River Huiches, east of Tamuin, *teste* Pessagno (1967, pp. 321–3), near Huiches, Hacienda El Limon, San Luis Potosí, Mexico), USNM; holotype refigured by Bolli *et al.* (1957, Pl. 11, Fig. 11), 'homeotypes' by Brönnimann and Brown (1956, Pl. 23, Figs 10–12) and by Pessagno (1967, Pls 79, 90). Globotruncanidae proposed by Brotzen (1924). (*Vide et* El-Naggar 1971a, Pl. 5; Berggren 1971a, Pl. 1, Figs 8–14; Postuma 1971, p. 18–19; Smith & Pessagno 1973, Pl. 18.) Figure 5.112: Key 321 11.

Globotruncanella Reiss 1957, p. 135; *Globotruncana citae* Bolli 1951, p. 197, O.D. (Maestrichtian marls/shales; Guaracara–Tabaquite Road, Central Range, Trinidad), USNM, subjective junior synonym of *Globotruncana havanensis* Voorwijk 1937, p. 195 ('Upper Cretaceous'; Habana, Cuba), MGIU, *teste* Brönnimann and Brown (1956, p. 552), 'homeotype' from late Maestrichtian, Habana, figured by Brönnimann and Brown (1956, Figs 4–5). Globotruncanellinae proposed by Maslakova (1964). (*Vide et* Corminboeuf 1961, p. 112; Berggren 1971a, Pl. 3, Figs 5–8; Marks 1972, Pl. 5; Masters 1977, pp. 569–72, Pl. 45; Blow 1979, Pl. 259.) Figure 5.95: Key 221 122.

Globotruncanita Reiss 1957, p. 136; *Rosalina stuarti* de Lapparent 1918, p.

Globorotaliinae (113 23): scale bar-line = 0.1 mm (100 μm)

Figure 5.42 (a–c) *Turborotalia centralis* (Cushman and Bermúdez); based on lectotype (113 231 111).

Figure 5.43 (a–d) *Neogloboquadrina dutertrei* (d'Orbigny) Bandy, Frerichs and Vincent; based on lectotype (a–c) and hypotype (d) (group 'A' of Srinavasan and Kennett (1976)) (113 231 112).

Figure 5.44 (a–c) *Clavatorella bermudezi* (Bolli) Blow, based on holotype and topotypes (113 231 12).

Figure 5.45 (a–d) *Menardella menardii* (Parker, Jones and Brady *ex* d'Orbigny) Bandy; based on neotype (a–c), compared to damaged lectotype (d) (113 231 211 1).

Figure 5.46 (a–c) *Globorotalia tumida* (Brady) Cushman; based on lectotype (113 231 211 2).

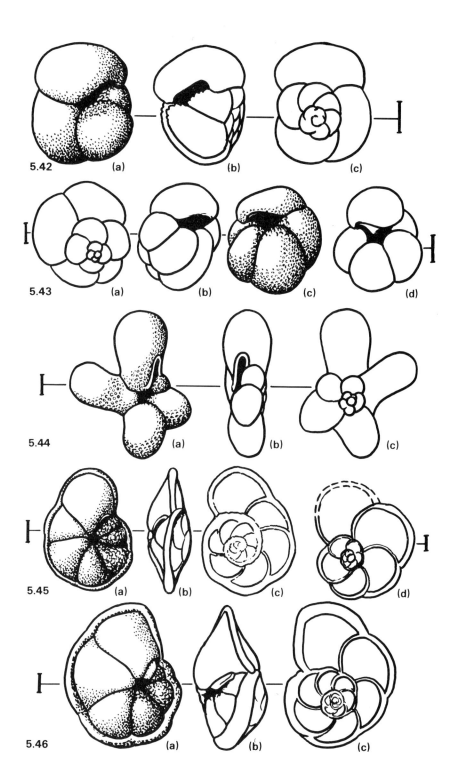

11, O.D. (Maestrichtian, chalk; Pointe Sainte-Anna, Hendaye region, western Pyrenees, France), possibly in DGUS, *teste* l'Abbé de Lapparent (verbal communication, 1959); lectotype figure selected by Pessagno (1967, p. 356) (*vide et* Reichel 1950, Pls 16, 17; Maslakova 1963, Pl. 7, Figs 1, 3; Pessagno 1967, Pls 81, 93; Postuma 1971, pp. 60, 61; Berggren 1971a, Pl. 1, Figs 1–7; Longoria & Gamper 1975, p. 91; Masters 1977, Pls 53, 54; Hofker 1978, Pl. 1, Figs 19, 20). Figure 5.115: Key 321 21.

Globoturborotalita Hofker 1976b, p. 52; *Globigerina rubescens* Hofker 1956b, p. 234, O.D. (Recent sediment; Virgin Islands), Hofker collection? (*vide et* Parker 1962, p. 226, Pl. 2, Figs 17, 18; Cifelli & Smith 1970, p. 35, Pl. 4, Fig. 1; Bé 1977, Pl. 9). Figure 5.12: Key 111 122 211.

Globuligerina Bignot and Guyader 1971, p. 79; *Globigerina oxfordiana* Grigelis 1958, p. 109, O.D. (Oxfordian; Lithuania). Topotype figured by Grigelis and Gorbatchik (1980, Pl. 1, Fig. 4). If Masters (1977, p. 462) is correct in the judgement that *G. oxfordiana* is a junior subjective synonym of *G. hauterivica* Subbotina 1953, then *Globuligerina* would become a senior subjective typonym of *Caucasella* Longoria 1974; Grigelis and Gorbatchik (1980) referred to *Globuligerina hauterivica* (Subbotina) and *G. oxfordiana* (Grigelis) as distinct species. (*Vide et* Premoli Silva 1966, p. 222, Fig. 2; Bignot & Guyader 1971, p. 83, Pl. 2, Fig. 3.) Figure 5.2: Key 111 111 12.

Guembelitrioides El-Naggar 1971, p. 431; '*Globigerinoides*' *higginsi* Bolli 1957b, p. 164, O.D. (early Middle Eocene, seabed core; Atlantic, 30°43'N, 62°28'W, 1554 m), USNM (*vide et* Stainforth *et al.* 1975, pp. 189–90; Saito *et al.* 1976, p. 287, Pl. 5; Blow 1979, pp. 862–4, Pl. 183). Figure 5.19: Key 111 122 221.

Hantkenina Cushman 1925a, p. 1; *H. alabamensis* Cushman 1925a, p. 3, O.D. (*Zeuglodon* bed, uppermost Eocene; Cocoa Post Office, Alabama), USNM. Topotype figured by Postuma (1971, pp. 224–5). A junior subjective synonym (by page priority) of *H. brevispina* Cushman 1925a, p. 2, *teste* Stainforth *et al.* (1975, p. 165) and Blow (1979, pp. 1153, 1163–4). Hantkeninidae proposed by Cushman (1927a, p. 64). (*Vide et* Blow & Banner 1962, Pl. 16; Hofker 1962, pp. 126–7; Blow 1979, pp. 1150–7, Pl. 189.) Figure 5.73: Key 131 221 12.

Globorotaliinae (113 23) (cont.): scale bar-line = 0.1 mm (100 μm)

Figure 5.47 (a–c) *Fohsella praefohsi* (Blow and Banner) Bandy; based on holotype (113 231 212).

Figure 5.48 (a–c) *Obandyella hirsuta* (d'Orbigny) Haman, Huddleston and Donahue: based on neotypes (113 231 213).

Figure 5.49 (a–c) *Truncorotalia truncatulinoides* (d'Orbigny) Cushman and Bermúdez; based on neotypes (113 231 22).

Figure 5.50 (a–c) *Turborotalita humilis* (Brady) Blow and Banner; based on lectotype (113 232).

Pulleniatininae (121): scale bar-line = 0.1 mm (100 μm)

Figure 5.51 (a–c) *Globigerinopsis aguasayensis* Bolli; based on holotype (121 1).

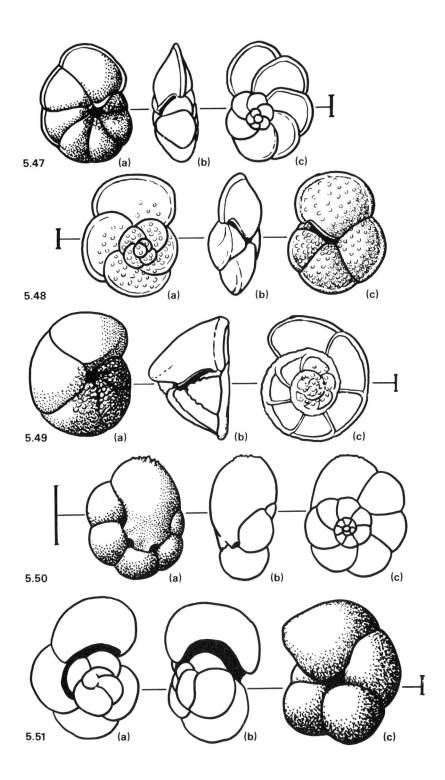

Hantkeninella Brönnimann 1950a, p. 399; *Hantkenina alabamensis* Cushman var. *primitiva* Cushman and Jarvis 1929, p. 16, O.D. (Mount Moriah Beds, San Fernando formation, Eocene; Vistabella Quarry, Trinidad), USNM, holotype refigured by Bolli *et al.* (1957, Pl. 2) (*vide et* Cushman & Cedestrom 1949, Pl. 5; Blow & Banner 1962, Pl. 16; Blow 1979, pp. 1153, 1161–2, Pl. 243). Figure 5.72: Key 131 221 11.

Hastigerina Thomson (in Murray) 1876, p. 534; *H. murrayi* Thomson (in Murray 1876, p. 534), O.D. (Recent, sediment; 'Challenger' 338, southern Atlantic, 1990 fathoms = 3639 m), BMNH, lectotype designated by Banner and Blow (1960b, p. 20), the same specimen as the neotype of its objective senior synonym, *Nonionina pelagica* d'Orbigny 1839b, p. 27, also selected by Banner and Blow (1960b, p. 20). Hastigerininae proposed by Bolli *et al.* (1957, p. 29). (For morphology, *vide et* Brady 1884, Pl. 83; Bé 1969, Pls 3, 4; Reiss *et al.* 1974, Pl. 82; Saito *et al.* 1976, Pls 2, 6, 8; Hemleben *et al.* 1979, Pl. 1. For biology, *vide et* Spindler *et al.* 1978; Anderson *et al.* 1979; Hemleben *et al.* 1979; Spindler *et al.*, 1979, *cum bibl.*) Figure 5.65: Key 131 112 1.

Hastigerinella Cushman 1927a, p. 87; *H. digitata* Rhumbler 1911, p. 202, O.D. and monotypy (Recent, northern Atlantic Deep Water, haul at 600 fathoms = 1097 m; 'Beebe' 1200, northern Atlantic Ocean), BMNH, neotype designated by Banner (1964, p. 115) *ex* Banner and Blow (1960b, p. 27, Fig. 8a–c). Objective senior synonym of *Hastigerinella rhumbleri* Galloway 1933, p. 333, *nom. van., nom. non necessarium*, and of *Hastigerinopsis digitiformans* Saito and Thompson (in Saito *et al.* 1976), type-species of *Hastigerinopsis* Saito and Thompson (in Saito *et al.* 1976), q.v., NOT *Hastigerinella eocanica* Nuttall 1928, as cited by Charmatz (1963) and by Saito and Thompson (in Saito *et al.* 1976), a citation neither necessary nor valid under the ICZN *Règles*, see Banner (1964) and Loeblich and Tappan (1964a,b). (*Vide et* Bé 1977, Pl. 6.) Figure 5.67: Key 131 112 2.

Hastigerinoides Brönnimann 1952b, p. 52; *Hastigerinella alexanderi* Cushman 1931, p. 87, O.D. (Austin Chalk, Senonian; Howe, Grayson County, Texas), USNM, holotype refigured by Bolli *et al.* (1957, Pl. 1) and

Pulleniatininae (121) (cont.): scale bar-line = 0.1 mm (100 μm)
Figure 5.52 (a–c) *Pulleniatina obliquiloculata* (Parker and Jones) Cushman; based on lectotype (121 2).

Globigerapsinae (122): scale bar-line = 0.1 mm (100 μm)
Figure 5.53 (a–c) *Orbulinoides beckmanni* (Saito) Cordey, based on holotype (a), hypotype (b), and axially sectioned hypotype (c) (122 1).
Figure 5.54 (a–c) *Globigerapsis kugleri* Bolli, Loeblich and Tappan; based on holotype (a), hypotype (b) and axially sectioned hypotype (c) (122 2).

Porticulasphaerinae (123 11): scale bar-line = 0.1 mm (100 μm).
Figure 5.55 (a–c) *Porticulasphaera mexicana* (Cushman) Bolli, Loeblich and Tappan; based on holotype (123 111 1).
Figure 5.56 (a–c) *Globigerinatheka barri* Brönnimann; based on holotype (a, b) and axially sectioned hypotype (c) (123 111 2).

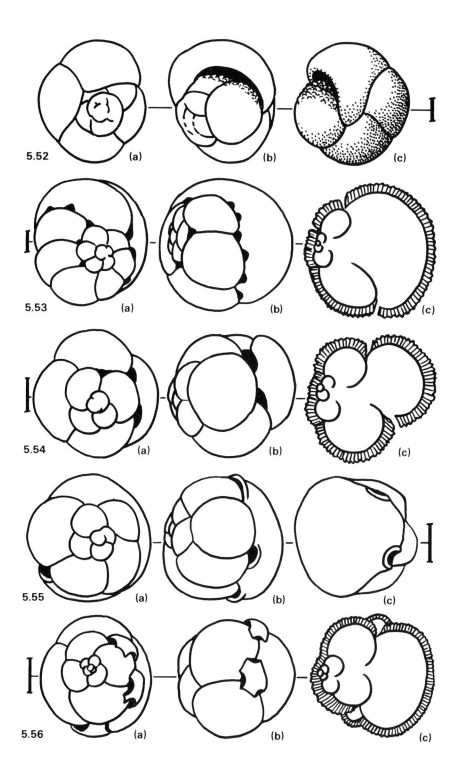

topotypes figured by Masters (1977, Pls 12, 13) (*vide et* Frerichs *et al.* 1977). Figure 5.84: Key 212 31.

Hastigerinopsis Saito and Thompson (in Saito *et al.*) 1976, pp. 284–6; *Hastigerinopsis digitiformans* Saito and Thompson (in Saito *et al.* 1976); objective junior synonym of *Hastigerina digitata* Rhumbler 1911, and therefore an objective typonym of *Hastigerinella* Cushman 1927a, q.v.

Hedbergella Brönnimann and Brown 1958, p. 15; *Anomalina lorneiana* d'Orbigny var. *trocoidea* Gandolfi 1942, p. 98, O.D. (Scaglia variegata; Aptian; Breggia River section, southern Switzerland), NHMB; lectotype selected (and figured) by Caron and Luterbacher (1969, p. 23, Pl. 7, Fig. 2) to replace the lost lectotype selected by Brönnimann and Brown (1958). Topotypes figured by Longoria (1974, Pl. 17) and Masters (1977, pl. 25). Hedbergellinae proposed by Loeblich and Tappan (1961b, p. 309); Hedbergelloidea proposed by Longoria and Gamper (1975). Figure 5.87: Key 221 111 21.

Helvetoglobotruncana Reiss 1957, p. 137; *Globotruncana helvetica* Bolli 1945, p. 226, O.D. (Turonian; Santis Section, Santis Range, St Gall canton, eastern Switzerland), NHMB. Ideotype figured by Bolli (1957c, Pl. 13, Fig. 1). Helvetoglobotruncaninae proposed by Lamolda (1976), (*Vide et* Banner & Blow 1959, Pl. 3; Reiss 1963, p. 74; Caron 1966, Pl. 3, Fig. 2; Postuma 1971, pp. 44–5; Longoria & Gamper 1975, pp. 89–90; Robaszynski & Caron 1979b, Pl. 46.) Figure 5.96: Key 221 211.

Hirsutella Bandy 1972, p. 298; *Globigerina hirsuta* d'Orbigny 1839c, p. 131, O.D.; homonym of *Hirsutella* Cooper and Muir-Wood 1951 (Brachiopoda); see *Obandyella* Haman, Huddleston and Donahue 1981, nom. nov.

Inordinatosphaera Mohan and Soodan 1967, p. 24; *I. indica* Mohan and Soodan 1967, p. 24, O.D. (Middle Eocene; between Beranda and Bernana, western Kutch, India), ONGC (*vide et* Mohan & Soodan 1970; El-Naggar 1971a, p. 441). Figure 5.57: Key 123 112.

Kuglerina Brönnimann and Brown 1956, p. 557; *Rugoglobigerina rugosa* (Plummer) var. *rotundata* Brönnimann 1952b, p. 34 (Guayaguayare Beds,

Porticulasphaerinae (123 11) (cont.): scale bar-line = 0.1 mm (100 μm)
Figure 5.57 (a–d) *Inordinatosphaera indica* Mohan and Soodan; based on paratype (a, b) and holotype (c, d) (123 112).
Orbulininae (123 12): scale bar-line = 0.1 mm (100 μm)
Figure 5.58 (a–c) *Praeorbulina glomerosa* (Blow) Olsson; based on axially-sectioned hypotype (a) and on holotype (b, c) (123 121 1).
Figure 5.59 (a–d) *Globigerinatella insueta* Cushman and Stainforth; based on paratype (a), holotype (b) and early growth stages dissected from hypotypes (c, d) (123 121 21).
Figure 5.60 (a&b) *Biorbulina bilobata* (d'Orbigny) Blow, based on axially-sectioned hypotype (a) and on lectotype and hypotypes (b) (123 121 221).
Figure 5.61 (a&b) *Candorbulina suturalis* (Brönnimann) = *C. universa* Jedlitschka; based on axially sectioned hypotype (a) and on lectotype and hypotypes (b) (123 121 222).
Figure 5.62 (a&b) *Orbulina universa* d'Orbigny; based on lectotype and hypotypes (a) and on broken hypotype with inner globigerinoidine whorls (b) (123 122).

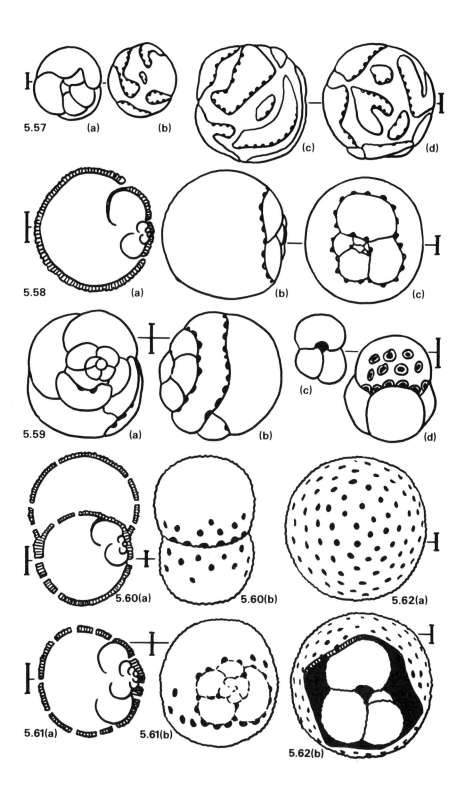

Maestrichtian; Trinidad), USNM (*vide et* El-Naggar 1971b, Pl. 10; Smith & Pessagno 1973, p. 58, Pl. 24; Hofker 1978, Pl. 1, Fig. 17). Figure 5.107: Key 312 111 2.

Labrobiglobigerinella Sigal 1966b, p. 22; *Globigerinelloides algerianum* Cushman and ten Dam 1948, p. 48, *nom. corr.* (designated as 'morphogénérotype: *Labroglobigerinella (Labrobiglobigerinella)* spectr.-*algerianum* (Cushman & ten Dam 1948) emend. Sigal, 1966 [b], Pl. 1, Figs 8, 10.') Junior typonym of *Globigerinelloides* Cushman and ten Dam 1948, q.v., under ICZN *Règles*.

Labroglobigerina Sigal 1966b, p. 23; *Globigerinelloides algerianum* Cushman and ten Dam 1948, p. 48, *nom. corr.* (designated as 'morphogénérotype: *Labroglobigerinella (Labroglobigerina)* spectr.-*algerianum* (Cushman & ten Dam 1948) emend. Sigal 1966[b], pl. 1, figs 3, 5, 6'). Junior typonym of *Globigerinelloides* Cushman and ten Dam 1948, q.v., under ICZN *Règles*.

Labroglobigerinella Sigal 1966b, pp. 21–2; *Globigerinelloides algerianum* Cushman and ten Dam 1948, p. 48, *nom. corr.* (designated as 'spectro-générotype et spectro-holotype: *Labroglobigerinella* spectrum-*algerianum* (Cushman & ten Dam 1948) Sigal, 1966[b]', subj. *Planulina cheniourensis* Sigal 1952, wrongly cited by Masters (1977, p. 400)). Junior typonym of *Globigerinelloides* Cushman and ten Dam 1948, q.v., under ICZN *Règles*.

Leupoldina Bolli 1958, p. 275; *L. protuberans* Bolli 1958, p. 275, O.D. (Aptian; Trinidad), USNM; metatypes figured by Bolli (1959, Pl. 20) and by Masters (1977, Pl. 14, 15). Subjective junior synonym of *Schackoina cabri* Sigal 1952, pp. 20–1 (Aptian; Djebel Rhazouane, Tunisia), Sigal collection; metatypes figured by Sigal (1959) (*vide et* Longoria 1974, pp. 89–90). Figure 5.82: Key 212 2.

Loeblichella Pessagno 1967, p. 288; *Praeglobotruncana hessi sensu stricto* Pessagno 1962, p. 358, O.D. (Rio Yauco formation; early Maestrichtian, *teste* Smith and Pessagno (1973, p. 42); near Hacienda Josefa, Barrio Quebrada Limon, Puerto Rico), USNM; ideotype refigured by Longoria (1974, Pl. 24). Loeblichellinae proposed by Pessagno (1967, p. 267). (*Vide et* Smith & Pessagno 1973, Pl. 16.) Figure 5.89: Key 221 111 222.

Orbulininae (123 12) (cont.): scale bar-line = 0.1 mm (100 μm)

Figure 5.63 (a–c) *Globigerinopsoides algerianus* Cita and Mazzola; based on holotype (a–c) and paratypes (c) (123 2).

Hastigerininae (131 1): scale bar-line 0.1 mm (100 μm)

Figure 5.64 (a&b) *Pseudohastigerina micra* (Cole) Banner and Blow; based on topotypes (131 111 1).

Figure 5.65 (a&b) *Hastigerina murrayi* Thomson; based on lectotype (a) and topotype (b) (= *H. pelagica* (d'Orbigny), neotype as (a)) (131 112 1).

Figure 5.66 (a–c) *Protentella prolixa* Lipps; based on holotype and topotypes (131 111 2).

Figure 5.67 (a–c) *Hastigerinella digitata* (Rhumbler) Cushman (= *Hastigerinopsis digitiformans* Saito and Thompson); based on neotype (note that the last chamber has broken off and a residual part of it remains around the aperture) (131 112 2).

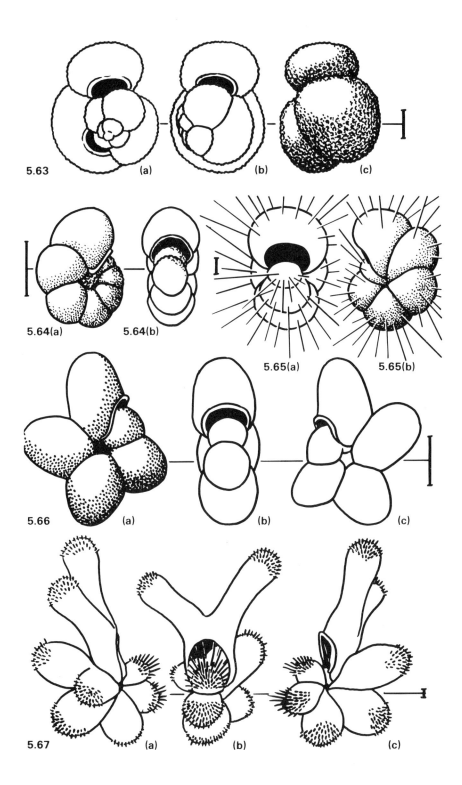

Marginotruncana Hofker 1956a, p. 319; *Rosalina marginata* Reuss 1846, p. 36, O.D. (lectotype figure selected by Bolli *et al.* (1957), Turonian, Plänermergel, Bohemia; type series lost, so neotype proposed and described by Jírová (1956): Emscherian marl, *'plänermergel'*, late Coniacian; Lužice, near Bilina, Chlomky Hills, northern Bohemia, Czechoslovakia), KPUK. Neotype photographed by Štemprokova-Jírová (1970, Pl. 1); topotype figured by Robaszynski and Caron (1979b, Pl. 63, Fig. 1). Marginotruncanidae proposed by Pessagno (1967, p. 298). (*Vide et* Porthault, in Donze *et al.* 1970, pp. 73–82; Masters 1977, p. 530 *et seq.*, Pl. 47; Hofker 1978, p. 59; Robaszynski & Caron 1979b, pp. 97–101.) Figure 5.114: Key 321 122.

Menardella Bandy 1972, p. 297; *Rotalia menardii* Parker, Jones and Brady 1865, p. 20 (*ex* d'Orbigny 1826, p. 273, *nom. nud.*). Lectotype proposed by Banner and Blow (1960a, p. 31, Pl. 6, Fig. 2) (Recent, sediment; off Laxey, Isle of Man, Irish Sea, 15 fathoms = 27 m), BMNH; ICZN proposal Z.N.(S)2145 by Stainforth *et al.* (1978) to suppress that lectotype and to designate a neotype (Late Miocene, 'Tortonian'; Senigallia section, south-east of Rimini), EPRC for USNM. Neotype described by Stainforth *et al.* (1978, pp. 260–1, Pl. 1). (*Vide et* Fleisher 1974, p. 1026; Stainforth *et al.* 1975, pp. 367–71, Fig. 177; Hemleben *et al.* 1977.) Figure 5.45: Key 113 231 211 1.

Morozovella McGowran (in Luterbacher) 1964, p. 636; *Pulvinulina velascoensis* Cushman 1925a, p. 19, O.D. (Velasco Shale, 'Upper Cretaceous', *recte* Palaeocene; Hacienda el Limon, San Luis Potosi State, Mexico), USNM; topotype figured by Postuma (1971, pp. 218–19), near-topotypes by Loeblich and Tappan (1957a, Pl. 64) and by Luterbacher (1964) (*vide et* McGowran 1968, p. 190; Fleisher 1974, p. 1029; Blow 1979, pp. 972–4, Pls 92, 94, 95, 99, 216, 217). Figure 5.41: Key 113 222.

Muricoglobigerina Blow 1979, pp. 1118–19; *Globigerina soldadoensis* Brönnimann 1952c, pp. 7, 9, O.D. (Lizard Springs Marl, Late Palaeocene

Hastigerininae (131 1) (cont.): scale bar-line = 0.1 mm (100 μm)
Figure 5.68 (a&b) *Globigerinella aequilateralis* (Brady) Cushman, based on lectotype (= *G. siphonifera* (d'Orbigny)) (131 113 1).
Figure 5.69 (a&b) *Berggrenia praepumilio* (Parker) Parker; based on holotype and paratypes (131 121).
Figure 5.70 (a&b) *Bolliella adamsi* Banner and Blow; based on holotype (131 113 2).
Hantkenininae (131 2): scale bar-line = 0.1 mm (100 μm)
Figure 5.71 (a&b) *Clavigerinella akersi* Bolli, Loeblich and Tappan; based on holotype and hypotypes (131 21).
Figure 5.72 (a&b) *Hantkeninella primitiva* (Cushman and Jarvis) Brönnimann; based on holotype and hypotypes (131 221 11).
Figure 5.73 (a&b) *Hantkenina alabamensis* Cushman; based on syntypes and hypotypes (= *Sporohantkenina brevispina* (Cushman) Bermúdez) (131 221 12).
Figure 5.74 (a&b) *Applinella dumblei* (Weinzierl and Applin) Thalmann; based on lectotype and hypotypes (a), paralectotype and hypotypes (b), (131 221 21).
Figure 5.75 (a&b) *Aragonella aragonensis* (Nuttall) Thalmann; based on lectotype and hypotypes (131 221 22).

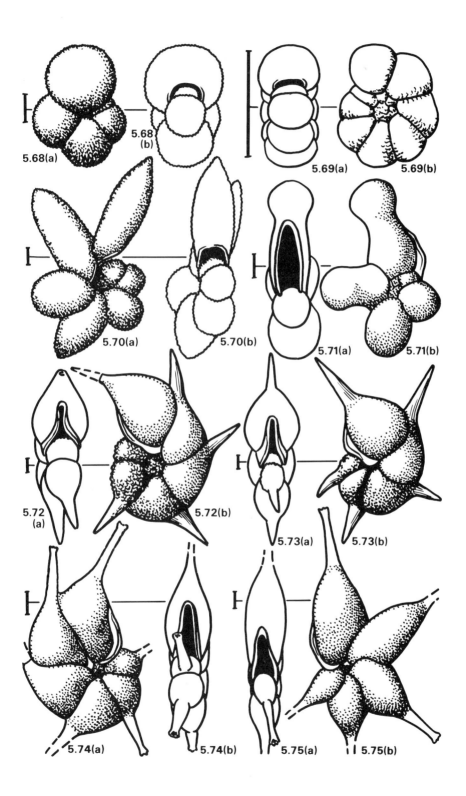

(zone P.5, type *Globorotalia velascoensis* zone of Bolli (1957d), *teste* Blow (1979, p. 1122)); Ampelu River ravine, Trinidad), USNM; topotype figured by Bolli (1957d, Pl. 16) (*vide et* Postuma 1971, pp. 158–9; Stainforth *et al.* 1975, pp. 229–30; Blow 1979, Pls 98, 107, 109, 110, 124, 131, 235). Figure 5.9: Key 111 122 11.

Neoacarinina Thompson 1973, p. 470; *N. blowi* Thompson 1973, p. 470, O.D. (Pleistocene, sediment; core RC8-39, Crozet Basin, southwestern Indian Ocean), USNM. Figure 5.11: Key 111 122 122.

Neogloboquadrina Bandy, Frerichs and Vincent 1967, p. 152; *Globigerina dutertrei* d'Orbigny 1839a, p. 84, O.D. (Recent, sediment; Cuba), MHN; lectotype selected and figured by Banner and Blow (1960a, p. 11, Pl. 2, Fig. 1); lectotype photographed by Le Calvez (1977, pp. 27–8, Pl. 4) (*vide et* Parker 1962, Pl. 7; Fleisher 1974, pp. 1034–5; Stainforth *et al.* 1975, pp. 347–8, Fig. 163; Maiya *et al.* 1976, pp. 409–12; Srinavasan & Kennett 1976, Pls 1, 3). Figure 5.43: Key 113 231 112.

Obandyella Haman, Huddleston and Donahue, 1981, p. 1265; *Globigerina hirsuta* d'Orbigny 1839c, p. 131, by substitution of O.D. (new name for *Hirsutella* Bandy 1972, homonym, *non Hirsutella* Cooper and Muir-Wood 1951, Brachiopoda); neotype proposed by Blow (1969, pp. 398–400, Pl. 8) from Recent sediment, 'Challenger' 8, off Gomera, Canary Islands, BMNH, and, independently, another neotype proposed by Le Calvez (1974, pp. 69–71, Pl. 16) from Recent sediment, d'Orbigny sample, Ile de Ténériffe, Canary Islands, MHN (*vide et* Fleisher 1974, p. 1027). Figure 5.48: Key 113 231 213.

Orbulina d'Orbigny 1839a, p. 2; *O. universa* d'Orbigny 1939a, p. 2, monotypy (Recent, sediment; Cuba), MHN; lectotype selected by Loeblich and Tappan (unpublished manuscript), recorded with figured topotype by Le Calvez (1977, p. 56–8, Pls 1, 2). Orbulinida proposed by Schultze

Hantkenininae (131 2) (cont.): scale bar-line = 0.1 mm (100 μm)

Figure 5.76 (a–c) *Cribrohantkenina bermudezi* Thalmann; based on lectotype and hypotypes (a), on paralectotype and hypotypes (b) and on hypotype with undivided aperture (c) (= *C. inflata* (Howe)) (131 222).

Planomalinidae (21): scale bar-line 0.1 mm (100 μm)

Planomalininae (211)

Figure 5.77 (a&b) *Globigerinelloides algerianus* Cushman and ten Dam; based on holotype, paratypes and hypotypes (211 111 1).

Figure 5.78 (a&b) *Blowiella blowi* (Bolli) Kretzschmar and Gorbachik; based on holotype and hypotypes (211 111 2).

Figure 5.79 (a–c) *Biglobigerinella multispina* Lalicker; based on holotype and hypotypes (a, b) (= *B. aspera* (Ehrenberg), based on strict topotype, Ehrenberg collection) (c) (211 112).

Figure 5.80 (a&b) *Planomalina apsidostroba* Loeblich and Tappan; based on holotype (= *P. buxtorfi* (Gandolfi)) (211 2).

Figure 5.81 (a–c) *Biticinella breggiensis* (Gandolfi) Sigal; based on holotype and topotypes (211 12).

Schackoininae (212): scale bar-line = 0.1 mm (100 μm)

Figure 5.82 (a&b) *Leupoldina protuberans* Bolli; based on holotype and metatypes (= *L. cabri* (Sigal)) (212 2).

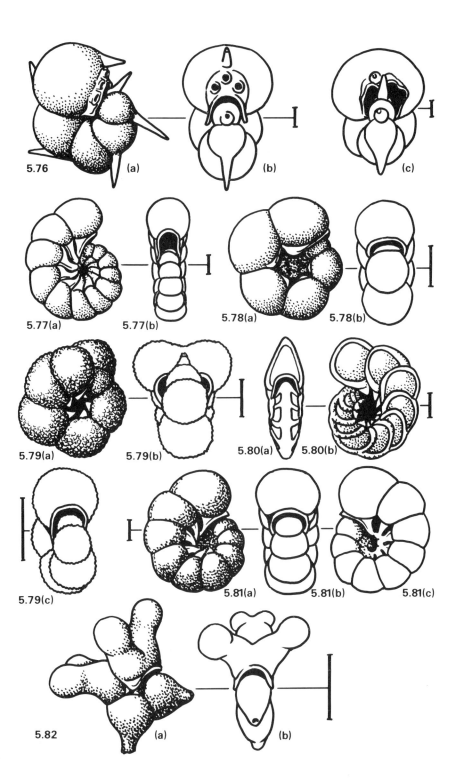

(1854, p. 52). (*Vide et* Blow 1956; Bé *et al.* 1973; Bé *et al.* 1976.) Figure 5.62: Key 123 122.

Orbulinoides Cordey (*ex* Blow and Saito) 1968, p. 371; *Porticulasphaera beckmanni* Saito 1962, pp. 221–2, O.D. and monotypy ('*Globigerina* tuff', Middle Eocene; Onion Beach, Okimura, Haha-jima), IGPS. Cordey's work was published in June 1968, inadvertently predating the publication of the genus by Blow and Saito (September 1968, for July issue), see Blow and Saito (1968b). (*Vide et* Proto Decima & Bolli 1970; Postuma 1971, pp. 230–1; Blow 1979, pp. 1137–40, Pl. 192). Figure 5.53: Key 122 1.

Parvularugoglobigerina Hofker 1978, p. 60; *Globigerina eugubina* Luterbacher and Premoli Silva 1964, p. 105, O.D. ('Danian', earliest Palaeocene, Scaglia; Ceselli, Valle della Nera, central Appenines, Italy), NHMB (*vide et* Stainforth *et al.* 1975, pp. 183–4, Fig. 47; Hofker 1978, Pl. 2, Fig. 6). Figure 5.32: Key 113 11.

Planogyrina Zakharova-Atabekyan 1961, p. 50; *Globigerina gaultina* Morozova 1948, p. 41, O.D. (Green marls, Albian; Bol'shaya Khota R., southwestern Caucasus, USSR), VNIGRI (*vide et* Subbotina 1953, Pl. 1, pp. 169–73, in Lees translation, 1971). Subjective junior synonym of *Globigerina planispira* Tappan 1940, p. 122 (Late Albian or early Cenomanian (*teste* Pessagno (1969, pp. 55–6)), Grayson formation, Washita Group; Denton County, Texas), USNM; topotypes figured by Loeblich and Tappan (1961a, p. 276, Pl. 5, Figs 7, 8), Longoria (1974, pp. 64–5, Pl. 23), Masters (1977, pp. 470–3, Pl. 24) and Robaszynski and Caron (1979a, p. 139–44, Pl. 27). (*Vide et* Bolli *et al.* 1957, pp. 39–40, Pl. 9, Fig. 3.) Figure 5.88: Key 221 111 221.

Planomalina Loeblich and Tappan 1946, p. 257; *P. apsidostroba* Loeblich and Tappan, 1946, p. 257, O.D. (Main Street formation, Late Albian; near Godley, Johnson County, Texas), USNM; holotype refigured by Bolli *et al.* (1957, Pl. 1, Fig. 2). Subjective junior synonym (*teste* Loeblich and Tappan (1961a, p. 269)) of *Planulina buxtorfi* Gandolfi 1942, p. 103 (Albian, Scaglia Bianca; Breggia River, northeast of Balerna, Canton

---

Schackoininae (212) (cont.): scale bar-line = 0.1 mm (100 μm)

Figure 5.83 (a–c) *Schackoina cenomana* (Schacko) Thalmann; based on hypotypes and near-topotypes (a, b) and equatorially sectioned hypotype (c) (comparable to original syntype figures) (212 1).

Figure 5.84 (a&b) *Hastigerinoides alexanderi* (Cushman) Brönnimann; based on holotype, paratype and topotypes (212 31).

Figure 5.85 (a&b) *Eohastigerinella watersi* (Cushman) Morozova; based on topotype and hypotypes (212 32).

Hedbergellidae (221): scale bar-line = 0.1 mm (100 μm)
Hedbergellinae (221 11)

Figure 5.86 (a–c) *Praehedbergella tuschepsensis* (Antonova) Gorbatchik and Moullade; based on holotype (221 111 1).

Figure 5.87 (a–c) *Hedbergella trocoidea* (Gandolfi) Brönnimann and Brown; based on lectotype and topotypes (221 111 21).

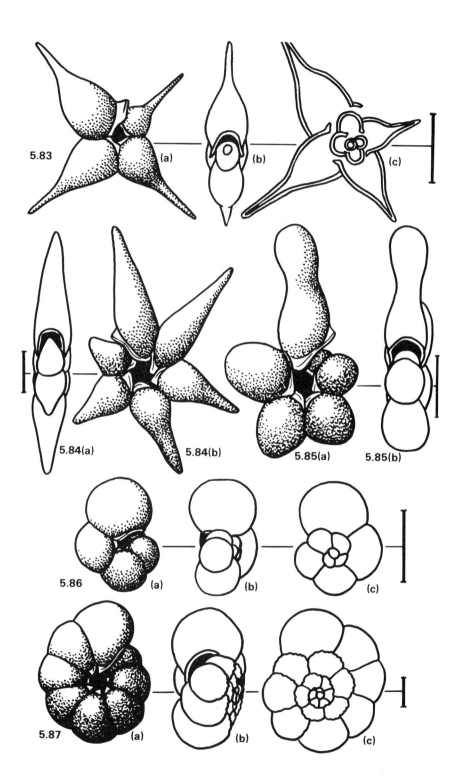

Ticino, Switzerland), MHN; holotype redrawn by Caron and Luterbacher (1969, Pl. 8, Fig. 5); topotypes figured by Robaszynski and Caron (1979a, pp. 41–3, Pl. 1, Figs 2–4). (*Vide et* Postuma 1971, pp. 68–9; Longoria 1974, Pls 8, 25; Blow 1979, p. 532, Text Fig. P.) Figure 5.80: Key 211 2.

Planorotalites Morozova 1957, p. 1112; *Globorotalia pseudoscitula* Glaessner 1937, p. 32, O.D. (Variegated Series, F_1, 'Lower Middle Eocene', *recte* Palaeocene–Early Eocene, *teste* Subbotina (1953); Il'sk station, northern Caucasus, USSR), MUPL? (*vide et* Subbotina 1953, Pl. 16, Figs 17, 18, Pl. 17, Fig. 1; McGowran 1968, pp. 188, 190; Poore & Brabb 1977, pp. 252, 264, Pl. 7, Figs 12, 13; Blow 1979, pp. 897–8, Pls 116, 173). Figure 5.36: Key 113 212 1.

Plummerita Brönnimann 1952d, p. 146 (*nom. nov. pro Plummerella* Brönnimann 1952b, p. 37, homonym *non Plummerella* de Long 1942, Insecta); *Rugoglobigerina (Plummerella) hantkeninoides hantkeninoides* Brönnimann 1952a, p. 146, O.D. (Maestrichtian, Guayaguayare Beds; Trinidad), USNM. Holotype refigured by Bolli *et al.* (1957, Pl. 11, Fig. 5). Figure 5.111: Key 312 2.

Polskanella Fuchs 1973, pp. 456–7; *Globigerina oxfordiana* Grigelis 1958. Objective junior typonym of *Globuligerina* Bignot and Guyader 1971, q.v. (*Vide et* Fuchs 1975, p. 230; Grigelis & Gorbatchik 1980, p. 184.)

Polyperibola Liska 1980, p. 136; *P. christiani* Liska 1980, p. 137, O.D. (Late Miocene, claystones; Lizard Road, Rio Claro to Guayaguare, Trinidad), USNM (holotype), BMNH and MHN (paratypes). Figure 5.18: Key 111 122 212 22.

Porticulasphaera Bolli, Loeblich and Tappan 1957, pp. 34–5; *Globigerina mexicana* Cushman 1925b, O.D. (Middle Eocene, probably Chapapote formation, reworked and redeposited into Oligo-Miocene, *teste* Blow and Saito (1968a), or into upper Eocene Tantoyuca formation, *teste* Cushman (1925a); Palacho Hacienda, Vera Cruz, Mexico), USNM; holotype redrawn by Blow and Saito (1968a, p. 358), who emended the genus (*vide et* Bolli 1972, Pl. 2, Figs 1–5; Blow 1979, pp. 868–74; Pl. 198). Figure 5.55: Key 123 111 1.

Praeglobotruncana Bermúdez 1952, p. 52; *Globorotalia delrioensis* Plummer 1931, p. 199, O.D. (Del Rio Shale, Washita Group, early Cenomanian; Shoal Creek, Austin, Travis County, Texas), PRII. Topotypes figured by

---

Hedbergellinae (221 11) (cont.): scale bar-line = 0.1 mm (100 μm)
Figure 5.88 (a–c) *Planogyrina gaultina* (Morozova) Zakharova-Atabekyan; based on hypotypes (= *P. planispira* (Tappan), holotype and topotypes) (221 111 221).
Figure 5.89 (a–c) *Loeblichella hessi* (Pessagno) Pessagno; based on holotype and ideotypes (221 111 222).
Figure 5.90 (a–c) *Clavihedbergella subcretacea* (Tappan) Banner and Blow; based on holotype, topotypes and hypotypes (221 112 1).
Figure 5.91 (a–c) *Asterohedbergella asterospinosa* Hamoui; based on holotype drawing. The umbilical and apertural details are not known, but are presumed to be similar to those of other Hedbergellinae (221 112 2).

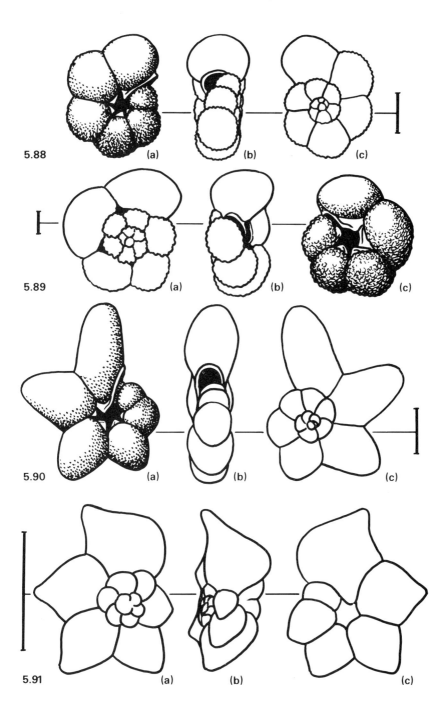

Brönnimann and Brown (1956, Pl. 21, Figs 8–10), by Bolli *et al.* (1957, Pl. 9, Fig. 1) and by Loeblich and Tappan (1961a, pp. 280, 282, Pl. 6, Figs 9, 10). (*Vide et* Masters 1977, pp. 486–9, Pl. 27, 28; Robaszynski & Caron 1979b, Pl. 43.) Figure 5.92: Key 221 121 11.

Praehedbergella Gorbatchik and Moullade 1973, p. 2662; *Globigerina tuschepsensis* Antonova (in Antonova *et al.*) 1964, p. 59, O.D. (Barremian; Tusheps River, Psheka–Ubin area, northwestern Caucasus), KFVNII. Figure 5.86: Key 221 111 1.

Praeorbulina Olsson 1964, p. 770; *Globigerinoides glomerosa glomerosa* Blow 1956, pp. 64–5, O.D. (Pozón formation, Early Miocene; El Mene–Pozón Road, eastern Falcón, Venezuela), USNM; near-topotypic metatype figured by Blow (1969, Pl. 23, Fig. 7). Figure 5.58: Key 123 121 1.

Prosphaeroidinella Ujiié 1976, pp. 9–11; *Sphaeroidinella disjuncta* Finlay 1940, p. 467, O.D. (Altonian, Early–Middle Miocene; Tangihanga, Waikohu, New Zealand), NZGS; holotype refigured by Jenkins (1971, pp. 171–2, Pl. 17, Figs 536–8) (*vide et* Hornibrook 1968, p. 86, Fig. 17; Ujiié 1976, Pls 11, 12). Figure 5.27: Key 111 211.

Protentella Lipps 1964, p. 122; *P. prolixa* Lipps 1964, p. 122, O.D. and monotypy (Monte Rey formation, Middle Miocene, Upper Luisian, zones N.14–N.15, *teste* Lipps (1967); Newport Bay, California), USNM; topotypes figured by Fleisher (1974, Pl. 5, Figs 3–6) and by Saito *et al.* (1976, pp. 292, 294, Pl. 7). Figure 5.66: Key 131 111 2.

Pseudogloboquadrina Jenkins 1966, p. 1122; *Globoquadrina primitiva* Finlay 1946, p. 291, O.D. (Bortonian, Middle Eocene; Hampden Beach, north of Kakaho Creek, Otago Province, New Zealand), NZGS; holotype and paratype refigured by Jenkins (1971, p. 170, Pl. 8, Figs 555–60); topotype figured by Postuma (1971, pp. 154–5) (*vide et* McGowran 1968, p. 186; Hornibrook 1968, p. 83, Fig. 16; Scott 1976; Blow 1979, pp. 902, 949–51). Figure 5.10: Key 111 122 121.

Pseudohastigerina Banner and Blow 1959, pp. 9, 20–1; *Nonion micrus* Cole 1927, p. 22, O.D. (Guayabal formation, Middle Eocene (zone P.12, *teste* Blow (1979, p. 1188)); Hacienda Tamatoco, Vera Cruz, Mexico), HCCU; topotypes figured by Berggren *et al.* (1967, Text Fig. 9), Postuma (1971,

Globotruncanellinae (221 12): scale bar-line = 0.1 mm (100 μm)

Figure 5.92 (a–c) *Praeglobotruncana delrioensis* (Plummer) Bermúdez; based on holotype and topotypes (221 121 11).

Figure 5.93 (a–c) *Rotundina stephani* (Gandolfi) Subbotina; based on holotype and topotypes (221 121 12).

Figure 5.94 (a–c) *Dicarinella indica* (Jacob and Sastry) Porthault = *D. imbricata* (Mornod); based on neotype and topotypes of the latter (221 121 21).

Figure 5.95 (a–c) *Globotruncanella citae* (Bolli) Reiss (= *G. havanensis* (Voorwijk)); based on holotype of the former and hypotypes of both. (221 122).

[For *Falsotruncana* see Figure 5.118 below]

Helvetoglobotruncaninae (221 2): scale bar-line = 0.1 mm (100 μm)

Figure 5.96 (a–d) *Helvetoglobotruncana helvetica* (Bolli) Reiss; based on hypotypes (a–c) and on axially sectioned holotype and hypotypes (d) (221 211).

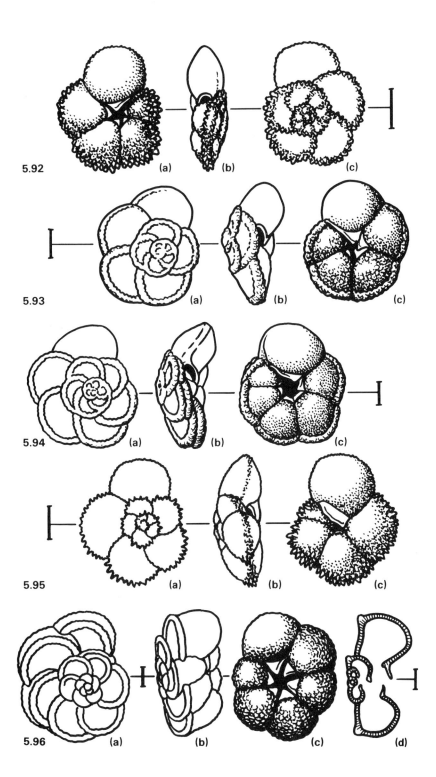

pp. 228–9) and Blow (1979, Pl. 253, Figs 7–9). Specimens referred to this species, figured by Banner and Blow (1959), Blow and Banner (1962) and Blow (1969) more probably belong to *Pseudohastigerina danvillensis* (Howe and Wallace) (*Nonion danvillensis* Howe and Wallace 1932, p. 51), *teste* Blow (1979, pp. 1181–5). (*Vide et* Cordey *et al.* 1970; Steineck 1971.) Figure 5.64: Key 131 111 1.

Pseudothalmanninella Wonders 1978, pp. 124–5; *Globotruncana ticinensis* var. *tipica* Gandolfi 1942, pp. 116–23, O.D. (corrected to *Globotruncana (Thalmanninella) ticinensis ticinensis* by Gandolfi (1957, p. 59)), (Scaglia Bianca, Albian; La Breggia River, near Balerna, Canton Ticino, Switzerland), MHN; holotype redrawn by Caron and Luterbacher (1969, p. 25, Pl. 8, Fig. 6); topotypes figured by Caron (1967, pp. 70–3, Text Fig. 21, Pls 1, 2), by Masters (1977, p. 516, Pl. 34) and by Robaszynski and Caron (1979a, pp. 111–2, Pl. 20, Fig. 1) (*vide et* Reichel 1950, pp. 603–4, Pl. 16, Fig. 3, Pl. 17, Fig. 3; Postuma 1971, pp. 86–7). Figure 5.100: Key 222 12.

Pseudoticinella Longoria (September) 1973, p. 418; *Globorotalia? multiloculata* Morrow 1934, p. 200, O.D. Objective junior typonym of *Anaticinella* Eicher 1973 (1972), q.v. (*vide et* Longoria & Gamper 1975, pp. 82–5).

Pulleniatina Cushman 1927a, p. 90; *Pullenia sphaeroides* (d'Orbigny) var. *obliquiloculata* Parker and Jones 1865, pp. 365, 368, O.D. (Recent sediment; Atlantic Ocean, off Abrolhos Bank, 22°54'S, 40°37'N at 260 fathoms = 475 m), BMNH, lectotype selected by Bolli *et al.* (1957, p. 33) (topotype figured in Bolli *et al.* (1957, Pl. 4, Fig. 3)), lectotype figured by Banner and Blow (1960a, pp. 25, 40, Pl. 7, Fig. 4); lectotype refigured, species and genus emended by Banner and Blow (1967, pp. 137–9, Pl. 3, Fig. 4). Pulleniatininae proposed by Cushman (1927a, p. 89). (*Vide et* Towe 1971, Pl. 3, Fig. 2; Lamb & Beard 1972, p. 58, Pl. 29, Figs 1–4; Blow 1979, Text Fig. K, p. 522). Figure 5.52: Key 121 2.

Radotruncana El-Naggar 1971a, p. 434; *Globotruncana calcarata* Cushman 1927b, p. 115, O.D. (proposed as subgenus of *Plummerita*, q.v.) (Pecan

Helvetoglobotruncaninae (221 21) (cont.): scale bar-line = 0.1 mm (100 μm)

Figure 5.97 (a–c) *Whiteinella archaeocretacea* Pessagno; based on holotype and topotypic paratypes (221 212).

Rotaliporidae (222): scale bar-line = 0.1 mm (100 μm)

Ticinellinae (222 1)

Figure 5.98 (a–c) *Ticinella roberti* (Gandolfi) Reichel; based on holotype and near-topotypic hypotypes (222 111).

Figure 5.99 (a–c) *Claviticinella digitalis* (Sigal) El-Naggar; based on holotype and hypotypes (222 112).

Figure 5.100 (a–c) *Pseudothalmanninella ticinensis sensu stricto* (Gandolfi) Wonders; based on holotype and topotypes (222 12).

Rotaliporinae (222 2): scale bar-line = 0.1 mm (100 μm)

Figure 5.101 (a–c) *Thalmanninella brotzeni* Sigal (= *T. greenhornensis* (Morrow)); based on holotypes and topotypes of both (222 211).

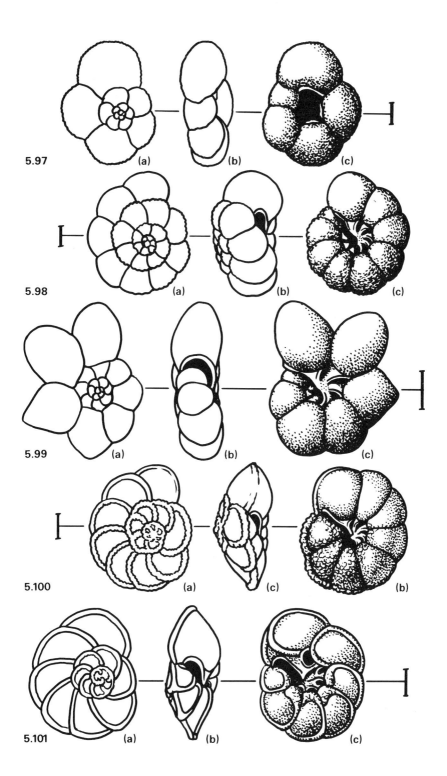

Gap Chalk, Mendez Shale, Campanian; Collin County, Texas), USNM. Topotypes figured by Masters (1977, pp. 540-1, Pl. 38, 39) (*vide et* Brönnimann & Brown 1956, pp. 548-9, Pl. 23, Figs 1-3; Pessagno 1967, pp. 326-8, Pl. 64, 93; Postuma 1971, pp. 22-3). Figure 5.116: Key 321 22.

Reticuloglobigerina Reiss 1963, p. 74; *Globigerina washitensis* Carsey 1926, p. 44, monotypy (Del Rio Clay, early Cenomanian; Shoal Creek, Austin, Travis County, Texas), MCUT? holotype lost so 'neoholotype' proposed and described by Plummer (1931, p. 193, Pl. 13, Fig. 12) (probably unnecessary replacement neotype proposed by Longoria (1974, pp. 74-6, Pl. 26), USNM). Topotypes figured by Loeblich and Tappan (1961a, p. 278, Pl. 4, Fig. 11), by Michael (1973, Pls 5, 7), by Longoria (1974, *loc. cit.*), and by Masters (1977, p. 479, Pls 25, 26). (*Vide et* Tappan 1940, p. 122, Pl. 19; Postuma 1971, pp. 66-7; Rösler *et al.* 1979, pp. 273-81.) Senior objective typonym of *Favusella* Michael 1973 (q.v.). Figure 5.5: Key 111 112 12.

Rosalinella Marie 1941, pp. 237, 256, 258; *Rosalina linneiana* d'Orbigny 1839a, p. 101 (Cretaceous, redeposited into Recent beach sediment; estuary of River Martin Pérez, Habana Bay, Cuba), original types destroyed, *teste* Le Calvez (1977, pp. 90-2), neotypified by Brönnimann and Brown (1956, pp. 540-2, Pl. 20, Figs 13-15), GOC? (*vide et* Longoria & Gamper 1975, Pl. 9, Fig. 1; Masters 1977, pp. 583-5, Pl. 46). Figure 5.113: Key 321 121.

Rotalipora Brotzen 1942, p. 32; *R. turonica* Brotzen 1942, p. 32, O.D. (Lower Turonian Chalk; Gristow in Pommern (Odermundung), Germany), PCNMS; topotypes figured by Reichel (1950, pp. 607-8, Pl. 16, Fig. 5, Pl. 17, Fig. 5) and by Loeblich and Tappan (1961a, Pl. 8, Fig. 1). Subjective junior synonym of *Globorotalia cushmani* Morrow 1934, p. 199 (late Cenomanian, Hartland shale, Greenhorn formation; Hodgeman County, Kansas), USNM, holotype redrawn by Brönnimann and Brown (1956, p. 538, Pl. 20, Figs 10-12); topotypes figured by Loeblich and Tappan (1961a, pp. 297-8, Pl. 8, Figs 2, 3, 6, 7, 9), by Masters (1977, pp. 501-6, Pls 30, 31) and by Robaszynski and Caron (1979a, pp. 51-8. Pls 7, 8). Rotaliporidae proposed by Sigal (1958, p. 264). (*Vide et* Bolli *et al.*

Rotaliporinae (222 2) (cont.): scale bar-line = 0.1 mm (100 μm)

Figure 5.102 (a-c) *Rotalipora turonica* Brotzen (= *R. cushmani* (Morrow)); based on primary types and topotypes (222 212).

Figure 5.103 (a-c) *Anaticinella multiloculata* (Morrow) Eicher (= *Pseudoticinella multiloculata* (Morrow) Longoria); based on holotype, topotypes and hypotypes (222 22).

Globotruncanidae (3): scale bar-line = 0.1 mm (100 μm)

Rugoglobigerininae (31)

Figure 5.104 (a-c) *Archaeoglobigerina blowi* Pessagno; based on holotype, topotypic paratypes and topotypes (311 11).

Figure 5.105 (a-c) *Fissoarchaeoglobigerina aegyptiaca* Abdel-Kireem; based on holotype (a, b) and paratype (c) (311 12).

Figure 5.106 (a-c) *Rugoglobigerina rugosa* (Plummer) Brönnimann; based on syntypes (a, b) topotypes (b, c) and hypotypes (a, b, c); (b) with tegillum lost; (c) with normal tegillum (312 111 1).

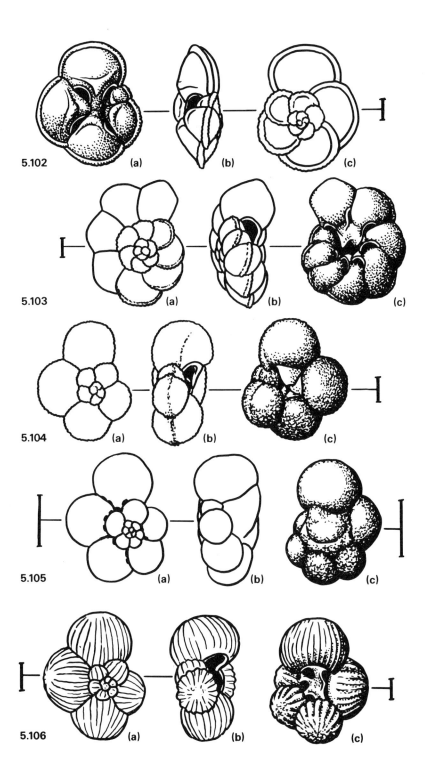

1957, Pl. 9, Fig. 6; Postuma 1971, pp. 78–9; Longoria 1974, Pl. 26, Figs 1–3.) Figure 5.102: Key 222 212.

Rotundina Subbotina 1953, p. 164; *Globotruncana stephani* Gandolfi 1942, O.D. (Scaglia Rossa, Cenomanian; Breggia River section, near Balerna, Canton Ticino, Switzerland), NHMB, holotype redrawn by Caron and Luterbacher (1969, p. 26, Pl. 8, Fig. 7); topotypes figured by Reichel (1950, Pls 16, 17), by Bolli *et al.* (1957, Pl. 9, Fig. 2) and by Robaszynski and Caron (1979b, p. 50, Pls 40, 41, 48). (*Vide et* Subbotina 1953, Pl. 3, Figs 1, 2; Hagn & Zeil 1954, Pl. 5, Figs 7, 8; Caron 1966, Pl. 2, Fig. 3; Postuma 1971, pp. 72–3.) Figure 5.93: Key 221 121 11.

Rugoglobigerina Brönnimann 1952b, p. 16; *Globigerina rugosa* Plummer 1927, p. 38, O.D. (Corsicana Marl, middle Maestrichtian; Walkers Creek, Milan County, Texas; *teste* Smith and Pessagno (1973)), PRII. Topotypes figured by Smith and Pessagno (1973, pp. 58–60, Pl. 25, Figs 1–4). Rugoglobigerininae proposed by Subbotina (in Rauser-Chernoussova & Fursenko, 1959, p. 303). (*Vide et* Bolli *et al.* 1957, pp. 42–3, Pl. 11, Figs 2–5; Postuma 1971, pp. 90–1; El-Naggar 1971a, p. 433, Pl. 5; El-Naggar 1971b, pp. 492–3, Pls 1, 4; Masters 1977, pp. 622–6, Pls 56, 57.) Figure 5.106: Key 312 111 1.

Rugotruncana Brönnimann and Brown 1956, p. 546; *R. tilevi* Brönnimann and Brown 1956, p. 547, O.D. (late Maestrichtian marl; construction pit, Habana, Cuba), USNM. Junior subjective synonym of *Globotruncana (Rugoglobigerina) circumnodifer subcircumnodifer* Gandolfi 1955, p. 44 (Maestrichtian, Colon formation; Colombia), USNM, *teste* Berggren (1962, pp. 67–9), Pessagno (1967, pp. 369–70) and Masters (1977, pp. 72–3). (*Vide et* Pessagno 1967, Pl. 62, Figs 14–16.) Figure 5.109: Key 312 121.

Schackoina Thalmann 1932, p. 288; *Siderolina cenomana* Schacko 1897, p. 166 (Cenomanian chalk; Moltzow, Mecklenburg, Germany), depository unknown. Schackoinidae proposed by Pokorný (1958, p. 348). (*Vide et* Reichel 1948; Bolli *et al.* 1957, Pl. 2; Loeblich & Tappan 1961a, p. 270, Pl. 1; Ayala-Castañares 1962; Masters 1977, pp. 426–8, 430–2, Pl. 16.) Figure 5.83: Key 212 1.

Sphaeroidinella Cushman 1927a, p. 90; *Sphaeroidina bulloides* d'Orbigny

Rugoglobigerininae (31) (cont.): scale bar-line = 0.1 mm (100 μm)

Figure 5.107 (a–c) *Kuglerina rotundata* (Brönnimann) Brönnimann and Brown; based on holotype (the last formed chamber is abnormal in comparison both to the previous growth mode and to complete paratypes) (312 111 2).

Figure 5.108 (a–d) *Trinitella scotti* Brönnimann; based on holotype (a, c) and metatype (b and c); tegillum removed from hypotype (d) (312 112).

Figure 5.109 (a–c) *Rugotruncana tilevi* Brönnimann and Brown; based on primary types (= *R. subcircumnodifer* (Gandolfi)), (b) with tegillum and (c) without (312 121).

Figure 5.110 (a–c) *Bucherina sandidgei* Brönnimann and Brown; based on primary types (umbilical tegillum in (a) apparently lost) (312 122).

Figure 5.111 (a–c) *Plummerita hantkeninoides* (Brönnimann) Brönnimann; based on holotype and paratypes (312 2).

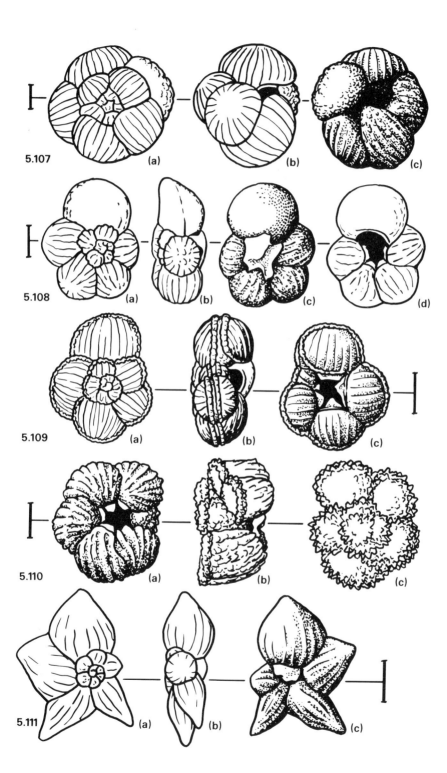

var. *dehiscens* Parker and Jones 1865, p. 369, O.D. (Recent sediment; Atlantic Ocean, 2°20'N, 28°44'W at 1080 fathoms = 1975 m), BMNH, lectotype selected by Bolli *et al.* (1957, p. 33), described by Banner and Blow (1960a, pp. 35–6, Pl. 7, Fig. 3). Sphaeroidinellinae proposed by Banner and Blow (1959, p. 5). (*Vide et* Bolli *et al.* 1957, Pl. 6; Bé 1965; Hofker 1972; Stainforth *et al.* 1975, pp. 344–7; Ujiié 1976; Blow 1969, Pl. 29; Blow 1979, Text Figs I, J.) Figure 5.29: Key 111 22.

Sphaeroidinellopsis Banner and Blow 1959, pp. 15–16; *Sphaeroidinella dehiscens* (Parker and Jones) *subdehiscens* Blow 1959, pp. 195–6, O.D. (Husito Clay member, Pozón formation, Middle Miocene; Pozón–El Mene Road, eastern Falcón, Venezuela), USNM. Ideotypes figured by Blow (1969, Pl. 30; 1979, p. 522, Text Fig. K(i)). (*Vide et* Mazzola 1971, Pl. 4, Figs 5, 7; Natori 1976, Pl. 1, Fig. 1; Ujiié 1976). Figure 5.28: Key 111 212.

Sporohantkenina Bermúdez 1937, p. 151; *Hantkenina brevispina* Cushman 1925a, p. 2, O.D. and monotypy (Tantoyuca formation, Late Eocene; Rio Pantepec, near Buena Vista, Mexico), USNM; a junior subjective typonym of *Hantkenina* Cushman 1925a, q.v. (*vide et* Bolli *et al.* 1957, pp. 28–9; Postuma 1971, pp. 226–7; Blow 1979, pp. 1153, 1163–4). Figure 5.73: Key 131 221 12.

Subbotina Brotzen and Pozaryska 1961, p. 160; *Globigerina triloculinoides* Plummer 1927, p. 134, O.D. (Wills Point formation, Palaeocene; Corsicana, Navarro County, Texas), WMUC. Topotypes figured by Loeblich and Tappan (1957a, pp. 183–4, Pl. 43, Fig. 9), by Postuma (1971, pp. 160–1) and by Blow (1979, pp. 1287–92, Pl. 248, Fig. 9). Genus emended by Blow (1979, pp. 1244–8). (*Vide et* Fleisher 1974, p. 1032; Blow 1979, Text Fig. M(i)–(vi), Pls 74, 80, 98, 248.) Figure 5.7: Key 111 121 1.

Tenuitella Fleisher 1974, p. 1033; *Globorotalia gemma* Jenkins 1966, O.D. (Lower Whaingaroan, Early Oligocene; Kakanui River section, New Zealand), NZGS, redescribed by Jenkins (1971, p. 115, Pl. 10, Figs 263–9). (*Vide et* Blow 1969, Pl. 34, Fig. 1; Blow 1979, pp. 1071–2, Pl. 245, Fig. 8; Poore & Brabb 1977, p. 260, Pl. 8, Figs 1–4.) Subjective junior synonym of *Globigerina postcretacea* Myatliuk 1950, p. 280 (Oligocene, Komachsk Series; western Ukraine, USSR), VNIGRI. (*Vide et* Subbotina 1953, Pl. 2, Figs 16–20; Blow & Banner 1962, Pl. 12, Figs G–J; Stainforth

Globotruncaninae (32): scale bar-line = 0.1 mm (100 μm)

Figure 5.112 (a–c) *Globotruncana arca* (Cushman) Cushman; based on holotype (umbilical structure from near-topotypes and hypotypes) (321 11).

Figure 5.113 (a–c) *Rosalinella linneiana* (d'Orbigny) Marie; based on neotype (tegillum after topotype) (321 121).

Figure 5.114 (a–c) *Marginotruncana marginata* (Reuss) Hofker; based on neotype (tegillum after hypotype) (321 122).

Figure 5.115 (a–c) *Globotruncanita stuarti* (de Lapparent) Reiss; based on lectotype and hypotypes (321 21).

Figure 5.116 (a–c) *Radotruncana calcarata* (Cushman) El-Naggar; based on holotype and topotypes (tegillum after hypotypes) (321 22).

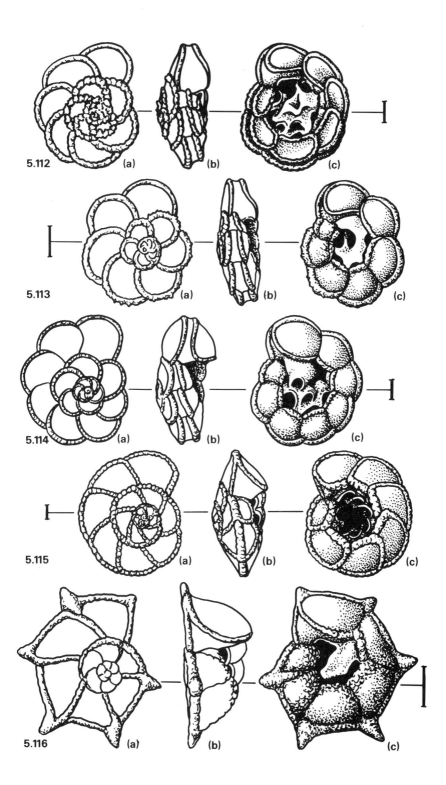

et al. 1975, pp. 300–2, Figs 4, 5. NOT Fleisher 1974, Pl. 17, or 1975, Pl. 3.) Figure 5.33; Key 113 12.

Testacarinata Jenkins 1971, pp. 110–11; *Globorotalia inconspicua* Howe 1939, p. 85, O.D. and monotypy (Cook Mountain formation, Middle Eocene; Winn Parish, Louisiana; *teste* Blow (1979), zones P. 11–12), HLSU; topotype figured by Blow (1979, pp. 930–1, Pl. 250, Figs 2–4). Figure 5.39: Key 113 221 12.

Thalmanninella Sigal 1948, p. 101; *T. brotzeni* Sigal 1948, p. 101, O.D. (middle Cenomanian; Sidi Aïssa, Algeria), Sigal collection; topotype figured by Bolli *et al.* (1957, Pl. 9, Fig. 7). Subjective junior synonym of *Globorotalia greenhornensis* Morrow 1934, p. 199 (Hartland shale, Greenhorn formation, late Cenomanian; Hodgeman County, Kansas), USNM; holotype redrawn by Brönnimann and Brown (1956, pp. 535–6, Pl. 20, Figs 7–9); topotypes refigured by Loeblich and Tappan (1961, pp. 299–301, Pl. 7, Fig. 9), by Masters (1977, pp. 508–11, Pl. 31) and by Robaszynski and Caron (1979a, pp. 85–90, Pl. 12). (*Vide et* Sigal 1952, p. 26, Fig. 25; Maslakova 1961, Pl. 4, Fig. 1; Maslakova 1963, Pl. 5, Fig. 2; Postuma 1971, pp. 80–1.) Figure 5.101: Key 222 211.

Ticinella Reichel 1950, p. 600; *Anomalina roberti* Gandolfi 1942, p. 100, O.D. (lower Scaglia Bianca, Albian; Breggia River section, near Balerna, Canton Ticino, Switzerland), NHMB; holotype redrawn by Caron and Luterbacher (1969, Pl. 7, Fig. 3); topotypes figured by Loeblich and Tappan (1961a, pp. 294–6, Pl. 6, Fig. 14) and by Longoria (1974, Pl. 12). Ticinellidae proposed by Longoria (1974, pp. 93, 98). (*Vide et* Sigal 1966a, pp. 187–9; Caron 1971, Figs 14–16; Postuma 1971, pp. 94–5; Masters 1977, pp. 521, 527–30, Pls 36, 37.) Figure 5.98: Key 222 111.

Tinophodella Loeblich and Tappan, 1957b, p. 112; *T. ambitacrena* Loeblich and Tappan, 1957b, pp. 113–14 (*Globigerina* ooze, Recent; 'Albatross' D2763, off eastern Brazil, 24°17′S, 42°48½′W, at 671 fathoms = 1227 m), USNM (*vide et* Parker 1962, Pl. 9; Blow 1979, p. 1320, Text Fig. H). Figure 5.17: Key 111 122 212 212 2.

Trinitella Brönnimann 1952b, p. 56; *T. scotti* Brönnimann 1952b, p. 56, O.D. and monotypy (Guayaguayare formation, Maestrichtian; Trinidad), USNM; holotype refigured by Bolli *et al.* (1957, Pl. 11, Fig. 4) (*vide et* Brönnimann & Brown 1956, p. 555, Pl. 23, Figs 13–15; Corminboeuf 1961, pp. 119–20, Pl. 2, Fig. 6; Smith & Pessagno 1973, pp. 60–1, Pl. 26; Masters 1977, pp. 629–30, Pl. 58, Figs 2, 4). Figure 5.108: Key 312 112.

Truncorotalia Cushman and Bermúdez 1949, p. 35; *Rotalina truncatulinoides* d'Orbigny 1839c, p. 86; neotype proposed by Blow (1969, pp. 370, 403–5, Pl. 5, Figs 10–11) (Recent, sediment; Atlantic Ocean, 'Challenger' 8, off Gomera, Canary Islands), BMNH, and another neotype proposed by Le Calvez (1974, p. 76) (Recent, sediment; Atlantic Ocean, off Ile de Ténériffe), MHN (d'Orbigny collection). Truncorotaliinae proposed by Subbotina (1971). (*Vide et* Fleisher 1974, p. 1028; Blow 1979, Text Fig. F, p. 512.) Figure 5.49: Key 113 231 22.

Truncorotaloides Brönnimann and Bermudez 1953, p. 817; *T. rohri* Brönnimann and Bermúdez 1953, p. 817 (Navet formation, marl pebble bed, Middle Eocene; Duff Road, 7 miles east of Point-à-Pierre, Trinidad), USNM; holotype refigured by Bolli *et al.* (1957, Pl. 10, Fig. 5). Truncorotaloidinae proposed by Loeblich and Tappan (1961b, p. 309) and by McGowran (1968, pp. 189–90). (*Vide et* Postuma 1971, pp. 232–3; Blow 1969, p. 372, Pl. 50, Figs 6–8; Blow 1979, pp. 1033–4, 1036–40, Pls 195, 196, 206, 231). Figure 5.40: Key 113 221 2.

Turborotalia Cushman and Bermúdez 1949, p. 42; *Globorotalia centralis* Cushman and Bermúdez 1937a, p. 26 (Eocene; Central Highway railroad bridge, Jicotea, Santa Clara Province, Cuba), USNM: original holotype refigured by Bolli *et al.* (1957, Pl. 10, Fig. 4), now lost; lectotype selected and figured by Cifelli and Belford (1977, Pl. 1, Figs 16–18) (senior subjective synonym of *G. cerroazulensis pomeroli* Toumarkine and Bolli 1970, 'replacement' taxon). (*Vide et* Beckmann 1954, p. 397, Pl. 26; Blow & Banner 1962, pp. 133–6, Pls 12, 16, 17; Postuma 1971, pp. 182–3; Fleisher 1974, p. 1034; Stainforth *et al.* 1975, pp. 258–61, Fig. 109; Blow

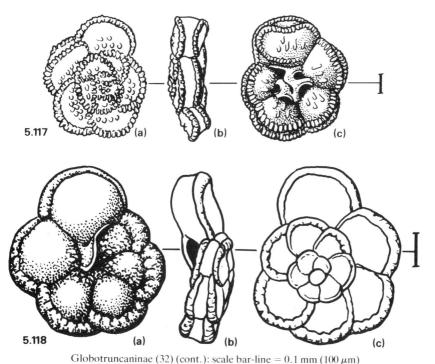

Globotruncaninae (32) (cont.): scale bar-line = 0.1 mm (100 μm)
Figure 5.117 (a–c) *Abathomphalus mayaroensis* (Bolli), Bolli, Loeblich and Tappan; based on holotype (a–c) and hypotypes (c, tegillum) (322).

Globotruncanellinae (222 12) (cont.): scale bar-line = 0.1 mm (100 μm)
Figure 5.118 (a–c) *Falsotruncana maslakovae* Caron; based on holotype and topotypic paratype (221 121 22).

1969, Pl. 36; Blow 1979, pp. 1045–9, 1052–4, Pls 173, 190, 261.) Figure 5.42: Key 113 231 111.

Turborotalita Blow and Banner 1962, p. 122; *Truncatulina humilis* Brady 1884, pp. 665–6, O.D. (Recent, sediment; Atlantic Ocean, 'Challenger' 5, southwest of Canary Islands, at 2470 fathoms = 4517 m), BMNH; lectotype selected and figured by Banner and Blow (1960a, pp. 36–7, Pl. 8, Fig. 1). Turborotalitidae proposed by Hofker (1976b). (*Vide et* Fleisher 1974, pp. 1037; Hofker 1976b, pp. 47–53; Blow 1979, pp. 175, 469–71.) Figure 5.50: Key 113 232.

Velapertina Popescu 1969, p. 105; *V. iorgulescui* Popescu 1969, p. 105, O.D. ('*Spirialis* marls horizon, Upper Tortonian', Miocene; Piatra, Bistriţa–Năsăud district, Subcarpathians, Romania), LPUB. Figure 5.24: Key 111 122 223 212.

Whiteinella Pessagno 1967, p. 298; *W. archaeocretacea* Pessagno 1967, p. 298, O.D. (South Bosque formation, Eagle Ford Group, late Turonian; Missouri–Pacific railroad, Austin, Travis County, Texas), USNM (*vide et* Robaszynski & Caron 1979a, pp. 151–6, 167, Pl. 33). Figure 5.97: Key 221 212.

Woletzina Fuchs 1973, pp. 460–1; *Globigerina jurassica* Hofman 1958, p. 125 (Bathonian–Lower Callovian; Crimea, USSR) (*vide et* Fuchs 1975, p. 230; Masters 1977, pp. 463–4; Grigelis & Gorbatchik 1980, p. 184). Figure 5.4: Key 111 111 112.

Appended notes: other generic names which have been applied

This short list excludes *nomina nuda* (e.g. *Eogloborotalia* Reiss 1957, *Pseudotruncorotalia* Reiss 1957, *Globalternina* Ivanova (in Subbotina *et al.*) 1955, etc.) as, strictly speaking, they have no meaning and are beyond comment.

A   Genera excluded from the superfamily Globigerinacea although once placed therein
A.1   Planktonic genera
A.1.1   Cassidulinacea d'Orbigny 1839a; Cassigerinellidae Bolli, Loeblich and Tappan 1957 (*nom. transl. ex* Cassigerinellinae): radial, hyaline walls; biserial throughout, enrolled in a low trochospire; high, laterally compressed, interiomarginal aperture (elongate comma-shaped), with fimbriate margin which may extend from septal foramina to form internal toothplate:
A.1.1.1   *Cassigerinella* Pokorny 1955; *C. boudecensis* Pokorny 1955, O.D. (subj. senr. syn.: *Cassidulina chipolensis* Cushman and Ponton (1932)) (Late Eocene–Early Miocene) (*vide et* Hofker 1963a; Steineck & Darrell 1971). Chambers and test inflated, primary aperture only.

A.1.1.2   *Cassigerinelloita* Stolk 1965; *C. amekiensis* Stolk 1965, O.D. (highest Early Eocene to Middle Eocene). Chambers inflated, with supplementary sutural aperture, sometimes covered by a bulla.

A.1.1.3   *Riveroinella* Bermúdez and Seiglie 1967; *R. martinezpicoi* Bermúdez and Seiglie 1967, O.D. (Early Miocene) (*vide et* Saito & Biscaye 1977). Chambers and test compressed; no supplementary apertures.

A.1.2   Heterohelicacea Cushman 1927a (*nom. transl. ex* Heterohelicidae): Guembelitriidae Montanaro-Gallitelli 1957 (*nom. transl. ex* Guembelitriinae, emend Blow 1979): radial, hyaline, hispid walls; triserial to quadriserial in high spire; umbilicus closed throughout growth; interiomarginal arched aperture with or without narrow rim; no toothplate or other internal or external apertural structures:

A.1.2.1   *Globoconusa* Khalilov 1956; *G. conusa* Khalilov 1956, O.D. (Danian) (*vide et* Khalilov 1964; Blow 1979): quadriserial.

A.1.2.2   *Gubkinella* Suleimanov 1955, p. 623; *G. asiatica* Suleimanov 1955, O.D. (late Senonian; Kizil-Kum, USSR) (*vide et* Rauser-Chernoussova & Fursenko 1959, p. 267, Fig. 470a–c): quinqueserial becoming quadriserial.

A.2   Benthonic genera:
A.2.1   Discorbacea Ehrenberg 1838:
A.2.1.1   Discorbidae Ehrenberg 1838:
A.2.1.1.1   *Hyderia* Haque 1962; *H. dubia* Haque 1962, O.D. (Early Eocene).
A.2.1.2   Oberhauserellidae Fuchs 1970:
A.2.1.2.1   *Praegubkinella* Fuchs 1968; *P. kryptumbilicata* Fuchs 1968, O.D. (late Triassic) (*vide et* Fuchs 1969, 1970; Hohenegger & Piller 1975).
A.2.1.3   Family uncertain:
A.2.1.3.1   *Jurassorotalia* Fuchs 1973; *J. grandis* Fuchs 1973, O.D. (Middle–Late Jurassic) (*vide et* Fuchs 1975; Grigelis & Gorbatchik 1980).
A.2.1.3.2   *Mariannenina* Fuchs 1973; *M. pulchra* Fuchs 1973, O.D. (Late Jurassic) (*vide et* Fuchs 1975; Grigelis & Gorbatchik 1980).
?A.2.1.3.3   *Tectoglobigerina* Fuchs 1973; *T. calloviana* Fuchs 1973, O.D. (Middle–Late Jurassic (*vide et* Fuchs 1975; Grigelis & Gorbatchik 1980): aberrant or deformed specimens?
A.2.2   Rotaliacea Ehrenberg 1839; Rotaliidae Ehrenberg 1839 (*nom. transl. ex* Rotalina); Indicolinae Singh and Kalia (in Singh) 1971 (*nom. transl. ex* Indicolidae):
A.2.2.1   *Indicola* Singh and Kalia 1970; *I. rajasthanensis* Singh and Kalia 1970, O.D. (Middle Eocene).

A.2.2.2   *Praeindicola* Singh and Kalia (in Singh 1971); *P. bikanerensis* Singh 1971, O.D. (Early Eocene).
B   Generic names which are dubious and should not be conserved:
B.1   *Coscinosphaera* Stuart 1866; *C. ciliata* Stuart 1866, monotypy. Originally supposed to be radiolarian; thought by Loeblich and Tappan (1964a) to be a junior synonym of *Orbulina* d'Orbigny 1839a; no primary types known.
B.2   *Hedbergina* Brönnimann and Brown 1956; *Globigerina seminolensis* Harlton 1927, *nomen dubium* (*vide et* Bolli *et al.* 1957, p. 39, 40; *Hedbergella* Brönnimann and Brown 1958).
B.3   *Planorotalia* Morozova 1957; *Planulina membranacea* Ehrenberg 1854, O.D. Proposed to accommodate Palaeocene species close to *Planorotalites* Morozova 1957 (q.v.), but the primary types of *Planulina membranacea*, once though to be Palaeocene, probably were Pliocene (Hay 1962). The lectotype is not known to exist, and has never been redescribed, although the type collection which may contain it is believed to exist in the Geologische-Paläontologische Institut der Humboldt Universität, East Berlin (the specimen represented by Ehrenberg (1854, Pl. 26, Fig. 43) was designated *non vide* by Loeblich and Tappan (1964a)); a supposed topotype (Loeblich & Tappan 1964a, Fig. 533.5) is subjectively synonymous with *Globorotalia margaritae* Bolli and Bermúdez 1965 (neotypified by Bolli & Bermúdez 1978; *vide et* metatypes of that species illustrated in Blow 1979). *G. margaritae* subjectively is referable strictly to *Menardella* (q.v.). Ehrenberg's figure of the lectotype is inadequate for firm determination, but, if the specimen exists, it is possible that *P. membranacea* is a senior subjective synonym of *G. margaritae*, a much-used zonal index, which would uselessly disturb current usage. It is equally possible that the lectotype, if it exists, is a senior synonym of another species (e.g. *G. scitula scitula* (Brady), as lectotypified by Banner and Blow (1960a), referable strictly to *Hirsutella*, q.v.). Many would consider the genus to be a junior subjective synonym of *Globorotalia* Cushman 1927a (e.g. Loeblich & Tappan 1964a). Because of its dubious identity, the species *P. membranacea* should be regarded as *nomen dubium, nomen non conservatum,* and the genus *Planorotalia* should be rejected in the interests of nomenclatural stability.
B.4   *Pylodexia* Ehrenberg 1858, p. 27; *Pylodexia pusilla* Ehrenberg 1858, pp. 27–8, cited by Stainforth *et al.* (1975, p. 35) as type by original indication and monotypy. The species was described briefly by its author but was not illustrated, and its name has never been subsequently used; the species and the genus based upon it are *nomina dubia* and should not be conserved. The

species called '*Pylodexia? tetratrias*' by Ehrenberg (1858) was illustrated by him and was designated type-species by Cushman (1927a) but, as it was only questionably assigned generically by the original author, it was not available for selection (ICZN Règles, 67(h)).

B.5 *Rhynchospira* Ehrenberg 1845; *R. indica* Ehrenberg 1845; *nomen dubium, non conservatum*.

Key to the discrimination of globigerinacean genera

1 Aperture or apertures without tegilla or arched, asymmetric portici; portical structures, if present at all, flat and confined to last chamber only: GLOBIGERINIDAE (Middle Jurassic–Recent):
11 Test trochospiral throughout (ventrally more involute, dorsally evolute):
111 Primary aperture wholly intraumbilically directed (throughout or in postnepionic whorls at least):
111 1 Wall not externally thickened by cortex, perforations not externally reduced by cortex formation:
111 11 Perforations minutely small, with diameters in order of 0.5 μm (invisible at magnifications below ×500): CAUCASELLINAE (Middle Jurassic–Late Cretaceous–Danian).
111 111 Initial very low trochospire of five or more chambers per whorl, followed by higher quadriserial spire:
111 111 1 Surface with scattered, low, rounded pustules, sporadically linking to form short ridges:
111 111 11 Quadriserial portion very high spired; aperture a low, narrow slit:
111 111 111 Chambers globular to depressed in quadriserial part: *Conoglobigerina* (Middle–Late Jurassic) (Fig. 5.1).
111 111 112 Chambers ovoid, high in axis of coiling: *Woletzina* (Middle–Late Jurassic) (Fig. 5.4) (subj. syn. *Conoglobigerina*).
111 111 12 Quadriserial portion low spired: *Globuligerina* (= *Polskanella*) (Late Jurassic–?Early Cretaceous) (Fig. 5.2).
111 112 Tests quadriserial throughout (sometimes reducing to three and a half in last whorl):
111 112 1 Test low spired, surface roughened by pustules or ridges or muricae; aperture a low intraumbilical arch:
111 112 11 Surface roughened by densely packed, low, rounded pustules: *Caucasella* (Early Cretaceous) (Fig. 5.3).
111 112 12 Surface covered by reticulation of ridges, each reticulation enclosing many minute perforations: *Reticuloglobigerina* (= *Favusella*) (Middle Cretaceous) (Fig. 5.5).
111 112 13 Surface with scattered muricae: *Globastica* (Danian) (Fig. 5.6).

111 12 Perforations large, with diameter in order of 5 μm (visible at ×200 magnification or less) when not secondarily closed: GLOBIGERININAE (Palaeocene–Recent):
111 121 Primary aperture with conspicuous portical structure (muricae, if present, weakly developed in immediate umbilical area only):
111 121 1 Aperture a low slit, with shelf-like porticus of breadth uniform along its length; wall surface pitted as perforations enlarge into 'pore' pits: *Subbotina* (Palaeocene–Miocene) (Fig. 5.7).
111 121 2 Aperture arched, with tooth-like, subtriangular, symmetrical porticus projecting into umbilicus; wall surface lacks 'pore' pits: *Dentoglobigerina* (Late Eocene–Pliocene) (Fig. 5.8).
111 122 Primary aperture rimless or with weak, narrow rim or lip:
111 122 1 Surface conspicuously muricate, ventrally or on both sides:
111 122 11 Aperture a high arch with no distinct lip: *Muricoglobigerina* (Palaeocene–Eocene) (Fig. 5.9).
111 122 12 Aperture a low slit with a distinct lip:
111 122 121 Test subquadrate; apertural face flattened to concave: *Pseudogloboquadrina* (Early–Late Eocene) (Fig. 5.10).
111 122 122 Test rounded to subtriangular; apertural face weakly convex: *Neoacarinina* (Quaternary) (Fig. 5.11).
111 122 2 Surface without muricae but with true spines and spine bases:
111 122 21 No dorsal, sutural supplementary apertures:
111 122 211 Perforation pits large, coalescent, producing favose surface: *Globoturborotalita* (Quaternary) (Fig. 5.12) (subj. syn. *Globigerina*).
111 122 212 Perforation pits not coalescent, surface not favose:
111 122 212 1 No bulla:
111 122 212 11 Chambers radially elongate: *Beella* (Quaternary) (Fig. 5.13).
111 122 212 12 Chambers not radially elongate: *Globigerina* (Eocene–Recent) (Fig. 5.14).
111 122 212 2 With umbilical bulla always regularly developed (Catapsydracinae):
111 122 212 21 With single bulla covering umbilicus:
111 122 212 211 Chamber walls with large perforation pits, macroperforate (~5 μm); bulla with few (one to four) large, arched infralaminal apertures: *Catapsydrax* (Eocene–Middle Miocene) (Fig. 5.15).
111 122 212 212 Chamber walls with no perforation pits, microperforate (~0.5 μm); bulla with small infralaminal apertures:
111 122 212 212 1 Infralaminal apertures few, sutural, no tunnel-like bulla extensions: *Globigerinita* (Early–Middle Miocene) (Fig. 5.16).
111 122 212 212 2 Infralaminal apertures many, intra- and extra-sutural,

	at ends of tunnel-like bulla extensions: *Tinophodella* (Quaternary) (Fig. 5.17).
111 122 212 22	With multiple bullae covering umbilicus and partly covering earlier bullae; many small infralaminal apertures; *Polyperibola* (Late Miocene) (Fig. 5.18).
111 122 22	With dorsal, sutural, supplementary apertures; primary aperture in umbilicus remains distinct (although it may possess a bulla):
111 122 221	Test high spired, loosely coiled; sutural apertures small, irregularly developed, in last one or two chambers only: *Guembelitrioides* (Middle Eocene) (Fig. 5.19) (subj. syn. *Globigerinoides*).
111 122 222	Test high spired, later chambers higher than broad, loosely coiled; sutural apertures regularly developed in last whorl or more: *Globicuniculus* (Middle Miocene) (Fig. 5.20) (subj. syn. *Globigerinoides*).
111 122 223	Test low or high spired, tightly coiled, chambers broader than high:
111 122 223 1	No bullae (or sporadic development only):
111 122 223 11	Chambers not extended into digitiform processes: *Globigerinoides* (Early Miocene–Recent) (Fig. 5.21).
111 122 223 12	Later chambers extended into digitiform processes: *Globigerinoidesella* (Late Miocene–Pliocene) (Fig. 5.22) (subj. syn. *Globigerinoides*).
111 122 223 2	Bullae regularly developed:
111 122 223 21	Bullae covering umbilicus only:
111 122 223 211	Infralaminal accessory apertures few and large: *Globigerinanus* (?Early–?Middle Miocene) (Fig. 5.23) (subj. syn. *Globigerinoita*).
111 122 223 212	Infralaminal accessory apertures many and small: *Velapertina* (Middle Miocene) (Fig. 5.24) (subj. syn. *Globigerinoita*).
111 122 223 22	Bullae covering primary and supplementary apertures: *Globigerinoita* (Middle Miocene) (Fig. 5.25).
111 122 23	Sutural supplementary apertures many: no primary aperture opens from last formed chamber (intraumbilical primary aperture preserved as a septal aperture only); wall microperforate (~0.5 μm): CANDEININAE (Late Miocene–Recent).
111 122 231	Test high spired, initial quadriseriality being rapidly replaced by regular triseriality; many supplementary apertures in each suture: *Candeina* (Late Miocene–Recent) (Fig. 5.26).
111 2	Wall thickened externally by cortex, reducing external perforation diameters: SPHAEROIDINELLINAE (Early Miocene–Recent).
111 21	No dorsal supplementary sutural apertures:

111 211 Cortex incompletely developed, especially weak on last formed chambers: *Prosphaeroidinella* (Early–Middle Miocene) (Fig. 5.27).

111 212 Cortex completely developed over whole test exterior: *Sphaeroidinellopsis* (Middle Miocene–Pliocene) (Fig. 5.28).

111 22 With dorsal supplementary sutural aperture (cortex developed over whole test exterior): *Sphaeroidinella* (Pliocene–Recent) (Fig. 5.29).

112 Primary aperture migrating from intraumbilical–extraumbilical position to a wholly intraumbilical position in ontogeny: aperture has portical tooth or shelf which may become covered by a bulla; true spine bases over whole test, but muricae may be present in umbilical area: GLOBOROTALOIDINAE (=Globoquadrinidae).

112 1 Test of rounded equatorial outline; umbilicus usually with bulla in adult; apertural face convex: *Globorotaloides* (Early–Middle Miocene, ?Recent) (Fig. 5.30).

112 2 Test of subquadrate equatorial outline; no bulla; apertural face flattened: *Globoquadrina* (Early–Middle Miocene) (Fig. 5.31).

113 Primary aperture intraumbilical–extraumbilical in position and/or direction.

113 1 Tests small; minutely perforate, diameters approximately 0.5 μm (invisible at magnifications below ×400); wall smooth or with small, scattered pustules (on early dorsal whorls and/or near aperture), typically with five to six chambers per whorl; aperture with narrow lip or rim: TENUITELLINAE nov. (Palaeocene–Oligocene–?Early Miocene).

113 11 Aperture arched extraumbilically (intraumbilical part much reduced) test very small, wall weakly muricate: *Parvularugoglobigerina* (Early Palaeocene) (Fig. 5.32).

113 12 Aperture low, equally open intraumbilically and extraumbilically; test small, wall weakly pustular: *Tenuitella* (Late Eocene?–Oligocene–?Early Miocene) (Fig. 5.33).

113 2 Tests with perforations visible at magnifications of ×100 or less (size comparable to the spine bases or muricae):

113 21 Surface pitted by perforations, otherwise essentially smooth; final aperture with small, asymmetric portical lip-like structure; tests small, typically low-spired and compressed:

113 211 Periphery rounded, not carinate; pustules or muricae rare, confined to umbilical area if present at all: EOGLOBIGERININAE (Danian–Middle Eocene).

113 211 1 Aperture not reaching periphery; test strongly lobulate, four to five chambers per whorl: *Eoglobigerina* (Danian–Palaeocene) (Fig. 5.34).

113 211 2 Aperture reaches periphery; test not strongly lobulate, five to

six chambers per whorl: *Globanomalina* (Palaeocene–?Early Eocene) (Fig. 5.35).

113 212 Periphery acute, carinate; pustules or weak muricae umbilically, and sometimes on early dorsal whorls also: PLANOROTALITINAE nov. (Palaeocene–Middle Eocene).

113 212 1 Chambers not radially elongate: *Planorotalites* (Late Palaeocene–Middle Eocene) (Fig. 5.36).

113 212 2 Chambers of last whorl radially elongate, each developing a radially directed spine: *Astrorotalia* (Middle Eocene) (Fig. 5.37).

113 22 Surface strongly and conspicuously muricate: TRUNCOROTALOIDINAE (=Acarinininae) (Late Palaeocene–Middle Eocene).

113 221 Periphery rounded to bluntly subangular, no carina, muricae developed over whole surface:

113 221 1 No supplementary apertures in dorsal spiral suture:

113 221 11 Muricae strongest around umbilicus: *Acarinina* (Late Palaeocene–Middle Eocene) (Fig. 5.38).

113 221 12 Muricae strongest at test periphery, giving appearance of a pseudospinose pseudocarina, but not fused there: *Testacarinata* (Middle–?Late Eocene) (Fig. 5.39).

113 221 2 With supplementary apertures in dorsal spiral suture: *Truncorotaloides* (Middle Eocene) (Fig. 5.40).

113 222 Periphery acutely angled, with peripheral muricae fused to form a peripheral muricocarina: *Morozovella* (Late Palaeocene–Middle Eocene) (Fig. 5.41).

113 23 Surface with true spines embedded in spine bases over all or part of the test: GLOBOROTALIINAE (Truncorotaliinae, Turborotalitinae) (Early Eocene–Recent).

113 231 Chambers of last whorl uniformly shaped, last not modified into an ampulla:

113 231 1 Periphery rounded or subangular but with no imperforate carina:

113 231 11 Chambers not radially elongate:

113 231 111 Aperture with narrow rim or lip of breadth uniform throughout its length: *Turborotalia* (Early Eocene–Recent) (Fig. 5.42).

113 231 112 Last one or two primary apertures sometimes with triangular lips which point into umbilicus: *Neogloboquadrina* (Quaternary) (Fig. 5.43) (subj. syn. *Turborotalia*).

113 231 12 Chambers radially elongate: *Clavatorella* (Early–Middle Miocene) (Fig. 5.44).

113 231 2 Periphery acutely angular, with an imperforate carina formed in part, at least, of the margin:

113 231 21 Tests almost equally biconvex:

113 231 211 Area of spinosity restricted to ventral surface, near aperture; dorsally, chambers approximately as broad as high:
113 231 211 1 Tests relatively compressed (the *scitula→archaeomenardii→praemenardii→menardii→cultrata* lineage): *Menardella* (Middle Miocene–Recent) (Fig. 5.45) (subj. syn. *Globorotalia*).
113 231 211 2 Tests relatively tumid (the *merotumida→plesiotumida→tumida* lineage): *Globorotalia* (gross homeomorphs in Late Eocene; strictly only Early Miocene–Recent) (Fig. 5.46).
113 231 212 Spinosity widespread over ventral surface; dorsally, chambers much higher than broad (the *praefohsi→fohsi* lineage): *Fohsella* (Middle Miocene) (Fig. 5.49) (subj. syn. *Globorotalia*).
113 231 213 Spinosity widespread ventrally and dorsally; dorsally, chambers much broader than high (reniform) (the *margaritae→praehirsuta→hirsuta* lineage): *Obandyella* (=*Hirsutella*) (Late Miocene–Recent) (Fig. 5.48) (subj. syn. *Globorotalia*).
113 231 22 Test strongly convex, subconical ventrally, flattened dorsally, chambers much broader than high (subquadrate) (the *crassaformis→tosaensis→truncatulinoides* lineage and related species): *Truncorotalia* (Pliocene–Recent) (Fig. 5.49) (subj. syn. *Globorotalia*).
113 232 Last chamber modified into an ampulla, normally with many 'infralaminal' apertures: *Turborotalita* (Late Miocene–Recent) (Fig. 5.50).
12 Test initially trochospiral, becoming streptospiral:
121 Test initially turborotalid (as 113 231 111), with true spine bases, trochospirality and umbilical–extraumbilical aperture; rate of chamber enlargement *not* changing with advent of streptospirality: PULLENIATININAE (Late Miocene–Recent).
121 1 Coiling becomes dorsally directed; umbilicus remains open; aperture extends into dorsal spiral suture: *Globigerinopsis* (Middle Miocene) (Fig. 5.51).
121 2 Coiling becomes (first) ventrally directed, with umbilical closure; then may become dorsally directed, with overlap of chambers on to dorsal side of test and development of a dorsally directed, wholly extraumbilical aperture: *Pulleniatina* (Late Miocene–Recent) (Fig. 5.52).
122 With muricate surface; rate of chamber enlargement increases with advent of ventrally directed streptospirality, so that later chambers cover umbilicus and supplementary apertures develop in dorsal sutures: test becoming globose: GLOBIGERAPSINAE.
122 1 With long early acarininid stage (as 113 221 11), with seven to five

chambers in last whorl; final test with many small apertures in spiral and in early and later intercameral sutures, sometimes with areal apertures also (all variably open or covered by small bullae): *Orbulinoides* (Middle Eocene) (Fig. 5.53).

122 2 With early muricoglobigerinid stage (as 111 221 1), with four to four and a half chambers in last whorl; final test with few, large supplementary apertures in spiral and later intercameral sutures (sometimes covered by small bullae): *Globigerapsis* (Middle–Late Eocene) (Fig. 5.54).

123 With spinose surface, lacking muricae:

123 1 Rate of chamber enlargement increases with advent of streptospirality, so that later chambers cover and enclose umbilicus; test becoming globose:

123 11 Early test globigerine (as 111 122 212 12); no sutural supplementary apertures in trochospiral part of test: PORTICULASPHAERINAE nov. (Middle–Late Eocene).

123 111 A few, large supplementary apertures in sutures of streptospiral final test; these may or may not be covered by (discrete) bullae:

123 111 1 Early coil with five to four chambers per whorl: supplementary apertures in suture of last chamber only: *Porticulasphaera* (Middle Eocene) (Fig. 5.55).

123 111 2 Early coil with four chambers per whorl; supplementary apertures in last and penultimate intercameral sutures: *Globigerinatheka* (Middle–Late Eocene) (Fig. 5.56).

123 112 Supplementary apertures of primary test completely covered by meandriform bullae formed in more than one series: *Inordinatosphaera* (Middle Eocene) (Fig. 5.57).

123 12 Early test globigerinoidine (as 111 122 223 11); final test with very small supplementary apertures in spiral suture; final chamber comprises more than half final test volume and has many small supplementary apertures in its suture: ORBULININAE (Early Miocene–Recent).

123 121 Spiral whorls always visible:

123 121 1 No areal apertures over surface of final chamber; no bullae: *Praeorbulina* (late Early Miocene) (Fig. 5.58).

123 121 2 With areal apertures over surface of final chamber:

123 121 21 With bullae covering supplementary and areal apertures (latter with raised rims): *Globigerinatella* (late Early Miocene) (Fig. 5.59).

123 121 22 No bullae, no raised rims around areal apertures:

123 121 221 Two large chambers comprise final whorl: *Biorbulina* (Middle Miocene–Recent) (Fig. 5.60) (subj. syn. *Orbulina*).

123 121 222 Single large chamber comprises final whorl: *Orbulina* (*Candorbulina* auctt.) microspheric form (Middle Miocene–Recent) (Fig. 5.61) (subj. syn. *Orbulina*).

123 122   Spiral whorls completely enclosed by final, last spherical chamber: *Orbulina* megalospheric form (Middle Miocene–Recent) (Fig. 5.62).
123 2   Rate of chamber enlargement remains unchanged with advent of streptospirality; early globigerinoidine test (as 111 122 223 11) becomes streptospiral with dorsal direction, closing umbilicus, with primary aperture becoming wholly extraumbilical: *Globigerinopsoides* (Middle Miocene) (Fig. 5.63).
13   Test biumbilicate, virtually equally involute in each umbilicus; coiling fundamentally planospiral, but may retain traces of initial trochospirality or may develop late streptospirality; aperture equatorial, extending into or towards each umbilicus:
131   Aperture with portical structures or lips visible on last chamber only:
131 1   Portical structures or lips small, narrow, never broad and flange-like; aperture a low arch, never elongate into apertural face; HASTIGERININAE (Early Eocene–Recent).
131 11   Perforations visible at magnifications of ×200 or less (diameters of order of 5 μm); true spines present, no muricae:
131 111   Test surface smooth: spine bases, if present at all, confined to area immediately adjacent to aperture; portical lip present:
131 111 1   Chambers not radially elongate; aperture extends fully into each umbilicus; portical lip broadest umbilically: *Pseudohastigerina* (Early Eocene–Oligocene) (Fig. 5.64).
131 111 2   Chambers radially elongate, clavate; aperture not extending fully into umbilici, lip strongest at midpoint, equatorially: *Protentella* (Middle Miocene–?Late Miocene) (Fig. 5.66).
131 112   Test surface smooth except for prominent triradiate spine bases scattered over surface; aperture with thin, narrow rim:
131 112 1   Chambers not radially elongate; adult coiling planospiral: *Hastigerina* (Late Miocene–Recent) (Fig. 5.65).
131 112 2   Chambers radially elongate, clavate; adult coiling streptospiral: *Hastigerinella* (=*Hastigerinopsis*) (Quaternary) (Fig. 5.67).
131 113   Test surface hispid, covered with small spine bases; aperture with thin rim, extending fully into each umbilicus:
131 113 1   Chambers not radially elongate: *Globigerinella* (Early Miocene–Recent) (Fig. 5.68).
131 113 2   Chambers radially elongate, clavate: *Bolliella* (Quaternary) (Fig. 5.70).
131 12   Perforations very small, diameter of order of 0.5 μm; surface with fine spine bases, heavier in sutures:
131 121   Aperture not extending deeply into either umbilicus; chambers not radially elongate: *Berggrenia* (Pliocene) (Fig. 5.69).
131 2   Portical structures strong, flat but flange-like; aperture elongate up into apertural face or replaced there by multiple, cribrate apertures: HANTKENININAE (Eocene).

131 21   Chambers radially elongate, clavate: *Clavigerinella* (late Early Eocene–top Middle Eocene–?earliest Late Eocene) (Fig. 5.71).
131 22   Chambers with radial spines; spines with thick, imperforate walls and narrow, central, longitudinal canal:
131 221   Aperture simple, an equatorial arch elongate narrowly up into apertural face:
131 221 1   Spines striate, proximally at least, not digitate terminally but with single, simple point:
131 221 11   Spines not developed on early chambers of last whorl: *Hantkeninella* (late Middle Eocene–latest Eocene) (Fig. 5.72) (subj. syn. *Hantkenina*).
131 221 12   Spines developed on all chambers of last whorl: *Hantkenina* (=*Sporohantkenina* subj.) (late Middle Eocene–latest Eocene) (Fig. 5.73).
131 221 2   Spines not striate, but terminally digitate:
131 221 21   Spines arise from anterior of chambers, at intercameral sutures: *Applinella* (early to middle Middle Eocene) (Fig. 5.74) (subj. syn. *Aragonella*)
131 221 22   Spines arise from midpoint of chamber periphery, away from sutures: *Aragonella* (early to middle Middle Eocene) (Fig. 5.75).
131 222   Multiple areal apertures form from division of broadly elongate primary aperture: *Cribrohantkenina* (Late Eocene) (Fig. 5.76).
2   Apertures with strong asymmetric portici on many or most of the relict apertures of the last whorl, as well as on the last aperture (no tegillum or convex symphysis):
21   Test biumbilicate, coiling planospiral (but may retain traces of initial trochospirality); aperture equatorial: PLANOMALINIDAE (Barremian–Maestrichtian):
211   Chambers not radially elongate: PLANOMALININAE (Globigerinelloidinae) (Aptian–Maestrichtian):
211 1   Periphery not carinate; test may be partly or wholly muricate:
211 11   Portici separated, not distally fused:
211 111   Planospiral, with chambers in single series throughout, final aperture undivided medially; surface mostly smooth:
211 111 1   Many chambers (ten or more) in increasingly evolute whorls; portici very long: *Globigerinelloides* (=*Labroglobigerinella*) (Aptian) (Fig. 5.77).
211 111 2   Fewer chambers (eight or less) per whorl, whorls of constant involution; portici relatively short: *Blowiella* (Aptian–Maestrichtian) (Fig. 5.78).
211 112   Last formed uniserial, planospiral chamber may have aperture divided medially sometimes to be followed by a biserial chamber pair, surface muricate: *Biglobigerinella* (?Albian–Cenomanian–Maestrichtian) (Fig. 5.79).

211 12   Portici fused distally, leaving intralaminal accessory apertures proximally: *Biticinella* (Albian) (Fig. 5.81).
211 2   Periphery carinate: *Planomalina* (Albian–Cenomanian) (Fig. 5.80).
212   Chambers radially elongate: SCHACKOININAE (Aptian–Maestrichtian).
212 1   Chamber elongations as slender, digitiform extensions, one or sometimes two or three per chamber: *Schackoina* (Albian–Maestrichtian) (Fig. 5.83).
212 2   Chamber elongations as bulbous lobes, usually two or more per chamber: *Leupoldina* (Aptian–Cenomanian) (Fig. 5.82).
212 3   Chambers clavate or as single, broad, phalange-like extensions; muricae may be present:
212 31   Chambers phalange-like, broadly tapering: *Hastigerinoides* (Aptian–early Campanian) (Fig. 5.84).
212 32   Chambers clavate: *Eohastigerinella* (early Santonian–early Campanian) (Fig. 5.85) (subj. syn. *Hastigerinoides*).
22   Test trochospiral, uniumbilicate, aperture ventral (surfaces partly or wholly muricate):
221   Portici separated, unfused (no intralaminal apertures; no supplementary sutural apertures): HEDBERGELLIDAE (Barremian–Maestrichtian):
221 1   Aperture umbilical, extending extraumbilically to periphery (if uncarinate) or to carina (if present):
221 11   Periphery without carina or imperforate peripheral band: HEDBERGELLINAE (Loeblichellinae) (Barremian–Maestrichtian).
221 111 Chambers not radially elongate:
221 111 1   Test thin walled, 'monolamellar', very finely perforate: *Praehedbergella* (Barremian) (Fig. 5.86).
221 111 2   Test wall of normal thickness, 'multilamellar' with growth, more coarsely perforate:
221 111 21 Chambers closely appressed, eight to ten chambers per whorl, muricate; umbilicus deep: *Hedbergella* (Aptian–Albian) (Fig. 5.87).
221 111 22 Chambers not closely appressed, but subglobular to depressed, five to seven per whorl; umbilicus relatively shallow and broad:
221 111 221 With no sutural apertures on dorsal side; dorsal muricae relatively weak: *Planogyrina* (Albian–Maestrichtian) (Fig. 5.88) (subj. syn. *Hedbergella*).
221 111 222 With small sutural apertures dorsally in last whorl; dorsal muricae strongly developed: *Loeblichella* (Campian–Maestrichtian) (Fig. 5.89).
221 112 Chambers radially elongate:
221 112 1   Chambers broadly elongate, clavate: *Clavihedbergella* (Aptian–Senonian–?Maestrichtian) (Fig. 5.90).

221 112 2   Chambers narrowly elongate into digitiform, slender extensions: *Asterohedbergella* (Cenomanian) (Fig. 5.91).
221 12   Chambers with imperforate peripheral band and carinae: GLOBOTRUNCANELLINAE (Albian–Maestrichtian).
221 121   Dorsal sutures limbate (at least in part) by carinal extensions; portici enlarge uniformly with chamber enlargement:
221 121 1   Carina single (a modified muricocarina):
221 121 11 Dorsal intercameral sutural limbations weak; muricae strong dorsally and ventrally: *Praeglobotruncana* (Albian–Turonian) (Fig. 5.92).
221 121 12 Dorsal intercameral sutural limbations strong; muricae weakly developed dorsally: *Rotundina* (Albian–Turonian) (Fig. 5.93) (subj. syn. *Praeglobotruncana*).
221 121 2   Carinae double:
221 121 21 Carinae closely spaced, appressed: *Dicarinella* (Turonian–Santonian) (Fig. 5.94).
221 121 22 Carinae widely separated: *Falsotruncana* (late Turonian–basal Coniacian) (Fig. 5.118).
221 122   Dorsal sutures not limbate; muricae dorsally and ventrally, strongest at periphery (muricocarina); last formed porticus large, covers most of umbilicus: *Globotruncanella* (Turonian–Maestrichtian) (Fig. 5.95).
221 2   Aperture intraumbilically directed (extending to full umbilical limit, sometimes to approach periphery): HELVETOGLOBOTRUNCANINAE.
221 21   Muricae fuse to form muricocarina on later-formed chambers:
221 211   Test dorsally flat, ventrally convex; muricocarina extends from angled periphery into dorsal intercameral sutures: *Helvetoglobotruncana* (Turonian) (Fig. 5.96).
221 212   Test almost equally biconvex: weak muricocarina on periphery of later chambers only, not extending into dorsal sutures: *Whiteinella* (Turonian–Santonian) (Fig. 5.97).
222   Portici laterally fused distally, producing intralaminal (as well as infralaminal) accessory apertures or supplementary sutural apertures or both; primary aperture umbilical–extraumbilical, reaching periphery: ROTALIPORIDAE (Aptian–Turonian).
222 1   Depressed portical structure, fused centrally (distally) but with accessory intralaminal apertures remaining open proximally (near chamber margins); no arched supplementary apertures in intercameral sutures: TICINELLINAE (Aptian–Albian).
222 11   Periphery rounded, not carinate:
222 111   Chambers not radially elongate (weak muricocarina may develop in spiral suture but not on periphery of last whorl): *Ticinella* (late Aptian–Albian) (Fig. 5.98).

222 112   Chambers elongate, clavate (no muricocarinate sutural thickening): *Claviticinella* (Late Albian) (Fig. 5.99).
222 12   Periphery acute, carinate (carina extending into early or all dorsal intercameral sutures): *Pseudothalmanninella* (Albian) (Fig. 5.100).
222 2   Portici fused distally (leaving intralaminal accessory apertures) or along their whole length; with arched supplementary apertures in the ventral intercameral sutures: ROTALIPORINAE (Albian–Turonian).
222 21   Periphery acutely angled, continuously carinate (carina extending into dorsal, and sometimes into ventral, intercameral sutures).
222 211   Supplementary apertures develop at umbilical shoulder, becoming sutural in later part of last whorl: *Thalmanninella* (Cenomanian) (Fig. 5.101).
222 212   Supplementary apertures fully sutural, extraumbilical in position: *Rotalipora* (late Cenomanian–Turonian) (Fig. 5.102).
222 22   Periphery rounded, with weak muricocarina on spiral suture and on periphery of last whorl: *Anaticinella* (= *Pseudoticinella*) (late Cenomanian–early Turonian) (Fig. 5.103).
3   Primary and relict primary apertures equipped with convex portical plates which extend across middle of umbilicus and fuse marginally there, one above the other, in whole or in part, to form a layered, complex tegillum covering umbilicus; aperture intraumbilically directed: GLOBOTRUNCANIDAE (Turonian–Maestrichtian).
31   Without continuously developed, solid, carinae (on imperforate peripheral band) and with no carinal extensions into intercameral dorsal sutures: RUGOGLOBIGERININAE (Turonian–Maestrichtian).
311   Wall muricate, but muricae do not align and fuse into costellae:
311 1   Chambers inflated, rounded, not elongate; weak peripheral muricocarina may or may not be present on earlier chambers:
311 11   Without dorsal supplementary apertures in spiral suture (suture completely closed), tegillum weak: *Archaeoglobigerina* (Turonian–early Maestrichtian) (Fig. 5.104).
311 12   With small supplementary apertures in dorsal sutures (sutures incompletely closed), tegillum massive: *Fissoarchaeoglobigerina* (Maestrichtian) (Fig. 5.105).
312   Muricae laterally fused to form meridionally aligned ridges (costellae) on part or all of test chamber surfaces:
312 1   Chambers rounded, not radially elongate:
312 11   Without peripheral muricocarinae:
312 111   Chambers subglobular, or ovoid throughout growth, periphery broadly rounded:
312 111 1 Test almost equally biconvex, dorso-ventrally; umbilicus relatively broad: *Rugoglobigerina* (Santonian–Maestrichtian) (Fig. 5.106).

312 111 2   Test much more convex ventrally than dorsally; umbilicus relatively narrow: *Kuglerina* (Maestrichtian) (Fig. 5.107) (subj. syn. *Rugoglobigerina*).
312 112   Last or later chambers of last whorl dorsally flattened and acquiring an acutely angled periphery: *Trinitella* (Maestrichtian) (Fig. 5.108).
312 12   With peripheral muricocarina:
312 121   Test almost equally biconvex, dorso-ventrally; imperforate peripheral band median to rounded periphery, with single or double muricocarina: *Rugotruncana* (Campanian–Maestrichtian) (Fig. 5.109).
312 122   Test flattened dorsally, convex ventrally; single muricocarina at broadly angled peripheral shoulder: *Bucherina* (Maestrichtian) (Fig. 5.110).
312 2   Chambers radially elongate: *Plummerita* (Maestrichtian) (Fig. 5.111).
32   With continuously developed imperforate peripheral band, which carries a single carina or pair of carinae; at anterior of periphery of each chamber, carinae diverge into the dorsal and ventral intercameral sutures (in last chamber, dorsally and ventrally to delimit the area of the terminal–apertural face, which will become a septum at next chamber addition): GLOBOTRUNCANINAE (Abathomphalinae, Marginotruncaninae) (Turonian–Coniacian–Maestrichtian).
321   Umbilicus relatively broad; aperture wholly intraumbilical:
321 1   Periphery bears two, separated or appressed continuous carinae; portici flat or only weakly convex below and within tegillum:
321 11   Ventral carina strong throughout its length, following ogival paths from the periphery along the ventral intercameral sutures, to reflex at the umbilicus; chamber sides flat or concave, not inflated, possessing a trapezoidal or rectangular axial profile: *Globotruncana* (Coniacian–Maestrichtian) Fig. 5.112).
321 12   Ventral carina extensions weak in the intercameral sutures, where they may be partly covered by overlap of the ventral wall of the succeeding chambers; chambers petaloid in equatorial view:
321 121   Chamber sides flat or weakly concave; chambers trapezoidal in axial profile; tegillum may fuse into continuous sheet: *Rosalinella* (?Turonian–Coniacian–Maestrichtian) (Fig. 5.113).
321 122   Chamber sides convex; chambers subtriangular to subovate in axial profile; tegillum clearly composite with strong intralaminal accessory apertures: *Marginotruncana* (Turonian–Campanian–?Maestrichtian) (Fig. 5.114) (subj. syn. *Rosalinella*).
321 2   Periphery bears single carina which diverges by dichotomy at each chamber anterior; portici highly arched below tegillum:
321 21   Carinate periphery not drawn out into radial spine-like extensions: *Globotruncanita* (Santonian–Maestrichtian) (Fig. 5.115).

321 22 Carinate periphery at posterior of each chamber elongate into radial, spine-like extension: *Radotruncana* (Campanian) (Fig. 5.116).
322 Umbilicus narrow; aperture umbilical–extraumbilical in extent (muricae may align and fuse into short, discontinuous ridges on chamber sides): *Abathomphalus* (Maestrichtian) (Fig. 5.117).

Review and discussion

These planktonic foraminifera with calcitic, perforate tests are morphologically characterised by their simplicity of skeletal structure (never possessing canal systems, plugs, pillars or internal toothplates), always being spinose, on part at least of the test surface (although the spinosity may be developed as murical or true spires or, occasionally, both, and may be preserved only on the 'juvenile', early chambers of the test), and in possessing, initially or throughout growth, a true umbilical cavity into which the successive chambers open through septal apertures which are still in direct communication with the exterior (as 'relict' primary apertures). In this last, the Globigerinacea are distinguishable from the Heterohelicacea (Fig. 5.121), where successive chambers communicate directly with each other by septal apertures, and where no primary chamber, except the last, has a primary aperture in direct communication with the exterior, because no true umbilical cavity is present (the only apparent exception to this rule is the heterohelicacean *Pseudotextularia* auctt. = *Racemiguembelina*, where a kind of secondary umbilicus develops in the last, extraordinary growth stage although it is *not* present in the earlier, normally coiled part of the test).

The origin and development of the Caucasellinae

The origin of the Globigerinacea is still uncertain, although Fuchs (1968, 1969, 1970, 1973, 1975) believes that the early Globigerinidae arose from discorbid Oberhauserellidae in the early and middle Jurassic; he has graphically displayed Triassic–Jurassic phylogenetic trees to illustrate this hypothesis (Fuchs 1975). It could well be that from the oberhauserellids the essentially intraumbilical aperture of the Globigerinidae developed, by

Morphological terms illustrated by diagrams:
Figure 5.119 (a&b) Oblique–axial sections across a late chamber showing a keel (a true carina) (K) and a muricocarina composed of fused muricae (MC), each superimposed upon an imperforate peripheral band (IPB); also showing the difference between a reflexed lip or rim (L) and a porticus (Pt) which blocks off the perforations (Pf).
Figure 5.120 Exemplifying the structure of a pitted wall with true spines (S) which are embedded in spine-bases (SB), distributed in this case between the externally widened ends of the perforations called 'pore-pits' or, more correctly, perforation-pits (PP). The wall is lamellate, successive lamellae (Lm) building the walls of the perforation-pits.

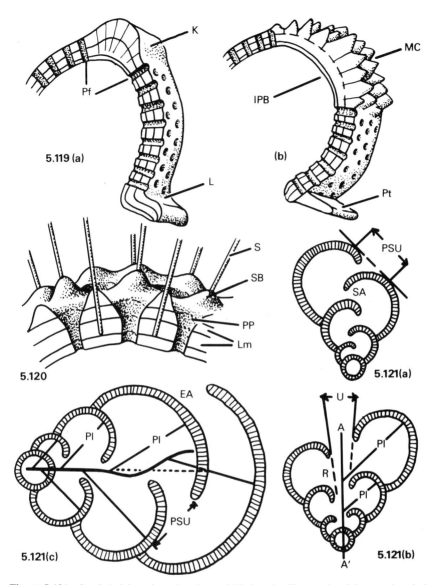

Figure 5.121 (a–c) Axial sections showing umbilical and coiling modes. (a) A trochospiral Heterohelicid, in which the chambers always overlap ventrally, *never* forming a true umbilicus, but, by their convexity, forming the central closed depression of a pseudumbilicus (PSU); earlier (septal) apertures (SA) open directly from one chamber to the next. (b) A high spired Globigerinid, its true trochospirality shown by the constant angle which the chamber-planes (Pl) make with the straight coiling axis (A–A′); the chambers do not overlap ventrally, leaving a true, open umbilicus (U) within the pseudumbilical depression; the last aperture and earlier, relict ones (R) open into this umbilicus. (c) A high-spired globigerinid in which the early trochospiral coiling (with chamber planes, Pl, at constant angles to a straight axis) becomes streptospiral, with chamber planes becoming set at angles which may be envisaged either as constantly changing relative to a projection of the straight axis or (which is the same thing) as remaining constant relative to a twisting coiling axis. The true umbilicus becomes closed with growth and apertures become extraumbilical (EA), with the formation of a pseudumbilicus and then its loss.

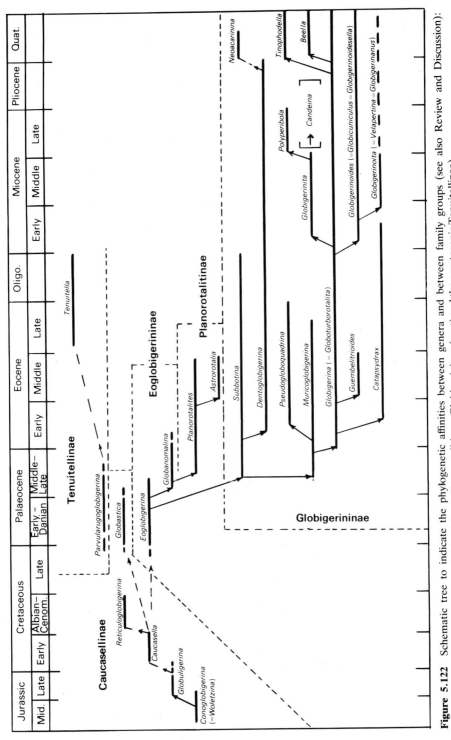

Figure 5.122 Schematic tree to indicate the phylogenetic affinities between genera and between family groups (see also Review and Discussion): Globigerinidae (Caucasellinae, Eoglobigerininae, Planorotalitinae, Globigerininae (part) and the cryptogenic Tenuitellinae).

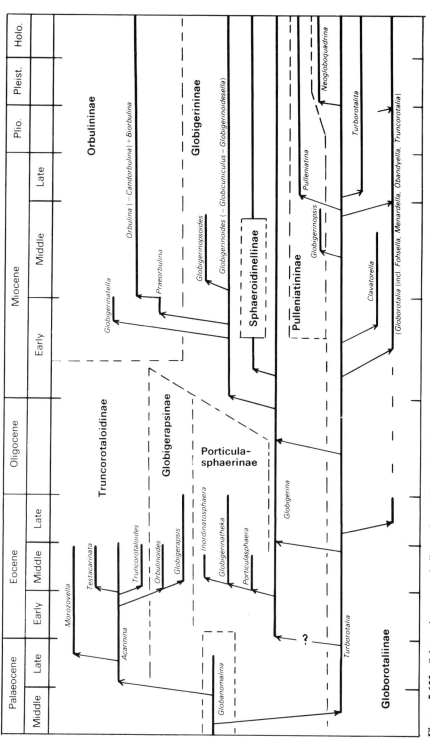

Figure 5.123 Schematic tree to indicate the phylogenetic affinities between genera and between family groups (see also Review and Discussion): Globigerinidae (Globigerininae (part), Globigerapsinae, Truncorotaloidinae, Porticulasphaerinae, Globorotaliinae, Pulleniatininae, Sphaeroidinellinae and Orbulininae).

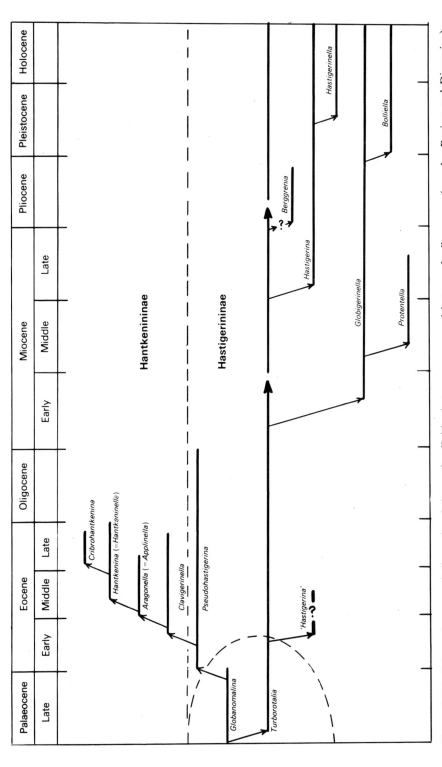

Figure 5.124 Schematic tree to indicate the phylogenetic affinities between genera and between family groups (see also Review and Discussion): Globigerinidae (Hantkenininae and Hastigerininae).

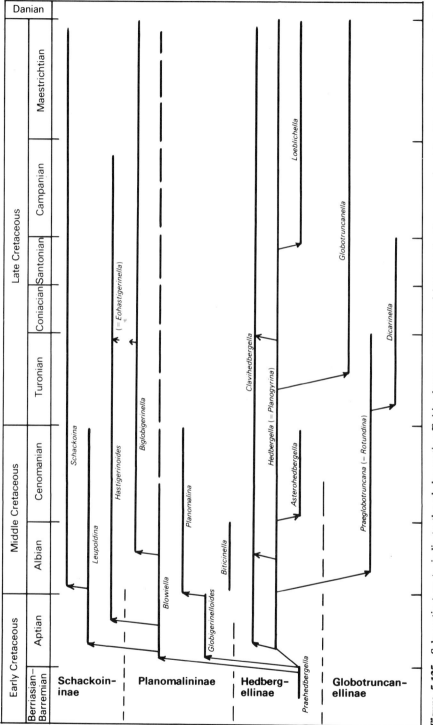

Figure 5.125 Schematic tree to indicate the phylogenetic affinities between genera and between family groups (see also Review and Discussion): Planomalinidae (Planomalininae, Schackoininae) and Hedbergellidae (Hedbergellinae, Globotruncanellinae).

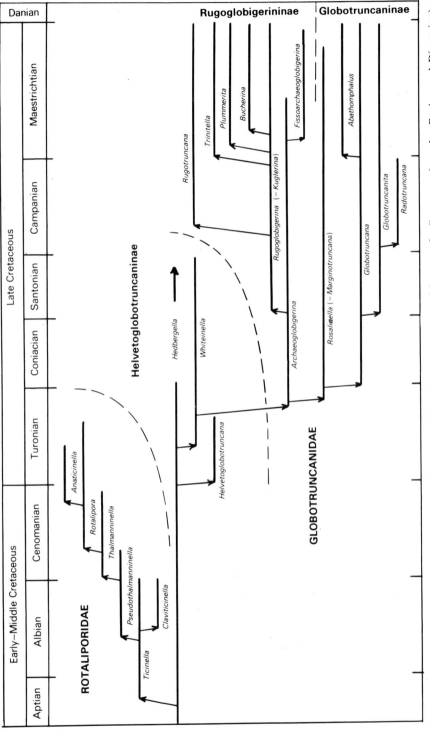

Figure 5.126 Schematic tree to indicate the phylogenetic affinities between genera and between family groups (see also Review and Discussion): Hedbergellidae (Helvetoglobotruncaninae), Rotaliporidae and Globotruncanidae (Globotruncaninae, Rugoglobigerininae).

simplification during adaptation to planktonic life. The earliest Globigerinidae (Caucasellinae) are all microperforate (perforations an order of magnitude smaller than in the typical, true Globigerininae of the Cenozoic), as, apparently are the oberhauserellids (e.g. the late Triassic *Praegubkinella*; see Hohenegger & Piller 1975). The earliest known (Jurassic) Caucasinellinae (*Conoglobigerina, Globuligerina*) have a 'discorbid' appearance to their early whorls (Figs 5.1, 2 & 4) and surfaces which are pustular, not yet truly spinose or muricate. Sometimes, the aperture of *Globuligerina* can be very highly arched and asymmetric (Bignot & Guyader 1971, Masters 1977), atypical of later Globigerinidae, but the pustules may fuse adjacently to form short ridges, heralding the surface of *Reticuloglobigerina* yet to come. It is curious that the final chamber may, in rare specimens, be reduced in volume and become sited directly over the umbilicus ('bulliform', shaped like a *bulla*), already like the 'Kummerform', 'abortive end-chambers' of some Cenozoic globigerine individuals and, perhaps, even with similarities to true bullae (see below). *Caucasella* of the Early Cretaceous is the first genus to become widely spread geographically and the first to have 'globigerine' coiling throughout its test (Fig. 5.3); perhaps these features suggest that a fully holoplanktonic existence, rather than a late-ontogeny meroplanktonic one, was acquired then, for the first time in the evolutionary history. Both *Caucasella* and its descendant *Reticuloglobigerina* commonly still display bulliform last chambers (Longoria's 'neotype' of *R. washitensis* has one). *Reticuloglobigerina*, with its characteristic surface reticulation of ridges (Fig. 5.5) is a worldwide index for Albian–Cenomanian strata, the shallow reticulations, occupied by many microperforations, only superficially resembling the deep perforation-pits of some Cenozoic *Globigerina* ('*Globoturborotalita*') (Figs 5.12 & 120). The late Cretaceous record of Caucasellinae is deficient, but tiny, spinose (muricate) globigerines occur (see, for example, Hofker 1966) which are the probable ancestors of *Globastica* (the last microperforate Caucaselline), a genus which, in its short geological existence, foreshadowed many truly globigerine characteristics (the erratic development of dorsal, sutural supplementary apertures, bullae on both primary and supplementary apertures, the Dano-Palaeocene *G. koslowskii* Brotzen and Pozaryska 1961 exaggerating features already seen in *G. daubjergensis*) but which apparently left no descendants.

Iterative homeomorphy and evolutionary radiation

There were three principal phases of evolutionary radiation in the Globigerinacea, each being marked by rapid morphological diversification and widening palaeoceanographic distribution (Cifelli 1969; Lipps 1970; Tappan & Loeblich 1972). The first, in the early to middle Cretaceous, was seen most spectacularly among the Hedbergellidae and its descendants, but the second (Palaeocene–Early Eocene) and third (Early Miocene) were within the Globigerinidae (Figs 5.122 & 123). The time-span during which these events took place is now well estimated (Berggren 1978; Hardenbol & Berggren 1978).

The assemblage of small, morphologically unspectacular species which characterises the Danian and its equivalents (Luterbacher & Silva 1964; Khalilov 1964; Hofker 1978; Blow 1979) contains the simple, almost archaetypical *Eoglobigerina* (Fig. 5.34) which is almost certainly directly ancestral to the characteristically Palaeogene genera *Subbotina* (Fig. 5.7) and *Globanomalina* (Fig. 5.35) and, through them, to the Cenozoic Globigerinidae as a whole. However, the origins of *Eoglobigerina* remain obscure. It is not impossible that it arose from unknown Caucasellinae, as suggested on Figure 5.122, but this would entail both a slightly extra-umbilical migration of the aperture and the acquisition of a distinct porticus (an attached apertural flange rather than the reflexed lip; Fig. 5.119), as well as the enlargement of its perforation diameters by an order of magnitude, the first two of these subsequently to be lost again in succeeding phylogeny. I believe it more likely that *Eoglobigerina* arose directly from small species of '*Hedbergella*' (Fig. 5.87 & 88) in the latest Maestrichtian, simply by loss of muricae and by an intraumbilical migration of the aperture – the latter trend occurred before (*Hedbergella* to the Helvetoglobotruncaninae; Fig. 5.126) and would occur again (*Turborotalia* to *Globigerina*; Fig. 5.123). This implies an origin for the Cenozoic Globigerininae (and their descendants) independent of that of the Caucasellinae, a polyphyletic Globigerinidae as here understood, and almost perfect heterochronous homeomorphy from separate ancestry for members of the two groups (compare *Caucasella* to *Globigerina* itself (Fig. 5.14), which differ only in the magnitude of their perforation diameters and the true spinosity of the latter, and *Caucasella* to *Muricoglobigerina* (Fig. 5.9), the latter being distinguished by its muricae and, again, by the magnitude of its perforation diameters).

Following the initiation of each phase of evolutionary upsurgence (Aptian, Middle Palaeocene, Early Miocene) morphological radiation produced forms which were grossly homeomorphic of others in the other phases of evolutionary radiation. This so-called 'iterative evolution' (and its resulting 'heterochronous homeomorphy') was the principal problem of early classifications and led to misleadingly broad interpretations of the stratigraphic ranges of the taxa. For example, the trochospiral, non-carinate tests of *Caucasella* (Fig. 5.3), *Reticuloglobigerina* (Fig. 5.5), *Whiteinella* (Fig. 5.97) and *Globigerina* (Fig. 5.14) led Cushman (1948) to suppose that *Globigerina* itself ranged from the Cretaceous to the Holocene. Similarly, the clavate chamber extensions of *Hastigerinella* (Fig. 5.67) had earlier homeomorphy in *Clavatorella* (Fig. 5.44), *Clavigerinella* (Fig. 5.71) and '*Eohastigerinella*' (Fig. 5.85), among others, so that, again, Cushman (1948) supposed *Hastigerinella* to range from Late Cretaceous to Recent, even though there were significant biostratigraphical gaps between each of these species groups. The strictly Neogene *Globorotalia* (Fig. 5.46) has a carinate, lensiform trochospiral test which was not distinguished either from *Plan-*

orotalites (Fig. 5.36) or *Morozovella* (Fig. 5.41) of the Palaeogene or from *Globotruncanella* (Fig. 5.95) or *Praeglobotruncana* (Fig. 5.92) of the Cretaceous, so that, again, the 'genus' (i.e. *'Globorotalia'*) was supposed to have a geological range as great as that known for the Globigerinacea itself, although the range of the 'genus' was far from being a continuous one. It was the importance placed upon apertural structures and umbilical characteristics by Bolli *et al.* (1957) and by Banner and Blow (1959) which led to valid taxonomic and biostratigraphical distinctions at supraspecific level.

A survey of Figures 5.1–118 and a comparison with Figures 5.122–6 will substantiate the principle behind the discriminatory key: that the Caucasellinae (Figs 5.1–6) and all the other subfamilies referred to the Globigerinidae (Figs 5.7–76) all have primary umbilici in which the primary apertural structures (lips, 'teeth', portici) are visible on the last one or two chambers only, and that, with the exception of the Caucasellinae, all the subfamilies are Cenozoic. In contrast, all the Mesozoic families and subfamilies (again, with the exception of the Caucasellinae) have umbilici which display the umbilical structures (portici, tegilla) associated with all of the relict apertures of the whole final whorl. Within these two major groups iterative evolution occurred, each time beginning with a simple, trochospiral, unkeeled form with an aperture extending from the true umbilicus to or near to the limit of the basal suture on the ventral side (an umbilical–extraumbilical aperture), as in *Praehedbergella* (Fig. 5.86) and *Hedbergella* (Fig. 5.87) in the Early Cretaceous and in their gross homeomorphs *Eoglobigerina* (Fig. 5.34) and *Globanomalina* (Fig. 5.35) of the early Palaeogene. From these rootstocks developed lineages which independently acquired the following:

(a) *Peripheral keels:* in the Globotruncanellinae (Figs 5.92–4 & 125) the later Rotaliporidae (Figs 5.100–2 & 126) and the Truncorotaloidinae (Figs 5.39–40 & 123) from the fusion at the periphery of the short, conical spine-like projections from the wall surface (called 'pseudospines' or 'muricae') to form a 'muricocarina' (Fig. 5.119b, where it is superimposed upon an imperforate peripheral band), and in the Globorotaliinae (Figs 5.45–9 & 123) by lamellar thickening of the imperforate narrow peripheral band to form a true keel or 'carina' (Fig. 5.119a).

(b) *Radially elongate chambers:* these occur in such species as *Clavihedbergella* (Figs 5.90 & 125), *Asterohedbergella* (Figs 5.91 & 125) and *Clavatorella* (Figs 5.44 & 123).

In each case, the umbilical characters remained distinct, betraying the separate origin of each homeomorph. The potential which the genotype with specialised umbilical characters held is revealed in the fact that those rotaliporids which acquired keels had also enlarged their portici so that separate accessory, 'intralaminal' apertures could form within them (accessory to the underlying primary relict apertures) (Figs 5.98–103 & 126), while the Truncorotaloidinae and Globorotaliinae, which had inherited no long hedbergelloid portici, could not.

Intraumbilical restriction of the primary aperture resulted in the evolutionary sequence *Eoglobigerina→Subbotina* (Figs 5.34, 7 & 122), *Turborotalia→Globigerina* (Figs 5.42, 14 & 123, repeatedly in the Palaeogene; see Blow & Banner 1962), *Hedbergella→Whiteinella→Archaeoglobigerina* (Figs 5.87, 97, 104 & 126), each time producing gross homeomorphs, the first two lines (Globigerininae) having very different potential from that of the last (Globotruncanidae). The Globigerininae, having narrow portici or lips only conspicuous on the last aperture, could develop them only to the extent of a so-called 'umbilical tooth' (*Dentoglobigerina*, Figs 5.8 & 122), a feature seen to appear again in *Globoquadrina* (Fig. 5.31, also derived from *Turborotalia*) and in geologically late specimens of some *Turborotalia* itself ('*Neogloboquadrina*', Figs 5.43d & 123). In contrast, the descendants of *Whiteinella* could enlarge their numerous portici into a complex of imperforate sheets spanning the whole umbilicus (the 'tegillum', Figs 5.104–17 & 126). Restriction of cytoplasmic exchange between the chamber interiors and exterior was achieved repeatedly in the Globigerinidae by the development of bullae (Figs 5.15–18 & 23–4) or of ventrally directed streptospirality (*Acarinina→Globigerapsis*, Figs 5.38, 53, 54 & 123; *Globigerina→Globigerinatheka*, Figs 5.14, 55, 56 & 123; *Globigerinoides→Orbulina*, Figs 5.21, 58, 61, 62 & 123), developments never possible with Hedbergellid–Globotruncanid genetic potentials. Also, the Globigerinidae could and did regularly develop true supplementary apertures in the dorsal sutures (*Truncorotaloides, Globigerinoides, Guembelitrioides*, Figs 5.40 & 19–21), a character which must have become a necessity in the Globigerapsinae, Porticulasphaerinae and Orbulininae, but this never became a significant feature in hedbergellid–globotruncanid stocks (the incomplete dorsal sutural closure in heavily muricate species, called *Loeblichella* (Figs 5.89 & 125) and *Fissoarchaeoglobigerina* (Figs 5.105 & 126) was insignificant phylogenetically and irregularly sporadic morphologically). As an example of phylogenetic differences of potential displayed in characters which would appear to be functionally unrelated, one can instance the fact that Globigerinidae with intraumbilical apertures never became carinate, but the Helvetoglobotruncaninae could develop a muricocarina (Figs 5.96 & 122) while the Globotruncanidae could develop double muricocarinae often of great strength and conspicuousness (Figs 5.109, 112–117 & 122).

Extension of the intraumbilical aperture towards the periphery of the test (to form an 'umbilical–extraumbilical interiomarginal primary aperture') resulted in the rapid development in Palaeocene time of *Globanomalina*, almost certainly from *Eoglobigerina*, and thence to the Planorotalitinae (Fig. 5.122) or the Truncorotaloidinae (Fig. 5.123). The latter strongly developed their muricae (*Acarinina*, Fig. 5.38, paralleling the independent globigerinine development of *Muricoglobigerina*, Fig. 5.9), which became particularly well developed peripherally (*Testacarinata*, Fig. 5.39), sometimes even so much so as to restrict closure of the dorsal spiral suture (*Truncorotaloides*, Fig. 5.40); the peripheral concentration of fused muricae in *Morozovella* (Fig. 5.41) produced the distinctively late Palaeocene–Middle Eocene 'conical globorotalids' (of Subbotina 1953) with their characteristically 'hispid' muricocarinae (Fig. 5.119b). In the Planorotalitinae, the muricae were never so conspicuous; the muricocarina of *Planorotalites* (Fig. 5.36) became somewhat smoothed by calcareous deposits between the muricae, and this keel became very similar indeed to those true carinae (Fig. 5.119a) which were to develop repeatedly in *Globorotalia* (from *Turborotalia*) in the later Late Eocene (*T. centralis→T. cerroazulensis→G. cunialis*, see Toumarkine & Bolli 1970; Blow 1979) and in the Miocene (*T. peripheroronda→G. praefohsi*, see Blow & Banner 1966; *T. praescitula→G. miozea*, see Walters 1965; Scott 1973; *T. scitula→G. margaritae*, see Blow 1969, 1979; etc.). In some cases, the homeomorphy between structures, and even species, of the Planorotalitinae and of the Globorotaliinae is almost perfect (compare *T. centralis*, Middle to Late Eocene, with *T. inflata*, Late Miocene–Recent, and *P. pseudomenardii*, Palaeocene, with *G.* ['*Obandyella*'] *hirsuta*, Quaternary) even though very great stratigraphical gaps, and quite different phylogenies, separate them. However, unlike their Cretaceous analogues (the Rotaliporidae and the Globotruncanellinae), the Cenozoic Truncorotaloidinae, the Planorotalitinae and the Globorotaliinae never could develop umbilical complexes of portical plates, with infralaminal and (Rotaliporidae) intralaminal accessory apertures, because they lacked both the genetic potential (and the genotypic umbilical skeletal structures) with which to do it; it is this umbilical–apertural morphology which, therefore, primarily must distinguish between the latter (Cenozoic, globigerinid) and former (Mesozoic, hedbergellid and rotaliporid) groups.

Again, once this distinction is made, other differences of potential become apparent. For example, the Globotruncanellinae could develop paired keels (*Dicarinella*, Figs 5.94 & 125), a distinctively Late Cretaceous phenomenon which never occurs in the Cenozoic, even though phylogenetically earlier forms (e.g. *Praeglobotruncana*, Figs 5.92 & 93) are grossly very similar (in all but umbilical structure) to Cenozoic analogues (Figs 5.41 & 45). Conversely, both Truncorotaloidinae and Globorotaliinae could develop streptospiral descendants (Fig. 121c), ventrally directed in the Globigerapsinae (Figs 5.53, 54 & 123) and in most Pulleniatininae (*Pulleniatina*, Figs 5.52 & 123), restricting and then losing their umbilical access to the exterior by this means in a manner analogous to that achieved by the porticulasphaerine and orbuline descendants of the Globigerininae. In some Pulleniatininae, however, the streptospirality was dorsally directed (*Globigerinopsis*, Fig. 5.51), maintaining an open true umbilicus. Such streptospirality appears to have been excluded from the genetic potential of the families and subfamilies which are excluded here from the Globigerinidae proper. The Globigerinidae had yet other ways to cover the umbilicus and restrict its access to the exterior, even without adopting changes of coiling mode: in the Globorotaliinae, the chambers themselves could be extended umbilically, to change into 'ampullae' with division of the primary umbilical–extraumbilical aperture (*Turborotalita*, Fig. 5.50), structures which were never seen in the Mesozoic or Palaeogene, and which were developments parallel to those of the iterative bullate forms of the Globigerininae itself (Figs 5.15–18), where, however, the primary aperture remained undivided and the infralaminal accessory apertures of the bullae probably were more strictly analogous to the infralaminal accessory apertures of the Mesozoic portici and tegilla.

If the primary aperture already extended from umbilicus to periphery, it was, apparently, a genetically simple matter to extend it further, to beyond the periphery, into the dorsal spiral suture, to be followed by the development of planospiral coiling and biumbilicate test form. This happened repeatedly throughout the whole history of the Globigerinacea, probably during every geological age and stage (except the Oligocene and the post-Miocene), and resulted in repeated gross homeomorphy, both within and between families. Compare *Globigerinella* (Fig. 5.68) with *Hastigerina* (Fig. 5.65), and both of these with *Pseudohastigerina* (Fig. 5.64) and *Berggrenia* (Fig. 5.69); compare all of these to *Globigerinelloides* (Fig. 5.77), *Blowiella* (Fig. 5.78), *Biglobigerinella* (Fig. 5.79) and *Biticinella* (Fig. 5.81). Figure 5.125, for example, is necessarily a simplification. *Globigerinelloides sensu stricto*, itself, is cryptogenic but must have arisen from *Praehedbergella* or *Hedbergella* near the beginning of Aptian time; from it descended the carinate *Planomalina* (Fig. 5.80, see also Sigal 1966b). *Blowiella* arose independently from *Praehedbergella* (Fig. 5.86) and *Planogyrina* (Fig. 5.88) (if one wishes to distinguish between these two genera), while the conspicuously, uniformly muricate *Biglobigerinella* (Fig. 5.79) arose repeatedly from *Planogyrina* throughout the Cretaceous (Hofker 1960a; Sigal 1966b). From true *Hedbergella* (Fig. 5.87) came first *Ticinella* (Fig. 5.98) and then, by late ontogenetic planospirality, *Biticinella* (Fig. 5.81). *Biglobigerinella* repeatedly produced individuals which, late in ontogeny, could medially divide the primary aperture (Figs 5.79a–b), a character never displayed in Cenozoic species (except, very rarely, in *Pseudohastigerina* (see Berggren *et al.* 1967), a phylogenetically primitive genus directly descended from *Globanomalina*, Fig. 5.122). For the other Cenozoic planospiral forms, *Hastigerina* (Fig. 5.65), with its characteristically sparsely distributed spines, triradiate in cross section, and *Globigerinella* (Fig. 5.68), with its densely distributed, triradiate to cylindrical spines (Bé 1969), arose separately from different species of *Turborotalia* (as did the Early to Middle Eocene homeomorph '*Hastigerina*' *bolivariana* (Petters), see Blow 1979, p. 1177); it may be considered practical to regard them (*Hastigerina* and *Globigerinella*) as synonyms or as subgenerically distinct, at most. They display not only evolution to genera with elongate, clavate chambers (*Bolliella*, Fig. 5.70; *Protentella*, Fig. 5.66), which grossly homeomorph the schackoinid *Hastigerinoides* (Fig. 5.84) and *Eohastigerinella* (Fig. 5.85), but also a streptospiral variant on the planospiral theme, *Hastigerinella* (Fig. 5.67), a genus with bifurcating or trifurcating clavate chambers which mimics the essentially planospiral schackoinid *Leupoldina* (Fig. 5.82).

These examples (and there are many others) show that the genetic potential iteratively to produce certain gross test shapes (based on changes of coiling mode, carinal acquisition, supplementation of the primary aperture, chamber elongation, etc.) was retained throughout the evolutionary history of the Globigerinacea since Early Cretaceous time, but that fundamental changes in the genetic potential occurred (leading both to new morphological possibilities and to new morphological impossibilities) at times of rapid evolutionary diversification following wide extinctions. The first of these was accompanied by the establishment of a holoplanktonic mode of life, and the development of *Caucasella* in the earliest Cretaceous, closely followed by the appearance of the hedbergellids and their descendants. The second was in the Dano-Palaeocene, with the evolution of the major subfamilies of the Globigerinidae. In each case, the renaissance was marked by a fundamental change in the architecture of the umbilicus and its primary apertural structures. The third event (Oligocene) was rather less dramatic in its rapidity and profundity, but the extinction of almost all the dominantly muricate Globigerininae (as well as of the Truncorotaloidinae and Globigerapsinae) and their subsequent dominance by genera bearing true spines (Fig. 5.120), was still of great biostratigraphic significance.

Nothing is known about the genetics of Globigerinacea, but I suspect that, in all of its existence, very few genes have determined fundamental test architecture and morphology and that the appearance of maybe only one allelomorphic mutant gene in the Jurassic, another in the Early Cretaceous, and another in the Dano-Palaeocene, could theoretically have given rise to the diversities of genotypes which created the rich evolutionary history of the Globigerinacea.

Acknowledgements

I sincerely thank John B. Saunders, of the Naturhistorisches Museum Basel, for the time he devoted to lengthy and helpful discussions during the early planning of this chapter. I am, as always, especially grateful to the late Walter Blow, who spent many years arguing interpretations of the Globigerinacea with me. This is also an opportunity to express sincere gratitude to Professor Tom Barnard, who initially taught all three of us.

References

Abdel-Kireem, M. R. 1978. Wall microstructure of *Fissoarchaeoglobigerina* n.g. (foraminifer). *Rev. Esp. Micropaleont.* **10**, 57–65.

Anderson, O. R., M. Spindler, A. W. H. Bé and C. Hemleben 1979. Trophic activity of planktonic foraminifera. *J. Mar. Biol. Assoc. UK* **59**, 791-9.

Antonova, Z. A., A. G. Shmigina, G. A. Gnedina and O. M. Kalugina 1964. Foraminiferi neokoma i apta mezhdurechnya pshekhayubin (severo-Zapadnii Kavkaz). In *Voprosi stratigrafii i litologii Mesozoiskikh i Kainozoiskikh otlozhenii Krasnodarskogo Kraya*, B. L. Egoyana (ed.), *Krasnodarskikii Filial, Vses. Neft. Nauchno-Issled. Inst., Trudi* **12**, 3-72. Moscow: NEDRA.

Ayala-Castañares, A. 1962. Morfología y estructura de algunas foraminiferos planctonicos del Cenomaniano de Cuba. *Soc. Geol. Mex., Bol.* **25** (1) 20-1.

Bandy, O. L. 1967. Cretaceous planktonic foraminiferal zonation. *Micropaleontology* **13**, 1-31.

Bandy, O. L. 1972. Origin and development of *Globorotalia (Turborotalia) pachyderma* (Ehrenberg). *Micropaleontology* **18**, 294-318.

Bandy, O. L., W. E. Frerichs and E. Vincent 1967. Origin, development and geological significance of *Neogloboquadrina* gen. nov. *Contr. Cushman Fdn Foramin. Res.* **18**, 152-7.

Banner, F. T., 1964. On *Hastigerinella digitata* (Rhumbler, 1911). *Micropaleontology* **10**, 114-16.

Banner, F. T. and W. H. Blow 1959. The classification and stratigraphical distribution of the Globigerinaceae. *Palaeontology* **2**, 1-27.

Banner, F. T. and W. H. Blow 1960a. Some primary types of species belonging to the superfamily Globigerinaceae, *Contr. Cushman Fdn Foramin. Res.* **11**, 1-41.

Banner, F. T. and W. H. Blow 1960b. The taxonomy, morphology and affinities of the genera included in the subfamily Hastigerininae. *Micropaleontology* **6**, 19-31.

Banner, F. T. and W. H. Blow 1965. *Globigerinoides quadrilobatus* (d'Orbigny) and related forms: their taxonomy, nomenclature and stratigraphy. *Contr. Cushman Fdn Foramin. Res.* **16**, 105-15.

Banner, F. T. and W. H. Blow 1967. The origin, evolution and taxonomy of the foraminiferal genus *Pulleniatina* Cushman, 1927. *Micropaleontology* **13**, 133-62.

Barnard, T. 1954. *Hantkenina alabamensis* Cushman and some related forms. *Geol. Mag.* **91**, 384-90.

Bé, A. W. H. 1965. The influence of depth on shell growth in *Globigerinoides sacculifer* (Brady). *Micropaleontology* **11**, 81-97.

Bé, A. W. H. 1969. Microstructural evidence of the close affinity of *Globigerinella* Cushman to *Hastigerina* Thomson. In *Proceedings of the 1st international conference on planktonic microfossils, Geneva, 1967*, Vol. 1, P. Brönnimann & H. H. Renz (eds), 89-91. Leiden: Brill.

Bé, A. W. H. 1977. An ecological, zoogeographic and taxonomic review of Recent planktonic foraminifera. In *Oceanic micropaleontology*, Vol. 1, A. T. S. Ramsay (ed.), 2-88. London: Academic Press.

Bé, A. W. H. 1980. Gametogenic calcification in a spinose foraminifer, *Globigerinoides sacculifer* (Brady). *Mar. Micropaleont.* **5**, 283-310.

Bé, A. W. H., S. M. Harrison and L. Lott 1973. *Orbulina universa* d'Orbigny in the Indian Ocean. *Micropaleontology* **19**, 150-92.

Bé, A. W. H., S. M. Harrison, W. E. Frerichs and M. E. Heiman 1976. Variability in test porosity of *Orbulina universa* d'Orbigny at two Indian Ocean localities. In *Progress in micropaleontology*, Y. Takayanagi & T. Saito (eds), 1-9. New York: Micropalaeontology Press.

Bé, A. W. H., C. Hemleben, O. R. Anderson and M. Spindler 1980. Pore structures in planktonic foraminifera. *J Foramin. Res.* **10**, 117-28.

Beckmann, J. P. 1954 (1953). Die foraminiferen der Oceanic formation (Eocaen-Oligocaen) von Barbados, Kl. Antillen. *Eclog. Geol. Helv.* **46**, 301-412.

Berggren, W. A. 1960. Paleogene biostratigraphy and planktonic foraminifera of the S.W. Soviet Union: an analysis of recent Soviet investigations. *Stockholm Contr. Geol.* **6** (5), 65-125.

Berggren, W. A. 1962. Some planktonic foraminifera from the Maestrichtian and type Danian stages of southern Scandinavia. *Stockholm Contr. Geol.* **9** (1), 1-106.

Berggren, W. A. 1966. Problemi taksonomii i filogeneticheskikh otnoshenii nekotorikh tretichnikh planktonnikh foraminifer. *Vopr. Mikropaleont.* **10,** 309–32.

Berggren, W. A. 1971a. Paleogene planktonic foraminiferal faunas on legs I–IV (Atlantic Ocean) Joides deep-sea drilling program – a synthesis. In *Proceedings of the 2nd Planktonic Conference, Roma, 1970,* Vol. 1, A. Farinacci (ed.), 57–77. Rome: Ed. Tecnoscienza.

Berggren, W. A. 1971b. Multiple phylogenetic zonations of the Cainozoic based on planktonic foraminifera. In *Proceedings of the 2nd Planktonic Conference, Roma, 1970,* Vol. 1, A. Farinacci (ed.), 41–56. Rome: Ed. Tecnoscienza.

Berggren, W. A. 1977. Atlas of Palaeogene planktonic foraminifera. Some species of the genera *Subbotina, Planoratalites, Morozovella, Acarinina* and *Truncorotaloides*. In *Oceanic micropalaeontology,* Vol 1, A. T. S. Ramsay (ed.), 205–99. London: Academic Press.

Berggren, W. A. 1978. Recent advances in Cenozoic planktonic foraminiferal biostratigraphy, biochronology and biogeography: Atlantic Ocean. *Micropaleontology* **24,** 337–70.

Berggren, W. A. and J. A. van Couvering 1974. The late Neogene. *Palaeogeogr. Palaeoclimat. Palaeoecol.* **16,** 1–216.

Berggren, W. A., R. K. Olsson and R. A. Reyment 1967. Origin and development of the foraminiferal genus *Pseudohastigerina* Banner & Blow, 1959. *Micropaleontology* **13,** 265–88.

Bermúdez, P. J. 1937. Notas sobre *Hantkenina brevispina* Cushman. *Soc. Cubaña Hist. Nat., Mem.* **11** (3), 151–2.

Bermúdez, P. J. 1952. Estudio sistemático de los Foraminiferos rotaliformes. *Venez. Minist. Minas y Hidrocarb., Bull. Geol.* **2** (4), 1–230.

Bermúdez, P. J. 1961. Contribución al estudio de los Globigerinidae de la région Caribe–Antillana (Paleoceno–Recente). *Congr. Geol. Venez. Mem.* **3,** *Soc. Geol. Venez., Bol. Geol., Publ. Espec.* **3,** 1119–393.

Bermúdez, P. J. and J. R. Farias 1977. Zonación del Cenozoico al Reciente basada en el estudio de los foraminiferos planctonicos. *Rev. Esp. Micropaleont.* **9,** 159–90.

Bermúdez, P. J. and G. A. Seiglie 1967. A new genus and species of foraminifera from the Early Miocene of Puerto Rico. *Tulane Stud. Geol.* **5,** 177–9.

Bignot, G. and J. Guyader 1971. Observations nouvelles sur *Globigerina oxfordiana* Grigelis. In *Proceedings of the 2nd Planktonic Conference, Roma, 1970,* Vol. 1, A. Farinacci (ed.), 79–83. Rome: Ed. Tecnoscienza.

Blow, W. H. 1956. Origin and evolution of the foraminiferal genus *Orbulina* d'Orbigny. *Micropaleontology* **2,** 57–70.

Blow, W. H. 1959. Age, correlation and biostratigraphy of the Upper Tocuyo (San Lorenzo) and Pozón formations, eastern Falcón, Venezuela. *Bull. Am. Paleont.* **39** (178), 67–251.

Blow, W. H. 1965. *Clavatorella,* a new genus of the Globorotaliidae. *Micropaleontology* **11,** 365–8.

Blow, W. H. 1969. Late Middle Eocene to Recent planktonic foraminiferal biostratigraphy. In *Proceedings of the 1st international conference on Planktonic Microfossils, Geneva, 1957,* Vol. 1, P. Brönnimann & H. H. Renz (eds), 199–422. Leiden: Brill.

Blow, W. H. 1979. *The Cainozoic Globigerinida,* Parts I and II. Leiden: Brill.

Blow, W. H. and F. T. Banner 1962. The mid-Tertiary (Upper Eocene to Aquitanian) Globigerinaceae. In *Fundamentals of mid-Tertiary stratigraphical correlation,* F. E. Eames, F. T. Banner, W. H. Blow & W. J. Clarke (eds) 61–151. Cambridge: University Press.

Blow, W. H. and F. T. Banner 1966. The morphology, taxonomy and biostratigraphy of *Globorotalia barisanensis* LeRoy, *G. fohsi* Cushman & Ellisor, and related taxa. *Micropaleontology* **12,** 286–302.

Blow, W. H. and T. Saito 1968a. The morphology and taxonomy of *Globigerina mexicana* Cushman, 1925. *Micropaleontology* **14,** 357–60.

Blow, W. H. and T. Saito 1968b. Comments and errata. *Micropaleontology* **14,** 505.

Bolli, H. M. 1945 (1944). Zur stratigraphie der oberen Kreide in den höheren helvetischen Decken. *Eclog. Geol. Helv.* **37,** 217–328.

Bolli, H. M. 1951. The genus *Globotruncana* in Trinidad, B.W.I. *J. Paleont.* **25,** 187–99.

Bolli, H. M. 1957a. Planktonic foraminifera from the Oligo-Miocene Cipero and Lengua formations of Trinidad, B.W.I. *US Natn Mus. Bull.* **215**, 97–123.
Bolli, H. M. 1957b. Planktonic foraminifera from the Eocene Navet and San Fernando formations of Trinidad, B.W.I. *US Natn Mus. Bull.* **215**, 155–72.
Bolli, H. M. 1957c. The genera *Praeglobotruncana, Rotalipora, Globotruncana* and *Abathomphalus* in the Upper Cretaceous of Trinidad. *US Natn Mus. Bull.* **215**, 51–60.
Bolli, H. M. 1957d. The genera *Globigerina* and *Globorotalia* in the Paleocene–Lower Eocene Lizard Springs formation of Trinidad, B.W.I. *US Natn Mus. Bull.* **215**, 61–82.
Bolli, H. M. 1958. The foraminiferal genera *Schackoina* Thalmann, emended, and *Leupoldina* n. gen. in the Cretaceous of Trinidad, B.W.I. *Eclog. Geol. Helv.* **50**, 271–8.
Bolli, H. M. 1959. Planktonic foraminifera from the Cretaceous of Trinidad, B.W.I. *Bull. Am. Paleontol.* **39** (179), 257–77.
Bolli, H. M. 1962. *Globigerinopsis*, a new genus of the foraminiferal family Globigerinidae. *Eclog. Geol. Helv.* **55**, 281–4.
Bolli, H. M. 1966. Zonation of Cretaceous to Pliocene marine sediments based on planktonic foraminifera. *Bol. Informat. Asoc. Venez. Geol. Min. Petrol.* **9** (1), 3–32.
Bolli, H. M. 1969. Zonación de sedimentos marinos del Cretáceo hasta el Plioceno basada en foraminiferos planctónicos. Inst. Mexicano del Petroleo, Public. 69, AE/047.
Bolli, H. M. 1970. The foraminifera of sites 23–31, leg 4. In R. G. Bader *et al.*, *Initial Reports, Deep Sea Drilling Project* **4**, 577–643.
Bolli, H. M. 1972. The genus *Globigerinatheka* Brönnimann. *J. Foramin. Res.* **2**, 109–36.
Bolli, H. M. and P. J. Bermúdez 1965. Zonation based on planktonic foraminifera of Middle Miocene to Pliocene warm-water sediments. *Bol. Informat. Assoc. Venez. Geol. Min. Petrol.* **8**, 119–49.
Bolli, H. M. and P. J. Bermúdez 1978. A neotype for *Globorotalia margaritae* Bolli & Bermúdez. *J. Foramin. Res.* **8**, 138–42.
Bolli, H. M., A. R. Loeblich Jr and H. Tappan 1957. Planktonic foraminiferal families Hantkeninidae, Orbulinidae, Globorotaliidae and Globotruncanidae. *US Natn Mus. Bull.* **215**, 3–50.
Brady, H. B. 1877. Supplementary note on the foraminifera of the chalk (?) of the New Britain Group. *Geol Mag.* **4**, 534–6.
Brady, H. B. 1879. Notes on some of the reticularian Rhizopoda of the Challenger Expedition, 2: additions to the knowledge of porcellaneous and hyaline types. *Q. J. Microsc. Soc.* **19**, 261–99.
Brady, H. B. 1884. *Report on the foraminifera dredged by H.M.S. 'Challenger' during the years 1873–1876.* Rept. Sci. Results Explor. Voyage H.M.S. Challenger, Zool. **9**, 1–814.
Brönnimann, P. 1950a. The genus *Hantkenina* Cushman in Trinidad and Barbados, B.W.I. *J. Paleont.* **24**, 397–420.
Brönnimann, P. 1950b. Occurrence and ontogeny of *Globigerinatella insueta* Cushman and Stainforth from the Oligocene of Trinidad, B.W.I. *Contr. Cushman Fdn Foramin. Res.* **1**, 80–2.
Brönnimann, P. 1951a. *Globigerinita naparimaensis*, n. gen., n. sp., from the Miocene of Trinidad, B.W.I. *Contr. Cushman Fdn Foramin. Res.* **2**, 16–18.
Brönnimann, P. 1951b. The genus *Orbulina* d'Orbigny in the Oligo-Miocene of Trinidad, B.W.I. *Contr. Cushman Fdn Foramin. Res.* **2**, 131–8.
Brönnimann, P. 1952a. *Globigerinoita* and *Globigerinatheka*, new genera from the Tertiary of Trinidad, B.W.I. *Contr. Cushman Fdn Foramin. Res.* **3**, 25–8.
Brönnimann, P. 1952b. Globigerinidae from the Upper Cretaceous (Cenomanian–Maestrichtian) of Trinidad, B.W.I. *Bull. Am. Paleont.* **34** (140), 1–61.
Brönnimann, P. 1952c. Trinidad Paleocene and Lower Eocene Globigerinidae. *Bull. Am. Paleont.* **34** (143), 1–34.
Brönnimann, P. 1952d. *Plummerita*, new name for *Plummerella* Brönnimann, 1942 (not de Long, 1942). *Contr. Cushman Fdn Foramin. Res.* **3**, 146.
Brönnimann, P. 1953. Note on planktonic foraminifera from Danian localities of Jutland, Denmark. *Eclog. Geol. Helv.* **45**, 339–41.

Brönnimann, P. and P. J. Bermúdez 1953. *Truncorotaloides*, a new foraminiferal genus from the Eocene of Trinidad. B.W.I. *J. Paleont.* **27**, 817–20.
Brönnimann, P. and N. K. Brown Jr 1956 (1955). Taxonomy of the Globotruncanidae. *Eclog. Geol. Helv.* **48**, 503–62.
Brönnimann, P. and N. K. Brown Jr 1958. *Hedbergella*, a new name for a Cretaceous planktonic foraminifer. *J. Wash. Acad. Sci.* **48**, 15–17.
Brotzen, F. 1942. Die foraminiferengattung *Gavelinella* nov. gen. und die systematik der Rotaliiformes. *Sver. Geol. Undersök.* **36** (8) C (451), 1–60.
Brotzen, F. and K. Pozaryska 1961. Foraminiferès du Paléocène et de l'Éocène inférieur en Pologne septentrionale: remarques paléogéographiques. *Rev. Micropaléont.* **4**, 155–66.

Caron, M. 1966. Globotruncanidae du Crétacé Supérieur du synclinal de la Gruyère (Préalpes medianes, Suisses). *Rev. Micropaléont.* **9**, 68–93.
Caron, M. 1967. Étude biométrique et statistique du plusieurs populations de Globotruncanidae 2. Le sous genre *Rotalipora (Thalmanninella)* dans l'Albien supérieur de la Breggia (Tessin). *Eclog. Geol. Helv.* **60**, 47–79.
Caron, M. 1971. Quelques cas d'instabilité des caractères génériques chez les foraminifères planctoniques de l'Albien. In *Proceedings of the 2nd Planktonic Conference, Roma, 1970*, Vol. 1, A. Farinacci (ed.), 145–57. Rome: Ed. Tecnoscienza.
Caron, M. 1976. Révision des types de foraminifères planctoniques décrits dans le région du Montsalvers (Préalpes fribourgeoises). *Eclog. Geol. Helv.* **69**, 327–33.
Caron, M. 1981. Un nouveau genre de foraminifères planktonique du Crétacé: *Falsotruncana* n. gen. *Eclog. Geol. Helv.* **74**, 65–73.
Caron, M. and H. P. Luterbacher 1969. On some type specimens of Cretaceous planktonic foraminifera. *Contr. Cushman Fdn Foramin. Res.* **20**, 23–9.
Carpenter, W. B., W. K. Parker and T. R. Jones 1862. *Introduction to the study of the Foraminifera*. London: Ray Society.
Carsey, D. O. 1926. *Foraminifera of the Cretaceous of central Texas*. Bull. Univ. Texas Bur. Econ. Geol Technol. 2612, 1–56
Cati, F. and A. M. Borsetti 1968 (1967). The accessory structures in Tertiary planktonic foraminifera. *Giornale Geol. (2)* **35**, 387–400.
Chapman, F., W. J. Parr and A. C. Collins 1934. Tertiary foraminifera in Victoria, Australia – the Balcombian deposits of Port Phillip, Part III. *J. Linn. Soc. Zool.* **38** (262), 553–77.
Charmatz, R. 1963. On '*Hastigerina digitata* Rhumbler, 1911'. *Micropaleontology* **9**, 228.
Cifelli, R. 1969. Radiation of Cenozoic planktonic foraminifera. *System. Zool.* **18**, 154–68.
Cifelli, F. and D. J. Belford. 1977. The types of several species of Tertiary planktonic foraminifera in the collections of the US National Museum of Natural History. *J. Foramin. Res.* **7**, 100–5.
Cifelli, R. and R. K. Smith 1970. Distribution of planktonic foraminifera in the vicinity of the North Atlantic current. *Smithson. Contr. Paleobiol.* **4**, 1–52.
Cita, M. B. and Mazzola, G. 1970. *Globigerinopsoides* n. gen. from the Miocene of Algeria. *Riv. Ital. Paleont.* **76**, 465–76.
Cole, W. S. 1927. A foraminiferal fauna from the Guayabal formation in Mexico. *Bull. Am. Paleont.* **14**, 1–46.
Cordey, W. G. 1967. The development of *Globigerinoides ruber* (d'Orbigny 1839) from the Miocene to Recent. *Palaeontology* **10**, 647–59.
Cordey, W. G. 1968. Morphology and phylogeny of *Orbulinoides beckmanni* (Saito 1962). *Palaeontology* **11**, 371–5.
Cordey, W. G., W. A. Berggren and R. K. Olsson 1970. Phylogenetic trends in the planktonic foraminiferal genus *Pseudohastigerina* Banner & Blow, 1959. *Micropaleontology* **16**, 235–42.
Corminboeuf, P. 1961. Tests isolés de *Globotruncana mayaroensis* Bolli, *Rugoglobigerina, Trinitella* et Heterohelicidae dans le Maestrichtian des Alpettes. *Eclog. Geol. Helv.* **54**, 107–122.
Cushman, J. A. 1925a (1924). A new genus of Eocene foraminifera. *US Natn Mus. Proc.* **66** (2567, Art. 30), 1–4.

Cushman, J. A. 1925b. New foraminifera from the Upper Eocene of Mexico. *Contr. Cushman Lab. Foramin. Res.* **1**, 4–9.
Cushman, J. A. 1926. Some foraminifera from the Mendez Shale of eastern Mexico. *Contr. Cushman Lab. Foramin. Res.* **2**, 16–26.
Cushman, J. A. 1927a. An outline of a re-classification of the foraminifera. *Contr. Cushman Lab. Foramin. Res.* **3**, 1–105.
Cushman, J. A. 1927b. New and interesting foraminifera from Mexico and Texas. *Contr. Cushman Lab. Foramin.* **3**, 111–19.
Cushman, J. A. 1931. *Hastigerinella* and other interesting foraminifera from the Upper Cretaceous of Texas. *Contr. Cushman Lab. Foramin. Res.* **7**, 83–90.
Cushman, J. A. 1948. *Foraminifera: their classification and economic use.* Cambridge, Mass: Harvard University Press.
Cushman, J. A. and P. J. Bermúdez 1937a. Further new species of foraminifera from the Eocene of Cuba. *Contr. Cushman Lab. Foramin. Res.* **13**, 1–29.
Cushman, J. A. and P. J. Bermúdez 1937b. Additional new species of Eocene foraminifera from Cuba. *Contr. Cushman Lab. Foramin. Res.* **13**, 106–10.
Cushman, J. A. and P. J. Bermúdez 1949. Some Cuban species of *Globorotalia*. *Contr. Cushman Lab. Foramin. Res.* **25**, 31–2.
Cushman, J. A. and D. Cedestrom 1949 (1945). *An Upper Eocene foraminiferal fauna from deep wells in York County, Virginia.* Virginia Geol Surv. Bull. **67**.
Cushman, J. A. and A. L. Dorsey 1940. Some notes on the genus *Candorbulina*. *Contr. Cushman Lab Foramin. Res.* **16**, 40–2.
Cushman, J. A. and P. W. Jarvis 1929. New foraminifera from Trinidad. *Contr. Cushman Lab. Foramin. Res.* **5**, 6–17.
Cushman, J. A. and G. M. Ponton 1932. Foraminifera of the Upper, Middle and part of the Lower Miocene of Florida. *Florida State Geol Surv. Bull.* **9**, 1–147.
Cushman, J. A. and R. M. Stainforth 1945. The foraminifera of the Cipero Marl formation of Trinidad, B.W.I. *Cushman Lab. Foramin. Res. Sp. Publ.* **14**, 1–74.
Cushman, J. A. and A. ten Dam 1948. *Globigerinelloides*, a new genus of the Globigerinidae. *Contr. Cushman Lab. Foramin. Res.* **24**, 42–3.

Donze, P., B. Porthault, G. Thornel and O. de Villoutreys 1970. Le Senonien inférieur de Puget-Theniers (Alpes Maritimes) et sa microfaune. *Geobios* **3**, 41–106.

Ehrenberg, C. G. 1845. Über das kleinste organische leben an mehreren bisher nicht untersuchten Erdpunkten, mikroskopische Lebensformen . . . . *K. Preuss. Akad. Wiss. Berlin, Bes.* 357–81.
Ehrenberg, C. G. 1854. *Mikrogeologie: das Erden und Felsen Schaffende Wirken des unsichtbar Kleinen Selbständigen Lebens auf der Erde*, Vols 1 and 2 and Atlas. Leipzig: C. H. Voss.
Ehrenberg, C. G. 1838. Ueber dem blossen Auge unsichtbare Kalkthierchen und Kieselthierchen als Huptbestandtheile der Kreidegebirge. *K. Preuss. Akad. Wiss. Berlin, Ber., Jahrg 1838* **3**, 192–200.
Ehrenberg, C. G. 1839. Ueber die Bildung der Kreidefelsen und des Kreidemergels durch unsichtbare organismen. *K. Preuss. Akad. Wiss. Berlin, Abh.* (1838), 59–147.
Ehrenberg, C. G. 1858. Kurze characteristik der 9 neuen Genera und 105 neuen species des ägäischen Meeres und des Tiefgrundes des Mittel-Meeres. *K. Preuss. Akad. Wiss. Berlin, Monatsber.* 10–40.
Eicher, D. L. 1973 (1972). Phylogeny of the late Cenomanian planktonic foraminifer *Anaticinella multiloculata* (Morrow). *J. Foramin. Res.* **2**, 184–90.
El-Naggar, Z. R. 1971a. On the classification, evolution and stratigraphical distribution of the Globigerinacaea. In *Proceedings of the 2nd Planktonic Conference, Roma, 1970*, Vol. 1, A. Farinacci (ed.), 421–76. Rome: Ed. Tecnoscienza.
El-Naggar, Z. R. 1971b. The genus *Rugoglobigerina* in the Maestrichtian Sharawna shale of Egypt. In *Proceedings of the 2nd Planktonic Conference, Roma, 1970*, Vol. 1, A. Farinacci (ed.), 477–537. Rome: Ed. Tecnoscienza.

Finlay, H. J. 1940. New Zealand foraminifera: key species in stratigraphy, no. 4. *Trans Proc. R. Soc. NZ* **69**, 448–72.

Finlay, H. J. 1946. New Zealand foraminifera: key species in stratigraphy. *N.Z. J. Sci. Technol.* **28**, 259–92.

Fleisher, R. L. 1974. Cenozoic planktonic foraminifera and biostratigraphy, Arabian Sea Deep Sea Drilling Project, leg 23A. *Initial Reports, Deep Sea Drilling Project* **23**, 1001–72.

Fleisher, R. L. 1975. Oligocene planktonic foraminiferal biostratigraphy, central North Pacific Ocean, DSDP leg 32. In R. L. Larson *et al.*, *Initial Reports, Deep Sea Drilling Project* **32**, 753–63.

Frerichs, W. E., E. M. Pokras and M. J. Evetts 1977. The genus *Hastigerinoides* and its significance in the biostratigraphy of the western interior. *J. Foramin. Res.* **7**, 149–56.

Fuchs, W. 1968 (1967). Über ursprung und phylogenie der Trias 'Globigerinen' und die Bedeutung dieses formenkreises für das echte Plankton. *Verhandl. Geol. Bundesanstalt, Wien* **1967**, 135–76.

Fuchs, W. 1969. Zur kenntnis des Schalenbaues der zu den trias-'Globigerinen' zählenden foraminiferen-gattung *Praegubkinella*. *Verhandl. Geol. Bundesanstalt, Wien* **1969**, 158–67.

Fuchs, W. 1970. Eine alpine, tielfliassische Foraminiferen-fauna von Hernstein in Niederösterreich. *Verhandl. Geol. Bundesanstalt, Wien* **1970**, 66–145.

Fuchs, W. 1973. Ein Beitrag zus kenntnis der Jura-Globigerinen und verwandten Formen an Hand polnischen Materials der Callovien und Oxfordien. *Verhandl. Geol. Bundesanstalt* **3**, 445–87.

Fuchs, W. 1975. Zur Stammesgeschichte der Planktonforaminiferen und verwandten Formen im Mesozoikum. *Jb. Geol. Bundesanstalt* **118**, 193–246.

Galloway, J. J. 1933. *A manual of foraminifera*. Bloomington, Ind.: Principia.

Gandolfi, R. 1942. Ricerche micropaleontologiche e stratigrafiche sulla Scaglia e sul Flysch Cretacici dei dintorni di Balerna (Canton Ticino). *Riv. Ital. Paleont.* **14** (14), 1–160.

Gandolfi, R. 1955. The genus *Globotruncana* in northeastern Columbia. *Bull. Am. Paleont.* **36** (155), 1–118.

Gandolfi, R. 1957. Notes on some species of *Globotruncana*. *Contr. Cushman Fdn. Foramin. Res.* **8**, 59.

Glaessner, M. F. 1937. Plankton foraminiferen aus der Kreide und dem Eozän und ihre Stratigraphische Bedentung. *Moscow Univ. Lab. Paleont., Stud. Micropaleont.* **1**, 27–46.

Glintzboeckel, C. and J. Magné 1955. Sur la répartition stratigraphique de *Globigerinelloides algeriana* Cushman et ten Dam, 1948. *Micropaleontology* **1**, 153–5.

Gorbachik, T. N. 1964. Variability and microstucture of the test wall in *Globigerinelloides algeriana* (in Russian). *Paleont. Zh.* **4**, 33–6.

Gorbatchik, T. N. 1971. On early Cretaceous foraminifera from the Crimea (in Russian). *Vopr. Mikropaleont.* **14**, 125–39.

Gorbatchik, T. N. and M. Moullade 1973. Caractères microstructuraux de le parvis du test des foraminifères planctoniques du Crétacé inférieur et leur signification sur le plan taxonomique. *C. r. Hebd. Séanc. Acad. Sci. Paris* D **277**, 2661–4.

Grigelis, A. A. 1958. Discovery of *Globigerina oxfordiana* n. sp. in the Upper Jurassic of Lithuania (in Russian). *Nauk Dokl. Vyss. Shkol. Geol.-Geogr.* **3**, 109–11.

Grigelis, A. and T. Gorbatchik 1980. Morphology and taxonomy of Jurassic and Early Cretaceous representatives of the superfamily Globigerinacea. (Favusellidae). *J. Foramin. Res.* **10**, 180–90.

Hagn, H. and W. Zeil 1954. Globotruncanen aus dem Obercenoman und Unterturon der Bayerischen Alpen. *Eclog. Geol. Helv.* **47**, 33–4.

Haman, D., R. W. Huddleston and J. P. Donahue 1981 (1980). *Obandyella*, a new name for *Hirsutella* Bandy, 1972 (Foraminiferida), *non* Cooper and Muir-Wood, 1951 (Brachiopoda). *Proc. Biol Soc. Wash.* **93**, 1264–5.

Hamoui, M. 1965 (1964). On a new subgenus of *Hedbergella*, *Israel J. Earth Sci.* **13**, 133–42.

Haque, A. F. M. M. 1956. The foraminifera of the Ranikot and the Laki of the Nammal Gorge, Salt Range. *Paleont. Pakistanica* **1**, 1–300.

Haque, A. F. M. M. 1962 (1959). The smaller foraminifera of the Meting Limestone (Lower Eocene), Meting, Hyderabad Division, West Pakistan. *Paleont. Pakistanica* **2**, 23.

Hardenbol, J. and W. A. Berggren 1978. A new Paleogene numerical timescale. In *The geologic time scale*, Studies in Geology Vol. 6, 213–34. American Association of Petroleum Geologists.

Harlton, B. H. 1927. Some Pennsylvanian foraminifera of the Glenn formation of southern Oklahoma. *J. Paleont.* **1**, 15–27.

Hay, W. W. 1962. The type level of some of Ehrenberg's foraminifera. *J. Paleont.* **36**, 1392–3.

Hemleben, C., A. W. H. Bé, O. R. Anderson and S. Tuntivaté 1977. Test morphology, organic layers and chamber formation of the planktonic foraminifer *Globorotalia menardii* (d'Orbigny). *J. Foramin. Res.* **7**, 1–25.

Hemleben, C., A. W. H. Bé, M. Spindler and O. R. Anderson 1979. 'Dissolution' effects induced by shell resorption during gametogenesis in *Hastigerina pelagica* (d'Orbigny). *J. Foramin. Res.* **9**, 118–24.

Hillebrandt, A. V. 1976. Los foraminiferos planctonicós, nummulitidos y coccolitofóridos de la zona de *Globorotalia palmerae* del Cuisiense (Eocene inferior) en el S.E. de España. *Rev. Esp. Micropaleont.* **7**, 323–94.

Hofker, J. 1956a. Die Globotruncanen von nordwest Deutschland und Holland. *Neues Jb. Geol. Paläontol. Abh.* **103**, 312–40.

Hofker, J. 1956b. *Foraminifera dentata: foraminifera of Santa Cruz and Thatch Island, Virginia Archipelago, West Indies*. Spol. Zool. Mus. Hauniensis no. 15.

Hofker, J. 1959. On the splitting of *Globigerina*. *Contr. Cushman Fdn Foramin. Res.* **10**, 1–9.

Hofker, J. 1960a. The taxonomic status of *Praeglobotrunca*, *Planomalina*, *Globigerinella* and *Biglobigerinella*. *Micropaleontology* **6**, 315–22.

Hofker, J. 1960b. Planktonic foraminifera in the Danian of Denmark. *Contr. Cushman Fdn Foramin. Res.* **11**, 73–86.

Hofker, J. 1962. Studien an planktonischen foraminiferen. *Neues Jb. Geol. Paläontol. Abh.* **114**.

Hofker, J. 1963a. *Cassigerinella* Pokorný, 1955, and *Islandiella* Nørvang, 1958. *Micropaleontology* **9**, 321–4.

Hofker, J. 1963b. Mise au point concernant les genres *Praeglobotruncana* Bermúdez, *Abathomphalus* Bolli, Loeblich & Tappan, *Rugoglobigerina* Brönnimann, et quelques espèces de *Globorotalia*. *Rev. Micropaléont.* **5**, 280–8.

Hofker, J. 1966. La position stratigraphique du Maestrichtien-type. *Rev. Micropaléont.* **8**, 258–64.

Hofker, J. 1968a. Studies of foraminifera, Part 1, general problems. *Overdruk Publ. Natuurhist. Genootschap Limburg*, **18**, 1–135.

Hofker, J. 1968b. Tertiary foraminifera of Coastal Ecuador, Lower Oligocene and Lower Miocene. *Palaeontographica A* **130**, 1–59.

Hofker, J. 1969. Have the genera *Porticulasphaera*, *Orbulina (Candorbulina)* and *Biorbulina* a biologic meaning? In *Proceedings of the 1st International Conference on Planktonic Microfossils, Geneva, 1967*, Vol. 2, P. Brönnimann & H. H. Renz (eds), 279–86. Leiden: Brill.

Hofker, J. 1972. The *Spaeroidinella* gens from Miocene till Recent. *Rev. Esp. Micropaleont.* **4**, 119–40.

Hofker, J. 1976a. The importance of the *Globoconusa daubjergensis-kozlowskii* gens as a time-marker in the Lower Tertiary. *Paläont. Z.* **50**, 34–9.

Hofker, J. 1976b. La famille Turborotalitidae n. fam. *Rev. Micropaléont.* **19**, 47–53.

Hofker, J. 1978. Analysis of a large succession of samples through the upper Maastrichtian and the Lower Tertiary of drill hole 47.2, Shatsky Rise, Pacific, Deep Sea Drilling Project. *J. Foramin. Res.* **8**, 46–75.

Hofman, E. A. 1958. Novie nakhodki yurskikh Globigerin. *Nauchni Dokl. Vish. Shkoli, Geol.-Geogr. Nauki* **2**, 125–6.

Hohenegger, J. and W. Piller 1975. Wandstruckturen und Grossgliederung der Foraminiferen. *Sitzungsber. Österr.-Akad. Wiss., Math.-Naturw. Kl.* **184,** 67–96.
Hornibrook, N. de B. 1968. *A handbook of New Zealand microfossils (foraminifera and ostracoda).* N.Z. Dep. Scient. Ind. Res. Inf. Ser. no. 62, 1–136.
Hornibrook, N. de B. 1978. *Globoquadrina dehiscens* (foram.) in the Otekaike limestone (Waitakian Stage), New Zealand. *NZ J. Geol. Geophys.* **21,** 657–9.
Howe, H. V. 1928. An observation on the range of the genus *Hantkenina*. *J. Paleont.* **2,** 13–14.
Howe, H. V. 1939. Louisiana Cook Mountain Eocene foraminifera. *Louisiana Conserv. Dept, Geol Surv., Bull.,* no. 14.
Howe, H. V. and W. E. Wallace 1932. Foraminifera from the Jackson Eocene at Danville Landing. *Louisiana Dept Conserv., Geol Surv., Geol Bull.,* **2,** 7–118.
Howe, H. V. and W. E. Wallace 1934. Apertural characteristics of the genus *Hantkenina*, with description of a new species. *J. Paleont.* **8,** 35–7.

Jacob, K. and M. Sastry 1950. On the occurrence of *Globotruncana* in Uttatur stage of the Trichinopoly Cretaceous, South India. *Sci. Cult.* **16** (6), 266–8.
Jedlitschka, H. 1934 (1933). Über *Candorbulina*, eine neue foraminiferen-Gattung und zwei neue *Candeina*-Arten. *Naturforsch. Ver Brünn, Verhandl.* **65,** 17–26.
Jenkins, D. G. 1966 (1965). Planktonic foraminiferal zones and new taxa from the Danian to Lower Miocene of New Zealand. *NZ J. Geol. Geophys.* **8,** 1088–126.
Jenkins, D. G. 1971. New Zealand Cenozoic planktonic foraminifera. *NZ Geol Surv. Paleont. Bull.* **42,** 1–288.
Jenkins, D. G. and W. N. Orr 1972. Planktonic foraminiferal biostratigraphy of the east equatorial Pacific, DSDP leg 9. In J. D. Hays *et al. Initial Reports, Deep Sea Drilling Project* **9,** 1059–193.
Jíróva, D. 1956. The genus *Globotruncana* in Upper Turonian and Emscherian of Bohemia (in Czech). *Univ. Carolina Geologica* **2,** 239–55.

Khalilov, D. M. 1956. O pelagicheskoi faune foraminifer paleogenovikh otlozhenii Azerbaidzhana. *Tr. In-Ta Geol. A.N. Azerbaidzhan SSR* **17,** 234–55.
Khalilov, D. M. 1964. Foraminiferi datskogo yarusa Azerbaidzhana v svete yutochneniya granitisi mezhdi melovoi i tretichnoi sistemami. *Ogerki po Geol. Azerbaidzhana, Izdat A.N. Azerbaidzhan SSR* 153–68.
King, K. and P. E. Hare 1972. Amino-acid composition of the test as a taxonomic character for living and fossil planktonic foraminifera. *Micropaleontology* **18,** 285–93.

Lalicker, C. G. 1948. A new genus of foraminifera from the Upper Cretaceous. *J. Paleont.* **22,** 624.
Lamb, J. L. and J. H. Beard 1972. *Late Neogene planktonic foraminifera in the Caribbean, Gulf of Mexico and Italian stratotypes.* Kansas Univ., Paleont. Contr., Art. 57, 1–67.
Lamolda, M. A. 1976. Helvetoglobotruncaninae subfam. nov. y consideraciones sobre los Globigeriniformes del Cretacico. *Rev. Esp. Micropaleont.* **7,** 395–400.
Lapparent, J. de 1918. *Étude lithologique des terrains Crétacés de la région d'Hendaye.* Serv. Carte Géol. France, Mém. 1–155.
Le Calvez, Y. 1974. *Révision des foraminiferès de la collection d'Orbigny. I – Foraminiferès des Îles Canaries.* Paris: Ed. CNRS.
Le Calvez, Y. 1977. *Révision des foraminiferès de la collection d'Orbigny. II – Foraminiferès de l'Île de Cuba,* Vol. 2. Paris: Ed. CNRS.
Lipps, J. H. 1964. Miocene planktonic foraminifera from Newport Bay, California. *Tulane Stud. Geol.* **2** (4), 109–33.
Lipps, J. H. 1966. Wall structure, systematics and phylogeny studies of Cenozoic planktonic foraminifera. *J. Paleont.* **40,** 1257–74.
Lipps, J. H. 1967. Planktonic foraminifera, intercontinental correlation and age of Californian mid-Cenozoic microfaunal stages. *J. Paleont.* **41,** 994–9.
Lipps, J. H. 1970. Plankton evolution. *Evolution* **24,** 1–22.

Liska, R. D. 1980. *Polyperibola*, a new planktonic foraminiferal genus from the late Miocene of Trinidad and Tobago. *J. Foramin. Res.* **10**, 136–42.

Loeblich, A. R., Jr and H. Tappan 1946. New Washita foraminifera. *J. Paleont.* **20**, 238–58.

Loeblich, A. R., Jr and H. Tappan 1957a. Planktonic foraminifera of Paleocene and Early Eocene age from the Gulf and Atlantic coastal plains. *US Natn Mus. Bull.* **215**, 173–98.

Loeblich, A. R., Jr and H. Tappan 1957b. The new planktonic foraminiferal genus *Tinophodella* and an emendation of *Globigerinita* Brönnimann. *J. Wash. Acad. Sci.* **47** (4), 112–16.

Loeblich, A. R., Jr and H. Tappan 1961a. Cretaceous planktonic foraminifera: Part 1 – Cenomanian. *Micropaleontology* **7**, 257–304.

Loeblich, A. R., Jr and H. Tappan 1961b. Suprageneric classification of the Rhizopodea. *J. Paleont.* **35**, 245–330.

Loeblich, A. R., Jr and H. Tappan 1964a. Part C, Protista 2, Sarcodina. Chiefly 'Thecamoebians' and Foraminiferida. In *Treatise on invertebrate paleontology*, R. C. Moore (ed.), Kansas City: University of Kansas Press and Geological Society of America.

Loeblich, A. R., Jr and H. Tappan 1964b. On '*Hastigerina digitata* Rhumbler, 1911': a comment. *Micropaleontology* **10**, 494–5.

Loeblich, A. R., H. Tappan, J. P. Beckmann, H. M. Bolli, E. M. Gallitelli and J. C. Troelsen 1957. Studies in foraminifera. *US Natn Mus. Bull.* **215**, 1–323.

Longoria, J. F. 1973. *Pseudoticinella*, a new genus of planktonic foraminifera from the Early Turonian of Texas. *Rev. Esp. Micropaleont.* **5**, 417–23.

Longoria, J. F. 1974. Stratigraphic, morphologic and taxonomic studies of Aptian planktonic foraminifera. *Rev. Esp. Micropaleont.* num. extraord., 1–107.

Longoria, J. F. and M. A. Gamper 1975. The classification and evolution of Cretaceous planktonic foraminifera. Pt. 1: the superfamily Hedbergelloidea. *Rev. Esp. Micropaleont.* num. especial, 61–96.

Luterbacher, H. 1964. Studies in some *Globorotalia* from the Palaeocene and Lower Eocene of the Central Apennines. *Eclog. Geol. Helv.* **57**, 631–730.

Luterbacher, H. P. and I. Premoli Silva 1964. Biostratigraphica del limite Cretaceo-Terziario nell'Appennino Centrale. *Riv. Ital. Paleont.* **70**, 67–128.

McGowran, B. 1964. Foraminiferal evidence for the Palaeocene age of the King's Park Shale (Perth Basin, Western Australia). *Jl R. Soc. W. Aust.* **47** (3), 85–6.

McGowran, B. 1968. Reclassification of Early Tertiary *Globorotalia*. *Micropaleontology* **14**, 179–98.

Maiya, S., T. Saito and T. Sato 1976. Late Cenozoic planktonic foraminiferal biostratigraphy of north-west Pacific sedimentary sequences. In *Progress in micropaleontology*, Y. Takayanagi & T. Saito (eds), 395–422. New York: Micropaleontology Press.

Marie, P. 1941. Les foraminiferès de la craie à *Belemnitella mucronata* du Bassin de Paris. *Mus. Natn Hist. Nat., Mem.* **12** (1), 1–296.

Marks, P. 1972. Late Cretaceous planktonic foraminifera from Prebetic tectonic elements near Jaen (southern Spain). *Rev. Esp. Micropaleont.* num. extraord., XXX. Ann. E. N. Adaro, 99–123.

Martin, L. T. 1943. Eocene foraminifera from the type Lodo formation, Fresno County, California. *Stanford Univ. Publ. Geol Sci.* **3**, 93–125.

Maslakova, N. I. 1961. Systematics and phylogeny of the genera *Thalmanninella* and *Rotalipora* (in Russian). *Paleont. Zh.* **1**.

Maslakova, N. I. 1963. Structure of the wall of the test of Globotruncanids (in Russian). *Vopr. Mikropaleont.* **7**, 138–49.

Maslakova, N. I. 1964. K sistematike i filogenii Globotruncanid. *Vopr. Mikropaleont.* **8**, 102–17.

Masters, B. A. 1977. Mesozoic planktonic foraminifera. In *Oceanic micropalaeontology*, Vol. 1, A. T. S. Ramsay (ed.), 301–731. London: Academic Press.

Masters, B. A. 1980. Re-evaluation of selected types of Ehrenberg's Cretaceous planktonic foraminifera. *Eclog. Geol. Helv.* **73**, 95–107.

Mazzola, G. 1971. Les foraminiferès planctoniques du Miocene–Pliocenè de l'Algerie nord-occidentale. In *Proceedings of the 2nd Planktonic Conference, Roma, 1970*, Vol. 1, A. Farinacci (ed.), 787–812. Rome: Ed. Tecnoscienza.

Michael, F. Y. 1973 (1972). Planktonic foraminifera from the Comanchean series (Cretaceous) of Texas. *J. Foramin. Res.* **2**, 200–20.

Mohan, M. and K. S. Soodan 1967. *Inordinatosphaera*, a new genus of Globigerinidae. *Bull. Geol Soc. India* **4** (1), 24.

Mohan, M. and K. S. Soodan 1970. Middle Eocene planktonic foraminiferal zonation of Kutch, India. *Micropaleontology* **16** (1).

Montanaro-Gallitelli, E. 1957. A revision of the foraminiferal family Heterohelicidae. *US Natn Mus. Bull.* **215**, 133–54.

Moorkens, T. L. 1971. Some Late Cretaceous and Early Tertiary foraminifera from the Maastrichtian type area. In *Proceedings of the 2nd Planktonic Conference, Roma, 1970*, Vol. 2, A. Farinacci (ed.), 847–77. Rome: Ed. Tecnoscienza.

Mornod, L. 1950 (1949). Les Globorotalidés du Crétacé supérieur du Montsalvens (Préalpes fribourgeoises). *Eclog. Geol. Helv.* **42**, 573–96.

Morozova, V. G. 1948. Foraminiferi nizhnemelovikh otlozhenii rayona g. Sochi (Yugozapadnoi Kavkaz). *Moskov. Obschch. Ispyt., Prirodi, Otd. Geol., Byul.* **23** (3), 23–43.

Morozova, V. G. 1957. Nadsemeistvo foraminifer Globigerinidea superfam. nova i nekotorie ego predstaviteli. *Dokl. Akad. Nauk SSSR* **114**, 1109–12.

Morozova, V. G. 1958. K sistematike i morfologii paleogenovikh predstavitelei nadsemeistva Globigerinidea. *Vopr. Mikropaleont.* **2**, 23–52.

Morozova, V. G. 1959. Stratigrafiya datsko-montskikh otlozhenii Kryma po foraminiferam. *Dokl. Akad. Nauk SSSR* **124**, 1113–6.

Morozova, V. G. and T. A. Moskalenko 1961. Planktonnie foraminiferi progranichnikh otlozhenii bayosskogo i batskogo Yarusov tsentralnogo Dagestana (severo-vostochnii Kavkaz). *Vopr. Mikropaleont.* **5**, 3–30.

Morrow, A. L. 1934. Foraminifera and ostracoda from the Upper Cretaceous of Kansas. *J. Paleont.* **8**, 186–205.

Murray, J. 1876. Preliminary reports to Prof. Wyville Thomson, FRS, director of the civilian staff, on work done on board the 'Challenger' *Proc. R. Soc.* **24**, 471–544.

Myatliuk, E. V. 1950. Stratigrafiya flishevikh osadkov severnikh Karpat v sveta dannikh fauni foraminifer. *VNNIGRI Trudi* **51**(*Mikrofauna SSSR* **4**), 225–87.

Nakkady, S. E. 1950. A new foraminiferal fauna from the Esna shales and Upper Cretaceous chalk of Egypt. *J. Paleont.* **24**, 675–92.

Nakkady, S. E. 1959. Biostratigraphy of the Um Elghanayem, Egypt. *Micropaleontology* **5**, 453–72.

Natori, H. 1976. Planktonic foraminiferal biostratigraphy and datum planes in late Cenozoic sedimentary sequence in Okinawa-jima, Japan. In *Progress in micropaleontology*, Y. Takayanagi & T. Saito (eds), 214–43. New York: Micropaleontology Press.

Netskaya, A. I. 1948. Foraminifera of upper Senonian deposits of W. Siberia (in Russian). *Trudi, VNIGRI* **31**, 211–29.

Nuttall, W. L. F. 1928. Notes on the Tertiary foraminifera of southern Mexico. *J. Palaeont.* **2**, 372–7.

Nuttall, W. L. F. 1930. Eocene foraminifera from Mexico. *J. Paleont.* **4**, 271–93.

Olsson, R. K. 1964. *Praeorbulina* Olsson, a new foraminiferal genus. *J. Paleont.* **38**, 770–1.

Orbigny, A. D. d' 1826. Tableau méthodique de la classe des Céphalopodes. *Ann. Sci. Nat. Paris (1)* **7**, 96–314.

Orbigny, A. D. d' 1839a. Foraminiferès. In R. de la Sagra, *Histoire physique, politique et naturelle de l'Isle de Cuba*, 1–225. Paris.

Orbigny, A. D. d' 1839b. *Voyage dans l'Amerique Meridionale – Foraminiferès*, Vol. 5 (5), 1–86. Paris: Pitois-Levrault.

Orbigny, A. D. d' 1839c. Foraminiferès des Iles Canaries. In Barker-Webb and Berthelot, *Histoire naturelle des Iles Canaries*, Vol. 2 (2) *Zoology*, 119–46. Paris: Béthune.
Orbigny, A. D. d' 1846. *Foraminiferès fossiles du Bassin Tertiaire de Vienne (Autriche)*, 1–312. Paris: Gide.
Orr, W. N. 1969. Variation and distribution of *Globigerinoides ruber* in the Gulf of Mexico. *Micropaleontology* **15**, 373–9.
Ouda, Kh. 1978. *Globigerinanus*, a new genus of the Globigerinidae from the Miocene of Egypt. *Rev. Esp. Micropaleont.* **10**, 355–78.

Parker, F. L. 1962. Planktonic foraminiferal species in Pacific sediments. *Micropaleontology* **8**, 219–54.
Parker, F. L. 1967. Late Tertiary biostratigraphy (planktonic foraminifera) of tropical Indo-Pacific deep sea cores. *Bull. Am. Paleont.* **52** (235), 115–203.
Parker, F. L. 1976. Taxonomic notes on some planktonic foraminifera. In *Progress in micropaleontology*, Y. Takayanagi & T. Saito (eds), 258–62. New York: Micropaleontology Press.
Parker, W. K. and T. R. Jones 1865. On some foraminifera from the North Atlantic and Arctic Oceans, including Davis Straits and Baffin's Bay. *Phil Trans R. Soc.* **155**, 325–441.
Parker, W. K., T. R. Jones and H. B. Brady 1865. On the nomenclature of the foraminifera. XII: the species enumerated by d'Orbigny in the *Annales des Sciences Naturelles*, Vol. 7, 1826. *Ann. Mag. Nat. Hist. (3)* **16**, 15–41.
Parr, W. J. 1938. Upper Eocene foraminifera from deep borings in King's Park, Perth, Western Australia. *J. R. Soc. W. Aust.* **24** (8), 69–101.
Pessagno, E. A., Jr 1962. The Upper Cretaceous stratigraphy and micropaleontology of south-central Puerto Rico. *Micropaleontology* **8**, 349–68.
Pessagno, E. A., Jr 1967. Upper Cretaceous planktonic foraminifera from the Western Gulf Coastal Plain. *Palaeontogr. Am.* **5** (37), 245–445.
Pessagno, E. A., Jr 1969. Upper Cretaceous stratigraphy of the Western Gulf Coast area of Mexico, Texas and Arkansas. *Mem. Geol Soc. Am.* **111**, 1–139.
Petters, S. W. 1977. Upper Cretaceous planktonic foraminifera from the subsurface of the Atlantic Coastal Plain of New Jersey. *J. Foramin. Res.* **7**, 165–87.
Plummer, H. J. 1927. *Foraminifera of the Midway formation in Texas*. Univ. Texas Bull. no. 2644, 1–206.
Plummer, H. J. 1931. *Some Cretaceous foraminifera in Texas*. Univ. Texas Bull. no. 3101, 109–203.
Pokorný, V. 1955. *Cassigerinella boudecensis* n. gen., n. sp., z oligocenú ždánickeho flyše. *Ustřed. ústavu Geol., Věstník* **30**, 136–40.
Pokorný, V. 1958. *Grundzüge der Zoologischen Mikropaläontologie*, Vol. 1, 1–582. Berlin: VEB Verlag der Wissenschaften.
Poore, R. Z. and E. E. Brabb 1977. Eocene and Oligocene planktonic foraminifera from the Upper Butano Sandstone and type San Lorenzo formation, Santa Cruz Mountains, California. *J. Foramin. Res.* **7**, 249–72.
Popescu, Gh. 1969. Some new *Globigerina* (foraminifera) from the Upper Tortonian of the Transylvanian Basin and the Subcarpathians. *Rev. Roumanie Géol., Géophys. Géogr. (sér. Géol.)* **13** (1), 103–6.
Postuma, J. A. 1971. *Manual of planktonic foraminifera*. Amsterdam: Elsevier.
Premoli Silva, I. 1966. La struttura della parete di alcuni foraminiferi planctonici. *Eclog. Geol. Helv.* **59**, 219–34.
Proto Decima, F. and H. M. Bolli 1970. Evolution and variability of *Orbulinoides beckmanni* (Saito). *Eclog. Geol. Helv.* **63**, 883–905.

Rauser-Chernoussova, D. M. and A. V. Fursenko 1959. *Osnovi Paleontologii*, Pt 1, 1–368. Moscow: Akademia Nauk SSSR. [Translated, 1962, as *Fundamentals of paleontology, general part, Protozoa*, 1–551. Jerusalem: Israel Program for Scientific Translations.]
Reichel, M. 1948 (1947). Les Hantkeninides de la Scaglia et des Couches rouges (Crétacé Supérieur). *Eclog. Geol. Helv.* **60**, 391–409.

Reichei, M. 1950. Observations sur les Globotruncana du gisement de la Breggia (Tessin). *Eclog. Geol. Helv.* **42**, 596–617.

Reiss, Z. 1957. The Bilamellidae, nov. superfam., and remarks on Cretaceous globorotaliids. *Contr. Cushman Fdn Foramin. Res.* **8**, 127–45.

Reiss, Z. 1963. Reclassification of perforate foraminifera. *Bull. Geol Surv. Israel* **35**, 1–111.

Reiss, Z. and Halicz, E. 1976. Phenotypy in planktonic foraminiferida from the Gulf of Elat. *Israel J. Earth Sci.* **25**, 27–39.

Reiss, Z., E. Halicz and L. Perelis 1974. Planktonic foraminiferida from Recent sediments in the Gulf of Elat. *Israel J. Earth Sci.* **23**, 69–105.

Renz, O., H. P. Luterbacher and A. Schneider 1963. Stratigraphisch-paläontologische Untersuchungen im Albien und Cenomanien des Neuenberger Jura. *Eclog. Geol. Helv.* **56**, 1073–116.

Reuss, A. E. 1846 (1845). *Die Versteinerungen der böhmischen Kreideformation*, Pt 2, 1–148. Stuttgart.

Rhumbler, L. 1911. *Die Foraminiferen (Thalamophoren) der Plankton-Expedition*. Ergnebisse der Plankton-Exped. Humboldt-Stiftung **3** (C), 1–331.

Robaszynski, F. and M. Caron (eds) 1979a. *Atlas de Foraminifères planctoniques du Crétacé Moyen (mer Boréale et Téthys)*, Vol. 1, 1–185. Paris: Ed. CNRS.

Robaszynski, F. and M. Caron (eds) 1979b. *Atlas de Foraminifères planctoniques du Crétacé Moyen (mer Boréale et Téthys)*, Vol. 2, 1–181. Paris: Ed. CNRS.

Rösler, W., G. F. Lutze and U. Pflaumann 1979. Some Cretaceous planktonic foraminifers (*Favusella*) of DSDP Site 397 (eastern North Atlantic). *Initial Reports, Deep Sea Drilling Project* **47** (1), 273–81.

Ruggieri, G. 1963. *Globigerinelloides algeriana* nell'Aptiano della Sicilia. *Soc. Palaeont. Ital. Bull.* **2** (2), 77–8.

Saint-Marc, P. 1970. Sur quelques foraminifères Cénomaniens et Turoniéns du Liban. *Rev. Micropaléont.* **13**, 85–94.

Saito, T. 1962. Eocene planktonic foraminifera from Hahajima (Hillsborough Island). *Trans Proc. Palaeont. Soc. Japan* **45**, 209–25.

Saito, T. and P. E. Biscaye 1977. Emendation of *Riveroinella martinezpicoi* Bermúdez & Seiglie, 1967 and synonymy of *Riveroinella* with *Cassigerinella* Pokorný, 1955. *Micropaleontology* **23**, 319–29.

Saito, T., P. R. Thompson and D. Breger 1976. Skeletal microstructure of some elongate-chambered planktonic foraminifera and related species. In *Progress in micropaleontology*, Y. Takayanagi & T. Saito (eds), 278–304. New York: Micropaleontology Press.

Sastry, M. V. A. and V. D. Mamgain 1972. Planktonic foraminifera of the Uttatur Group, Upper Cretaceous, South India. *Rec. Geol Surv. India* **99** (2), 145–56.

Schacko, G. 1897 (1896). Beitrag über Foraminiferen aus dem Cenoman-Kreide von Voltzow in Mecklenburg. *Verh. Freunde Naturg. Mecklenburg, Archiv* **50**, 161–8.

Scheibnerova, V. 1962. Stratigrafiya strednei a verkhnei kreidi tetudnei oblasti na zaklade globotrunkanid. *Geol. Sb.*, Bratislava **13** (2), 197–226.

Schubert, R. J. 1910. Über foraminiferen un einen fischotolithen aus dem fossilen Globigerinenschlamm von Neu Guinea. *Geol Reichsanst., Verh., Vienna* 318–28.

Schubert, R. J. 1911. Die fossilen foraminiferen des Bismarkarchipels und einiger angrezender Insel. *Neues Jb. Min. Geol. Palaeont., Stuttgart* **2**, 318–20.

Schultze, M.S. 1854. *Ueber den organismus der Polythalamien (Foraminiferen) nebst Bemerkungen über die Rhizopoden im Allgemeinen*. Leipzig: W. Engelmann.

Scott, G. H. 1973. Peripheral structure in chambers of *Globorotalia scitula praescitula* and some descendants. *Rev. Esp. Micropaleont.* **5**, 235–46.

Scott, G. H. 1976. Estimation of ancestry in planktonic foraminifera: *Globoquadrina dehiscens*. *NZ J. Geol. Geophys.* **19**, 311–25.

Sigal, J. 1948. Notes sur les genres de Foraminifères *Rotalipora* Brotzen, 1942, et *Thalmanninella*; famille des Globorotaliidae. *Rev. Inst. Fr. Pétrole Ann. Combus. Liquides* **3** (4), 95–103.

Sigal, J. 1952. Aperçu Stratigraphique sur la micropaléontologie du Crétacé Alger. *19ᵉ Int. Geol Congr., Monogr. Régionales, Ser. 1, Algerie* **26,** 1–45.

Sigal, J. 1956. Notes micropaléontologiques nord-africains. 4. *Biticinella breggiensis* (Gand.), nouveau morphogenre. *Soc. Géol. France, C. r. Somm. Séanc.* nos 3–4, 35–57.

Sigal, J. 1958. La classification actuelle des familles de foraminiferès planctoniques du Crétacé. *Soc Géol. France, C. r. Somm. Séanc.* fasc. 11–12, 262–5.

Sigal, J. 1959. Notes micropaléontologiques alpines. Les genres *Schackoina* et *Leupoldina* dans le Gargasien Vocontien. Étude de morphogénèse. *Rev. Micropaléont.* **2** (2), 68–79.

Sigal, J. 1966a. Contribution à une monographie des Rosalines. 1. Le genre *Ticinella,* souche des Rotalipores. *Eclog. Geol. Helv.* **59,** 184–217.

Sigal, J. 1966b. Le concept taxinomique du spectre. Exemples d'application chez les foraminifères. Propositions de règles de nomenclature. *Soc. Géol. France, Mém. Hors-Serie* **3,** 1–126.

Singh, S. N. 1971. Planktonic foraminifera in the Eocene stratigraphy of Rajasthan, India. In *Proceedings of the 2nd Planktonic Conference, Roma, 1970,* Vol. 1, A. Farinacci (ed.), 1169–81. Rome: Ed. Tecnoscienza.

Singh, S. N. and P. Kalia 1970. A new planktonic foraminifer from the Middle Eocene of India. *Micropaleontology* **16,** 76–82.

Smith, C. G. and E. A. Pessagno Jr 1973. Planktonic foraminifera and stratigraphy of the Corsicana formation (Maestrichtian), north-central Texas. *Cushman Fdn Foramin. Res., Sp. Publ.* **12,** 1–68.

Spindler, M., O. R. Anderson, C. Hemleben and A. W. H. Bé 1978. Light and electron microscopic observations of gametogenesis in *Hastigerina pelagica* (foraminifera). *J. Protozool.* **25,** 427–33.

Spindler, M., C. Hemleben, U. Bayer, A. W. H. Bé and O. R. Anderson 1979. Lunar periodicity of reproduction in the planktonic foraminifer *Hastigerina pelagica. Mar. Ecol. Progress Ser.* **1,** 61–4.

Spraul, G. L. 1963. Current status of the Upper Eocene guide fossil, *Cribrohantkenina. J. Paleont.* **37,** 366–70.

Srinavasan, M. S. and J. P. Kennett 1976. Evolution and phenotypic variation in the Late Cenozoic *Neogloboquadrina dutertrei* plexus. In *Progress in micropaleontology,* Y. Takayanagi and T. Saito (eds), 329–55. New York: Micropaleontology Press.

Stainforth, R. M., J. L. Lamb, H. Luterbacher, J. H. Beard and R. M. Jeffords 1975. *Cenozoic planktonic foraminiferal zonation and characteristics of index forms.* Univ. Kansas paleont. Contr., Art. **62,** 1–425.

Stainforth, R. M., J. L. Lamb and R. M. Jeffords 1978. *Rotalia menardii* Parker, Jones & Brady, 1865 (Foraminiferida): proposed suppression of lectotype and designation of neotype. ZN(S)2145. *Bull. Zool Nomencl.* **34,** 252–62.

Steineck, P. L. 1971. Phylogenetic reclassification of Paleogene planktonic foraminifera. *Texas J Sci.* **23,** 167–78.

Steineck, P. L. and J. H. Darrell III 1971. *Cassigerinella winniana* from the Cook Mountain Eocene, Louisiana. *Micropaleontology* **17,** 357–60.

Štemproková-Jírová, D. 1970. Variation of *Globotruncana marginata* (Reuss, 1845) at its type locality. *Acta Univ. Carolinae, Geol.* **4,** 303–18.

Stolk, J. 1965. Contribution a l'étude des corrélations microfauniques du Tertiaire inférieur de la Nigeria méridionale. In *Colloque internationale de micropaleontologie (Dakar, 6–11 mai 1963).* 247–75.

Stuart, A. 1866. Ueber *Coscinosphaera ciliosa,* eine neue Radiolarie. *Z. Wiss. Zool.* **16,** 328–45.

Subbotina, N. N. 1953. Globigerinidi, Hantkeninidi i Globorotaliidi: Iskopaemie Foraminiferi SSSR. *Trudi, VNIGRI* **76,** 1–296. (Translations: M. Sigal 1960. BRGM Service d'Inf. Géol., Traduction 2239; E. Lees 1971. London & Wellingborough: Collets)

Subbotina, N. N. 1971. Novoe v sistematike mikrofauni. *Trudi Vses. Neft. Nauchno-issled. Geol.-razved. Inst. (VNIGRI)* **291,** 63–9.

Subbotina, N. N., V. V. Glushko and L. S. Pishvanova 1955. O vozraste nizhnei vorotishchenskoi sviti predkarpatskogo kraevogo progiba. *Dokl. Akad. Nauk SSSR* **104**, 605–7.
Subbòtina, N. N., L. S. Pishvanova and L. V. Ivanova 1960. Stratigrafiya Oligotsenovikh i Miotsenovikh otlozhenii predkarpatiya po foraminiferam. *Trudi, VNIGRI, Mikrofauna SSSR* **153**, 5–156.
Suleimanov, I. S. 1955. Novi rod *Gubkinella* i dva novikh vida semeistva Heterohelicidae iz verkhnego senona yugo-zapadnikh Kizil-kumov. *Dokl. Akad. Nauk SSSR* **102**, 623–4.

Tappan, H. 1940. Foraminifera from the Grayson formation of northern Texas. *J. Paleont.* **14**, 93–126.
Tappan, H. 1943. Foraminifera from the Duck Creek formation of Oklahoma and Texas. *J. Paleont.* **317**, 476–517.
Tappan, H. and A. R. Loeblich 1972. Fluctuating rates of protistan evolution, diversification and extinction. *Proc. 24th int. Geol. Congr.* Sect. 7, 205–13.
Thalmann, H. E. 1932. Die foraminiferen-Gattung *Hantkenina* Cushman, 1924, und ihre regional-stratigraphische Verbreitung. *Eclog. Geol. Helv.* **25**, 287–92.
Thalmann, H. E. 1942. Foraminiferal genus *Hantkenina* and its subgenera. *Am. J. Sci.* **240**, 809–20.
Thompson, P. R. 1973. Two new Late Pleistocene planktonic foraminifera from a core in the S.W. Indian Ocean. *Micropaleontology* **19**, 469–74.
Thompson, P. R., A. W. H. Bé, J. C. Duplessy and N. J. Shackleton 1979. Disappearance of pink-pigmented *Globigerinoides ruber* at 120,000 yr BP in the Indian and Pacific Oceans. *Nature, Lond.* **280**, 554–8.
Todd, R. 1957. *Geology of Saipan, Mariana Islands, Pt 3, Paleontology: smaller foraminifera.* US Geol Surv. prof. Paper 280-H, 265–320.
Toumarkine, M. and H. M. Bolli 1970. Evolution de *Globorotalia cerroazulensis* (Cole) dans l'Éocène moyen et supérieur de Possagno (Italie). *Rev. Micropaléont.* **13**, 131–45.
Towe, K. M. 1971. Lamellar wall construction in planktonic foraminifera. In *Proceedings of the 2nd Planktonic Conference, Roma, 1970*, Vol. 1, A. Farinacci (ed.) 1213–24. Rome: Ed. Tecnoscienza.
Turnovsky, K. 1958. Eine neue Art von *Globorotalia* Cushman aus dem Eozaen Anatoliens und ihre Zuordnung zu einer neuen Untergattung. *Bull. Geol Soc. Turkey* **6**, 80–6.

Ujiié, H. 1976. *Prosphaeroidinella* n.gen.: probable ancestral taxon of *Sphaeroidinellopsis* (foraminifera). *Bull. Natn. Sci. Mus., Tokyo (C, Geol.)* **2**, 9–26.

Voorwijk, G. H. 1937. Foraminifera from the Upper Cretaceous of Habana, Cuba. *R. Acad. Amsterdam, Proc.* **40**, 190–8.

Walters, R. 1965. The *Globorotalia zealandica* and *G. miozea* lineages. *NZ J. Geol. Geophys.* **8**, 109–27.
Weinzierl, L. L. and E. R. Applin 1929. The Claiborne formation on the coastal domes. *J. Paleont.* **3**, 384–410.
Wonders, A. A. H. 1978. Phylogeny, classification and biostratigraphic distribution of keeled Rotaliporinae. *K. Ned. Akad. Wetensch., Proc. (B)* **81**, 113–25.

Zakharova-Atabekyan, L. V. 1961. K revizii sistematiki Globotruncanid i predlozhenii novogo roda *Planogyrina* nov. gen. *Dokl. Akad. Nauk Armyanskoi SSR* **32** (1), 49–53.

6 Differential preservation of foraminiferids in the English Upper Cretaceous – consequential observations

Dennis Curry

In the English Upper Chalk, planktonic foraminiferids have been found to be much more common in flint meal than in the associated chalk, although benthonic forms are common in both. In some marl bands, foraminiferids are much rarer than in equal volumes of the chalk that lies above and below them, although agglutinating species remain relatively common. These two phenomena and others like them are believed to give evidence of selective preservation, which has arisen in various ways and on differing time scales. The new data on the relative abundance of planktonic foraminiferids now support evidence from other sources relating to the depth of the Chalk sea.

Introduction

Powder from cavities in the interior of Chalk flints ('flint meal') has long interested microscopists because of the beauty and variety of the fossils which it contains and because of the ease with which slide preparations can be made from it (Eley 1859, Wright 1875). However, such material has been little used in systematic scientific investigations of microfossils because, first, comparable preparations can be obtained from the Chalk itself, and, secondly, Chalk flints with suitable cavities are rare, as will be discussed later. One important exception to this general rule, however, is the work of Barr (1962, 1966), who described the faunas of planktonic foraminiferids and of the benthonic genus *Bolivinoides* from Senonian sequences at Culver Cliff, Isle of Wight, England (SZ639856). In the sea-cliffs at and near this locality there is a continuous succession (although with hiatuses) from Weald Clay (early Cretaceous) to Bembridge Marls (earliest Oligocene) (White 1921). The sequence is accessible for its whole length (although with some difficulty over a short stretch) and forms quite the finest continuous and accessible exposure in the Upper Cretaceous Chalk of Britain. It was no doubt for this reason that Barr chose the locality for his studies. However, the Chalk itself is hard, probably due to diagenesis in relation to the formation of the Isle of Wight monocline, and microfossil preparation from it is relatively difficult and may result in breakage of the more fragile specimens. Barr's sample material was taken mostly from the interiors of flints, fractured and exposed on the cliff face. He noted (Barr 1962, p. 552) that (in comparison with the associated chalk and marly chalk) such material yielded 'by far the most abundant and best-preserved Foraminifera'.

At about the same time, the present author had taken a set of serial samples of chalk from the succession at Culver Cliff and had processed them to provide a standard faunal sequence for use in dating the large number of gravity-core samples of chalk which were becoming available from the floor of the English Channel in the 1960s and early 1970s. Comparison of the faunas obtained at Culver Cliff with those recorded by Barr showed an important discrepancy, in that Barr's faunas were many times richer in planktonic foraminiferids than those collected by the author. Additional collecting established that this difference was associated in some way with the difference in source material in the two cases – flint instead of chalk. A systematic study was therefore carried out to explore this situation, and is reported on in the present note.

Collection and preparation of materials

The composition and possible origin of flint have been discussed many times (see Hancock 1976 for references) and will not be further commented on here. In the English Upper Cretaceous sequences, flints are absent or very rare in Cenomanian and low Turonian levels. They are present and, at some levels, abundant and large in the high Turonian and Coniacian. Relatively rare in the Santonian, they become common once more in the Campanian and are particularly abundant in the high Campanian and low Maastrichtian of Norfolk. The proportion of flints with internal cavities is very variable. At some levels, notably in the *Micraster* zones of southern England and the Trimingham Chalk, some 5% or so contain cavities, but in other areas and levels the percentage is much lower (0.5–2%). Indeed at one locality (high Campanian near Portsmouth, SU064666), the writer examined many hundreds of flints without finding a cavity.

The aim during collection was to locate a flint *in situ* within chalk, and showing a cavity on its fractured face. When such a cavity was located, its contents were removed and a sample of chalk was taken alongside, as nearly as could be judged at the same level, to act as a control. Part of the collected material was retained and the balance was processed by standard micropalaeontological techniques. Those used involve gentle crushing if necessary to reduce the sample to small (<5 mm) pieces and its transfer to the centre of a square of finely woven (75 μm equivalent) nylon or terylene cloth, which is then gathered up into a bag. The whole is then subjected to gentle friction under a running tap. The fine fraction of the sample passes outwards through the cloth and clouds the water. As the disintegration of the sample proceeds, the water becomes less cloudy and finally runs clear, signalling the end of the operation. The aim is to reduce the bulk of undisaggregated rock to, say, less than 2% of the original with the minimum of mechanical effort. After drying, the preparation is examined to check on the success of the operation, in reducing the bulk rock and cleaning up the microfossils on the one hand, and in avoiding damage to the latter on the other. The sample is now ready for further study.

Samples of chalk prepared in this way yield a variable proportion of organic remains, prominent amongst which are foraminiferids and *Inoceramus* prisms, with lesser quantities of ostracod valves and echinoderm and bryozoan fragments, and rare brachiopods and serpulids. Samples of flint meal yield, in most cases, a similar suite of organisms but commonly include abundant sponge spicules in addition. Radiolaria, ebridians and diatoms have not been observed. Originally calcareous fossils may be more or less silicified, the silicification commonly preserving specific details but infrequently being associated with solution and/or overgrowth to such an extent that even generic characters of the fossils are masked. In a small minority of cases, notably from very irregular cavities, the only recognisable material of organic origin turned out to be sponge spicules. It is assumed that this group of cavities are fundamentally different from those with an abundant (albeit perhaps diagenetically altered) fauna, and that they result from the silicification of more or less complete sponge skeletons. Samples of flint meal yielding either sponge spicules only or a varied fauna which was not identifiable as a whole to generic level were discarded. The examination of many thousands of flints from more than a score of localities resulted in the assembly of some fifty flint meal preparations, together with their associated chalk controls, spread over the range high Turonian to low Maastrichtian.

Comparative analysis of chalk and flint meal preparations

Foraminiferids were by far the most diverse group of organisms in the preparations and a preliminary inspection revealed that, in general, preparations from flint meal did indeed contain obviously higher proportions of planktonic foraminiferids than the corresponding chalk controls. A 300-specimen count was made from the foraminiferid population, using the whole size fraction retained by a 110 mesh ($=130\ \mu$m) sieve. Much statistical work on foraminiferids concentrates on a narrower dimensional band (commonly 30–60 mesh=500–250 μm), which may be justified on the grounds that material in this band is easier to identify. However, it excludes or seriously under-represents important parts of the adult fauna (for example, the heterohelicids and buliminaceans). The wider size band has therefore been adopted in the present work as a compromise which admits a more diverse adult fauna while inescapably allowing in a proportion of juveniles and fragments of larger species which may be difficult to identify. The absence of any size restriction at the upper end is not of great importance. Few individual foraminiferal tests are retained by a 30 mesh sieve, though these might of course represent an appreciable proportion of the original biomass. The selected fraction of the sample was divided to produce a split estimated to yield about 500 specimens and this split was spread on a picking surface. In total, 300 specimens were then picked, precautions being taken at all stages to avoid, in so far as possible, any non-random procedure. A similar pick was made from the corresponding chalk control and the two picks were mounted one above another on a 100-square slide mount in systematic order to generic and, if possible, specific level. An immediate visual comparison could then be made between the two faunas, and the material was ready for statistical analysis, if required.

A preliminary analysis was carried out of the results of replicating samples and, for this purpose, material was examined from the interiors of three flints and their associated chalk controls, all taken from about the same horizon and within a few metres of each other (locality 11 below), and its results are shown in Table 6.1. The inter-sample difference observed

Table 6.1 Chalton Down, Hampshire, *Micraster cortestudinarium* zone.

	Chalk			Flint meal		
	A	B	C	A	B	C
Arenobulimina presli (Reuss)	17	31	12	8	9	3
Ataxophragmium variabile (d'Orbigny)	4	0	1	3	0	0
Gaudryina parallela (Reuss)	1	1	0	3	0	1
Gaudryina rugosa d'Orbigny	7	3	1	1	5	1
Gavelinella costata Brotzen	68	0	10	42	3	25
Gavelinella pertusa (Marsson)	3	16	1	4	2	3
Gavelinella stelligera (Marie)	8	29	38	0	9	4
Gavelinella thalmanni (Brotzen)	0	2	7	0	4	4
Globigerinelloides sp.	0	0	0	0	1	0
Globorotalites michelinianus (d'Orbigny)	61	120	136	60	59	69
Globotruncana bulloides Vogler	0	0	1	40	29	21
Globotruncana pseudolinneiana Pessagno	1	0	0	7	6	6
Gyroidinoides nitidus (Reuss)	52	53	41	17	33	19
Hagenowella elevata (d'Orbigny)	16	7	7	1	3	3
Hedbergella brittonensis L. & T.	0	0	6	50	79	86
Hedbergella sp.	0	0	0	3	5	3
Heterohelix globulosa (Ehr.)	0	0	0	7	3	2
Lenticulina rotulata (Lamk.)	10	5	3	3	3	5
Marssonella trochus (d'Orbigny)	7	3	5	3	8	10
Osangularia whitei (Brotzen)	3	3	0	2	0	0
Praebulimina reussi (Morrow)	3	0	0	9	3	2
Reussella kelleri Vasilenko	21	9	7	17	9	10
Valvulineria lenticula (Reuss)	14	15	23	17	19	16
specimens of other species	4	3	1	3	8	7
number of planktonic specimens	1	0	7	107	123	118
number of benthonic specimens	299	300	293	193	177	182
number of benthonic species	20	16	15	18	19	19
α index (benthonic species)	4.83	3.61	3.35	4.86	5.41	5.34
similarity index (%)						
Chalk	A–B 62.7, B–C 80.7, C–A 57.3					
flint meal	A–B 72.7, B–C 83.3, C–A 76.3					

*Planktonic species.

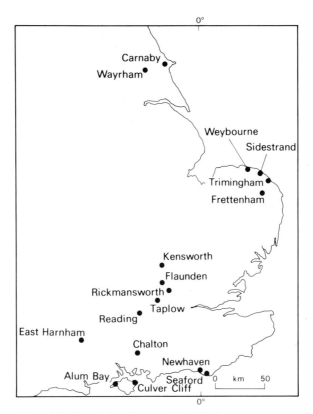

Figure 6.1 Sample localities mentioned in the text.

(57–83%) (similarity index of Murray & Wright 1974) is larger than might have been expected and notably greater than that (74–88%) reported by these authors for the London Clay of the Isle of Wight. However, the content of planktonic foraminiferids (within flint meal and chalk sample groups respectively) shows little variation.

Twelve sample-pairs (of flint meal and associated chalk) were then chosen from a wide range of horizons to investigate variation with time. The localities of the samples are listed below (see also Figure 6.1) and their approximate ages are indicated in Figure 6.2.

(1) Trimingham, Norfolk, TG301379. Lunata Chalk, Bed 4.
(2) Sidestrand, Norfolk, western bluff, TG258405.
(3) Frettenham, Norfolk, TG245173. Probably basal Paramoudra Chalk.
(4) Weybourne, Norfolk, TG112438. Weybourne Chalk.

(For details of the above localities, see Peake & Hancock 1961)

(5) Alum Bay, Isle of Wight, SZ305852. Highest Chalk exposed.
(6) East Harnham, Wiltshire, SU151284. Top of *quadrata* zone.

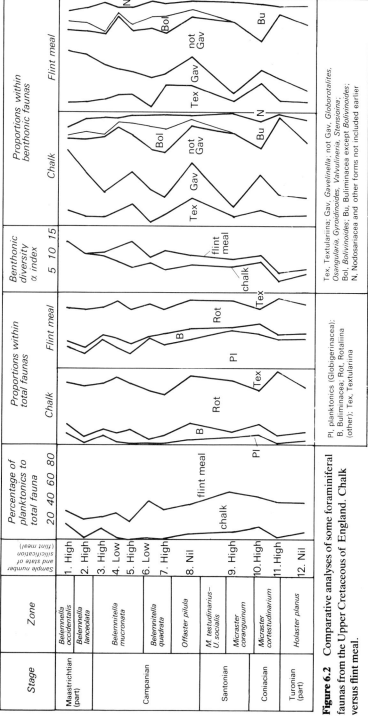

Figure 6.2 Comparative analyses of some foraminiferal faunas from the Upper Cretaceous of England. Chalk versus flint meal.

(7) Culver Cliff, Isle of Wight, SZ639856. Twenty-five metres below top of the *quadrata* zone.
(8) Newhaven, East Sussex, TQ447000. Ten metres above base of cliff.
(9) Rickmansworth, Hertfordshire, TQ065936. Road cutting, two metres below the base of the Reading Beds.
(10) Flaunden, Hertfordshire, TL005008. Small pit.
(11) Chalton Down, Hampshire, SU717172. Road cutting.
(12) Kensworth, Bedfordshire, TL020195. Cement works, two metres above top of Chalk Rock.

Differential loss of planktonic foraminiferids

Table 6.2 and Figure 6.2 present the observed faunal associations of foraminiferids, subdivided into major taxonomic groupings. They demonstrate the consistent presence of abundant planktonic foraminiferids in the flint meal samples and their relative rarity, or absence, in samples of the associated chalk. The pattern of occurrences of benthonic taxa is seen to be very varied but in this case no differences are immediately apparent between the two sets of samples.

The difference between the proportions of planktonic and benthonic forms in the sample-pairs could be due to (a) a difference in the original associations fossilised within the flint meal and chalk, or (b) subsequent selective preservation (or to some combination of the two). A possible mechanism for proposal (a) might be selective trapping of planktonic foraminiferids in burrowing structures or within sponges, both of which are known to be associated with the formation of flint (cf. Bromley 1967, Håkansson *et al.* 1974). This proposal fails to account for the cases where planktonic foraminiferids are abundant within flints, but absent from the associated chalk, nor is it easily reconciled with the lack of evidence for strong segregation within the benthonic taxa. No other machinery for segregation in association with the formation of flint has occurred to the author, so proposal (a) is rejected, and the phenomenon is ascribed to subsequent selective preservation. If this is the case, it is clear that planktonic foraminiferids have been selectively lost from the chalk controls. The alternative hypothesis, that there has been massive selective loss of benthonic taxa from the flint meal (which has failed to disturb significantly the proportions of the various benthonic taxa within the original assemblages), is clearly untenable on the figures presented.

Evidence was sought for selective loss within planktonic taxa as a group. The number of samples available to provide comparisons was too small to provide reliable indications, but *Globotruncana* appeared to survive rather better than *Hedbergella*, whilst *Globigerinelloides* and *Heterohelix* were particularly subject to loss. This result agrees with an estimate of the order of robustness of the tests of the genera concerned.

Differential loss of benthonic foraminiferids

If selective (and in some cases massive) loss of planktonic foraminiferids has occurred, is it possible that there has been some loss of benthonic forms also? To be detected by statistical means, such a loss would of course need to be to some extent selective. The total benthonic faunas picked were analysed by taxa to compare percentage abundances in the chalk controls with those of the flint meal (assumed to be less affected by loss). It was found that there was indeed a pattern in that, at one extreme, *Stensioina* was some 40% more common in the benthonic fauna of the chalk population as a whole than in the flint meal, while the reverse was true for certain members of the Textulariina. An order of relative survival was established as follows: *Stensioina* – *Gyroidinoides* – *Bolivinoides* – *Arenobulimina, Hagenowella* and other spirally coiled Textulariina – *Gavelinella* and related genera – *Globorotalites* – nodosariaceans – other buliminaceans – other Textulariina.

Although insufficient individuals were counted to provide statistically reliable evidence for this pattern, and especially for its middle part, it is clear that a pattern exists and that there has been appreciable differential loss of benthonic forms from the chalk samples also. It may be noted in passing that the test of *Stensioina* resembles that of *Globotruncana*, the most successful survivor of the planktonic genera, in that it is keeled and its sutures are limbate; both structures no doubt providing unusual strength to the test. Differential loss has reduced the species diversity of the benthonic population of the chalk in relation to that of the associated flint meal. This is shown by the records of the α index (Fisher *et al.* 1943) in each case, presented in Figure 6.1. The mean difference between pairs of diversity indices is 2.3, a figure which is highly significant statistically in the light of the fact that the standard deviation of individual records averages less than half this figure. This mean difference represents an expected loss of four species in a pick of 300 specimens or of about twelve species in a general search of 3000 specimens.

Those who have examined large numbers of microfossil preparations of Chalk will probably recognise the pattern which now emerges. This is, at one extreme, of samples with a diverse and well preserved benthonic fauna and, commonly, abundant planktonics and, at the other, of samples with a poorly preserved fauna of rugged individuals of a limited number of benthonic species only.

Other evidence in relation to selective preservation

The phosphatic chalk of Taplow (Buckinghamshire)

The sedimentary succession, macrofauna and age of this unit, exposed only in a pit in a private garden (SU905819), were described by White and

Table 6.2 Composition of foraminiferal faunas picked. The top line of figures against each locality relates to chalk; the second line relates to the associated flint meal (where * indicates marl, † indicates phosphatised forms, and ‡ indicates Palaeogene); the third line relates to either Chalk Rock (⁺) or Palaeogene (‡); the fourth line indicates a cavity in Chalk Rock (§).

Locality	Zone	Textularina			Nodosariacea	Buliminacea			Rotaliforms					Globigerinacea				Various others	Total picked	Total benthonics picked	Number of benthonic species	α index (benthonic species)
		Arenobulimina, Ataxophragmium	Marssonella	Others		Bolivinoides	Reussella	Others	Globorotalites	Osangularia	Gyroidinoides, Valvulineria	Stensioina	Gavelinella and allied forms	Globotruncana	Heterohelix	Globigerinelloides	Hedbergella					
Trimingham	occidentalis	21	6	8	30	3	2	16	0	10	16	2	133	0	0	0	48	5	300	252	41	13.89
		13	1	2	21	15	6	14	0	15	20	7	99	0	0	8	75	4	300	217	38	13.33
Sidestrand	lanceolata	15	2	8	17	18	14	20	0	28	12	47	114	1	0	2	1	2	300	297	35	10.37
		11	2	7	17	18	4	28	0	25	15	6	103	1	0	17	43	3	300	239	33	10.38
Frettenham	mucronata	4	1	15	10	1	3	42	46	5	24	54	60	0	0	13	21	1	300	266	32	9.51
		5	2	14	7	4	10	39	6	4	13	24	55	0	17	34	66	0	300	183	31	10.71
Weybourne	mucronata	23	3	13	10	11	4	7	11	37	36	54	81	1	0	1	4	4	300	294	30	8.36
		15	4	14	13	5	2	11	10	37	16	36	50	5	6	32	41	3	300	216	30	12.31
Alum Bay		41	1	17	7	35	0	22	20	1	39	27	90	0	0	0	0	0	300	298	26	6.84
		19	1	20	19	15	6	29	22	14	30	22	49	9	11	20	13	0	300	247	40	13.52
East Harnham	quadrata	11	0	3	7	58	16	14	14	10	18	65	81	2	0	0	0	1	300	298	28	7.57
		9	0	2	3	18	15	10	19	10	7	42	44	22	5	79	14	1	300	180	28	9.29
Culver Cliff	quadrata	25	6	6	5	61	3	34	86	12	17	1	43	0	0	0	1	0	300	299	24	6.15
		18	4	27	10	30	1	22	62	3	10	0	21	47	15	16	14	0	300	208	34	11.45
Newhaven	pilula	33	10	24	6	13	1	8	53	12	23	7	90	15	2	0	2	1	300	281	27	7.36
		21	5	19	3	10	2	14	34	11	13	8	49	18	59	0	33	1	300	190	30	10.02

Location	Zone																								
Rickmansworth	coranguinum	14	3	7	4	0	19	24	20	44	34	84	24	9	7	0	7	0	300	277	26	7.03
		5	2	7	3	0	14	15	20	29	13	31	11	42	28	0	80	0	300	150	26	9.08
Flaunden		17	5	16	17	0	21	44	36	0	50	0	55	11	5	0	22	1	300	262	29	8.35
	cortestudinarium	12	4	24	5	0	23	32	10	0	32	0	22	37	25	0	72	2	300	166	29	10.17
Chalton (C)		13	5	8	3	0	7	0	136	0	64	0	56	1	0	0	6	1	300	293	15	3.35
		3	10	5	5	0	10	2	69	0	35	0	36	27	2	0	89	7	300	182	19	5.34
Kensworth	planus	15	7	6	4	0	2	54	86	8	43	0	42	5	16	0	0	12	300	279	20	4.93
		8	7	2	2	0	5	27	46	14	46	0	17	15	90	0	13	8	300	182	21	6.14
		†7	0	1	0	0	2	18	35	0	17	0	33	18	0	0	44	0	175	113	14	4.21
		§4	1	0	4	0	9	16	28	0	18	0	11	18	19	16	156	0	300	91	18	6.72
									(comparison of normal with phosphatised specimens)													
Taplow	testudinarius	6	4	3	4	1	17	19	16	41	25	46	100	8	5	3	0	1	300	283	28	7.72
		†1	4	25	1	1	6	5	16	8	6	29	51	26	74	20	26	1	300	154	29	10.50
									(comparison of chalk with marl)													
Seaford	pilula	11	3	10	3	0	24	0	22	24	10	33	57	3	0	0	1	0	201	197	16	4.11
		*59	1	20	1	0	0	0	6	22	8	25	63	0	0	0	0	0	205	205	12	2.78
Seaford	coranguinum	11	2	9	11	0	1	0	190	0	16	0	42	12	0	0	3	0	297	282	12	2.54
		*12	0	3	2	0	0	0	11	0	2	0	5	0	0	0	0	0	35	35	6	2.08
Kensworth	lata	18	43	24	13	0	0	0	16	0	20	0	104	140	0	0	124	1	503	239	17	4.21
		*18	4	2	1	0	0	0	3	0	0	0	16	19	0	0	1	0	64	44	9	3.42
									(Yorkshire chalks)													
Carnaby	pilula	3	10	5	5	3	1	22	4	31	12	46	19	60	25	0	53	1	300	162	26	8.75
Wayrham	cortestudinarium	4	9	6	5	0	2	6	23	3	14	9	24	130	16	0	45	4	300	109	22	8.31
									(comparison of chalk with Palaeogene)													
Pincent's kiln, Reading	coranguinum	18	1	3	7	0	10	9	53	69	53	40	22	3	1	0	9	2	300	287	19	4.57
		‡6	0	2	1	0	0	14	11	18	3	6	14	2	10	0	10	3	100	78	19	8.00
		‡6	0	7	5	0	14	58	50	5	21	7	10	39	38	0	39	1	300	184	25	7.81

Treacher (1905). Chapman (1892) listed its rich foraminiferal fauna of ninety-eight species and recorded that planktonic species were abundant. Cayeux (1897), in a study of the Chalk of France, had observed that, as a rule, its contained foraminiferids were rather rare and of few species. At many levels they showed evidence of mechanical and chemical damage, part of which he thought to have occurred prior to burial (Cayeux 1897, p. 302), and part to have occurred in place (Cayeux 1897, pp. 250, 323). He noted that in certain phosphatic chalks, on the other hand (Cayeux 1897, pp. 241, 322, 323), foraminiferids were abundant, in considerable variety and mostly complete, although in associated unphosphatised levels they were rare and damaged. He concluded that 'La craie ne montre donc sa composition organique originelle que là où elle est phosphatée'. White and Treacher (1905, pp. 484, 485) recognised the close resemblance of the Taplow chalk to the phosphatic chalks of the Senonian of northern France and commented on 'the marked difference presented by the phosphatised and the calcareous [unphosphatised] foraminifera', the former being better preserved, in greater variety and including 'in abundance, certain families (notably the Textulariidae [in which they included the planktonic genus *Heterohelix*] and the Globigerinidae) which are rare in the calcareous examples'.

A sample from the highest part of the Taplow succession (bed D or E of White & Treacher 1905, *testudinarius* zone) was examined for foraminiferids. These were found to be abundant and about two-thirds of the individuals present were apparently phosphatised, being pale brown in colour and having a shiny surface. Most of the remainder of the specimens were white, with the matt surface typical of specimens from ordinary chalk. These appeared to be neither infilled nor phosphatised as they dissolved in acetic acid, leaving almost no trace behind. A few specimens presented an intermediate appearance, no doubt due to partial phosphatisation.

Specimens were picked at random from the >110 mesh fraction to a total of 300 of each of the shiny brown and matt white forms, specimens not clearly assignable to either group being disregarded (see Table 6.2 for details). It was found that planktonic species were much more abundant amongst the brown specimens (49% of the total, compared with 6% in the white specimens). No significant pattern was observed amongst the benthonic species except, possibly, in the Textulariina, in which *Textularia baudouiniana* d'Orbigny and ?*Tritaxia* sp. were much commoner in the phosphatised group. This pattern occurs also in the sample from Flaunden (the species mentioned being much commoner in the flint meal) but is not apparent in any of the other preparations studied, so these observations may be the result of chance. It was noticed that the pick of white specimens contained more large individuals than that of brown ones, in spite of the fact that the same overall size-fraction and the same split were picked in each case. This is attributed to a difference in density between the two groups, brought out by sorting by bottom-currents in the Chalk sea.

The α indices for the benthonic species are (brown) 10.50 and (white)

7.72. It will be seen that there is a close similarity between the results of investigating flint meal and those obtained on the Taplow sample. Cayeux (1897) exaggerated somewhat in suggesting that only phosphatised chalks displayed the original fossil content of the Chalk, but no doubt many such chalks, and some samples of flint meal, are much more representative in this respect than the average sample of chalk from the British Upper Cretaceous.

Yorkshire chalks and the Chalk Rock

The relatively hard chalks of Yorkshire are known to yield much higher proportions in general of planktonic foraminiferids than the softer chalks of the south of England and conclusions of a palaeoecological or palaeogeographical nature have been drawn from this fact. Statistical investigation of this phenomenon is, however, greatly hampered by the difficulty or impracticability of breaking down such hard chalks without destroying their contained foraminiferids. Two samples with sufficiently well preserved faunas were available to the author. They are from Wayrham (SE836567) (probably *cortestudinarium* zone) and Carnaby (TA144659) (*pilula* zone). Their faunal compositions were found to resemble those of flint meal from corresponding levels in southern England, planktonic percentages and α indices being 63.7% and 8.31 and 46.0% and 8.75 respectively.

Amongst sequences in the south of England the Chalk Rock is an especially indurated unit, and this induration was clearly of an early date because locally the unit contains borings filled with chalk which is not indurated at all (Bromley 1967), and it is preceded and immediately succeeded by soft chalk. The Chalk Rock of Kensworth (locality 12, already discussed) also yields relatively high percentages of planktonic foraminiferids (35%) as compared with the flint meal and chalk taken 2 m higher in the same section (with figures of 39.3% and 7.0% respectively). The soft material from borings in the Chalk Rock at this locality yielded no less than 69.7% of planktonics, the highest figure recorded in the present investigations, and a sample of soft chalk taken from the *lata* zone below the Chalk Rock yielded a figure of 52.5% (see subsection entitled 'Marl bands' below). The tentative conclusion that may be drawn from this and the preceding paragraph is that the survival conditions of the particular hard chalks discussed have something in common with those of the flint meal, and are quite different from those undergone by the hard Upper Chalk of the Isle of Wight succession, in which proportions of planktonics are typically very low.

Interiors of large fossils

It was speculated that fossils with sheltered interiors might protect the more vulnerable microfossils from destruction as happens, for instance, with pteropods trapped within the interiors of gastropod or paired bivalve shells

(Curry 1965). Material from the interior of several echinoids was therefore compared with control material from their exterior. No differences were observed between the foraminiferal populations in the two cases, however, indicating that the special conditions associated with the flint meal were not present in the case of the echinoids. In one case only was any difference observed, an *Echinocorys* from Trimingham yielded a great abundance of calcispheres, no doubt because its interior formed a sediment trap for this minute skeleton.

Assemblages of derived fossils

It is known that derived fossils may occur in a state of preservation as good as or better than that of the original rock; an example being the Campanian foraminiferids in the Thanetian Sables de Châlons-sur-Vesle, near Rheims, France. A similar occurrence, at the base of the Reading Beds at Pincent's Kiln, near Reading, Berkshire (SU650719), was investigated statistically. The lowermost unit of the Reading Beds there is a glauconitic sandy clay about 1 m thick, with *Ostrea bellovacina* and sharks' teeth, which overlies *coranguinum* zone Chalk. It yields foraminiferids derived from the Chalk, together with very rare indigenous specimens. A pick of 300 was carried out on the >110 mesh fraction of the chalk and on levitated material from the Reading Beds. A check pick was carried out on the corresponding sand fraction of the latter (which was very tedious because of the low proportion of foraminiferids to sand grains) and isolated 100 specimens before being halted. Material from the two picks from the Reading Beds presented characters reminiscent of the flint meal samples discussed above. Proportions of planktonics were 39 and 22% respectively and α indices for the benthonics were 8.00 and 7.81. Corresponding figures for the underlying chalk were 4.3% and 4.57. While the results of this single enquiry are clearly not definitive, and the *in situ* and *remanié* faunas compared are obviously not from precisely the same horizon, the result (that the reworked fauna is more diverse and contains a much higher proportion of planktonics than that of the source rock) appears clear cut and is thought to be important for a general discussion. This investigation cannot be repeated elsewhere at the moment because no other English locality presents the right combination of conditions; a Palaeogene basement bed, which is accessible, has a large derived fauna and rests directly on the Chalk.

Marl bands

In his earliest investigations into Chalk microfaunas, the author collected preferentially from marl bands, rather than the associated chalk, on the hypothesis that their higher clay content would help to protect the contained microfossils from solution, and that preparation would be easier and less likely to destroy the fauna. He was disappointed to find that intercalated marl bands contained relatively poor faunas, sometimes enclosing little but *Inoceramus* prisms and rare agglutinating foraminiferids, and so abandoned that procedure, subsequently collecting preferentially from chalk.

Three paired analyses have been carried out to quantify this contrast between marl bands and the associated chalk, and the results of these are reported in Tables 6.2 and 6.3. It should be noted that in this case the picks related to a standard weight of rock and the fraction counted was that retained by a 70 mesh ($=210\,\mu$m) sieve. Because a different size-fraction was used, the α indices derived are not directly comparable with others quoted herein. The oldest pair of samples was from Kensworth (locality 12), at 6 m below the Chalk Rock (which is about 2 m thick). The others were from Seaford, East Sussex; from the base of the *coranguinum* zone (TV507973) and the lower part of the *pilula* zone (TV490980), 21 m below the top of the local succession. The samples from the *lata* and *pilula* zones were from continuous bands several centimetres thick, that from the *coranguinum* zone formed wispy surrounds to nodules of normal chalk (cf. Hancock 1976, p. 515, '*flaserkalk*').

Table 6.3 demonstrates clearly that selective factors are present in these cases also, but in the reverse sense, the marls showing impoverishment in relation to the chalks. Once more the planktonic foraminiferids are reduced, but now the Textulariina survive markedly better in the marls than the benthonic members of the Rotaliina. In this respect the pattern of selective loss is different from that shown by the chalk in relation to the flint meal, which suggests that a different mechanism operated in the two cases. Assemblages of foraminiferal tests, both Recent and fossil, always include some broken specimens. However, the proportion of damaged forms and the extent of damage to individual tests is rather high in most samples from the Chalk (cf. Cayeux 1897) and is particularly high in the three marls examined. Some of this damage clearly points to solution, as when the whole

Table 6.3 Composition of foraminiferal faunas: chalk and marl.

	Original sample and weight (g)	Percentage insoluble in HCl	Total picked; benthonics in brackets	Species retained by 70 mesh sieve (percentage present)				
				Textulariina	*Nodosariacea*	*Buliminacea*	*Globigerinacea*	*Other Rotaliina*
Seaford, *pilula* zone, TQ490980	chalk 2.5	1.7	201 (197)	11.9	1.5	11.9	2.0	72.7
	marl 2.5	7.7	205 (205)	39.0	0.5	0	0	60.5
Seaford, base of *coranguinum* zone, TQ507973	chalk 10	not known	297 (282)	7.4	3.7	0.3	5.1	83.5
	marl 10		35 (35)	42.9	5.7	0	0	51.4
Kensworth, top of *lata* zone, 6 m below Chalk Rock, TL020195	chalk 10	8.1	503 (239)	15.9	2.6	0	52.5	28.9
	marl 10	36.6	64 (44)	37.5	1.6	0	31.2	29.7

of one side of a planispiral form is missing or when a strongly keeled *Globotruncana* is reduced to little but the keels.

The proportion of rock insoluble in dilute HCl is recorded in the case of two pairs of samples, in both of which the marls are seen to be enriched in insolubles by a factor of about four. This could be due to the influx of a higher than usual proportion of the clay fraction (by the mechanism envisaged by Jefferies (1963), for example) or by some reduction in the fraction of $CaCO_3$, probably by solution as is more widely believed (Scholle 1974, Hancock 1976) today. Of these two possibilities the pattern of the foraminiferids strongly suggests the latter hypothesis. If there were an influx of, say, 20% of clay, the abundance of foraminiferids might be somewhat reduced, but no change in relative proportions would be expected. On the other hand, a depletion of $CaCO_3$ sufficient to push up the insolubles to the extent demonstrated would be of the order of 75%, and the foraminiferid remains could be expected to disappear massively and selectively, as the analyses indicate. Indeed it may be no coincidence that the Textulariina in the marl samples are enhanced by about the same factor as the insolubles.

Discussion

It seems that no single explanation can account in detail for the various examples of selective preservation which have been recorded here. The overriding cause in each case seems to be solution but the phenomenon described in relation to the marls differs from those of the flint meal, the phosphatic chalk and the derived fossils in (at least) two respects. First the pattern of selective dissolution is different and secondly it seems that the timing must have been different also.

Timing of dissolution

The case of the base of the Reading Beds indicates that there was selective loss from the underlying Chalk *after* the Reading Bottom Bed had been deposited, that is, some 15–20 Ma after the Chalk itself was deposited. The case of the flint meal also involves a relatively long period of time, it seems. Selective loss occurred *after* the flint meal was enclosed in the flint. In both cases, of course, earlier selective loss could have occurred between the original time of burial in the Chalk and incorporation in the Reading Beds and flint respectively. The precise mode and depth of genesis of flint in rock sequences are not known but it has long been observed that 'flints have never served as surfaces of attachment of faunas' (Hancock 1976) and that they never occur as accumulations of drifted pebbles (rounded or not) on Chalk hardgrounds. It is now recognised that hardgrounds are common in the Chalk and it would be surprising if some of these omission surfaces were not associated with the contemporaneous removal of many metres of rock. With these facts in mind one might propose that flint does not form under less than (say) 10 m of sediment, equivalent to about 0.5 Ma, to provide an estimate for the time of sealing of the flint meal. This time could be substantially reduced if the flint resulted, for instance, from the accumulation of silica gel near the surface of the sediment, which could provide the necessary seal yet could not at that stage be reworked without destruction. The two cases discussed in this paragraph thus involve a long, and perhaps extremely long, time-scale for the dissolution observed. It will be remembered that these cases are associated with a selective removal of planktonic foraminiferids, but no well marked selective removals within benthonic forms. The lapse of time to the sealing date suggested reflects a situation during which abundant planktonics still survived. The transition from that situation to the one demonstrated in the associated chalks would, it seems, have taken a period several times longer. Preferential removal of planktonic foraminiferids is well documented in DSDP cores, as, for instance, by Schlanger and Douglas (1974, Fig. 2). The rate at which it occurs there cannot be deduced precisely from their curves, and was in any case probably very variable. However, these curves do suggest that the extent of preferential removal displayed by the contrast between flint meal and chalk may indicate a lapse of millions of years.

Selective loss in marls appears to have a much shorter time scale. Before discussing this it is necessary to make clear that the alternations of more or less marly units so common in parts of the Lower Chalk of England are not considered by the author to be solution-controlled, but to be due to differences in the rate of influx of insoluble material in relation to carbonate accumulation. A significant difference between these and the marl layers discussed below is the essentially gradual nature of the transitions between more or less marly units, by contrast with the typically abrupt transitions of the latter. To the author's knowledge, however, no analyses have been carried out on the foraminiferid populations of the Lower Chalk with the specific purpose of identifying solution patterns. Certainly there is no evidence for important solution effects within the Plenus Marl, to judge from the analyses published by Jefferies (1962). In that paper, amongst sixty-five levels in a total of six localities, the mean planktonic percentage recorded is about 50%, and no record is less than 20%.

Marl occurs in the Middle and Upper Chalk in (at least) three different sedimentary styles. It may be present in well defined bands, typically of centimetric thickness and of considerable extent, which have flat and rather sharp lower and upper boundaries. The samples examined from the *lata* and *pilula* zones are of this type. Alternatively it may occur as wavy sheets filling the interstices between layers of lenticular to rounded kernels of normal chalk (*'flaserkalk'*) or, especially in Yorkshire, as centimetric bands of marl with irregular lower and upper surfaces, sparsely intercalated between massive beds of very indurated but otherwise normal chalk. Hancock (1976) postulates early post-depositional solution for the *flaserkalk* and contrasts it with the late-stage solution proposed by him for the irregular marls of Yorkshire. The present author agrees with these propositions.

The marls examined from the *lata* and *pilula* zones, with their regular lower and upper surfaces, included within sequences of soft chalk, seem not to fit into the above pattern. Solution in these appears to have taken place at or very near to the contemporary seafloor. A hypothesis which involved dissolution at depth would have to explain how it was possible to remove 75% of the original bulk of the marl band beneath a moderate thickness of unaffected, porous and, at that stage, probably unconsolidated sediment over a wide area, and to do this in such a way that the band of dissolution was of uniform thickness and had flat lower and upper surfaces. No mechanism capable of yielding the above result has been envisaged, and so the hypothesis of dissolution at depth is rejected for such marl bands. The presence of a typical (though selectively depleted) fauna indicates that the overlying bottom water was not hostile, and the obvious conclusion must be that such marl bands are the result of destructive diagenesis in the top few centimetres of the accumulating sediment (cf. Alexandersson 1978), no doubt due to acid conditions associated with unusually large accumulations of decaying organisms. This dissolution would, in a geological sense, be effectively contemporaneous with the deposition of the marls concerned. It is concluded that, so far as marls within the Upper Chalk of southern England are concerned, the evidence points to early or very early and rapid selective destruction which contrasts with that discussed in the first part of this section both in timing and results.

Selective destruction of planktonic foraminiferids

It is widely accepted that planktonic foraminiferids may (in some cases) be destroyed selectively in relation to benthonics within sedimentary rocks, and this has been demonstrated to occur also within samples suspended at great depths in the ocean. The generally accepted reason for this is that 'benthonic species usually have thicker walls and fewer pores than planktonic Foraminifera, which tend to make the former more resistant to solution and, in some cases, to mechanical damage' (Schlanger & Douglas 1974). Because of this structural difference the preferential removal of planktonic foraminiferids is believed to follow a path of more rapid corrosion, followed by fragmentation and final complete dissolution over a period during which benthonic forms in general suffer relatively little loss or damage.

However, some planktonic forms have tests which are notably thicker than those of some comparable benthonic species, and some have extremely fine pores as reference to the details (including many cross-sections) produced by Hofker (e.g. Hofker 1957, 1968) confirms. The mean thickness of the test of planktonic species is perhaps one-half to one-third of that of benthonic species of the same size and such a difference seems inadequate to explain the extent of differential survival commonly encountered. Some additional and as yet unidentified factor may thus be suspected.

It might be supposed that this relative proneness to mechanical damage could account for the rarity or absence of planktonics in samples prepared from the Upper Chalk of the Isle of Wight, for example. Some damage certainly does occur during the processing of such hard rocks, but this is normally recognisable by the presence of fragments of fossils of especially vulnerable shape, e.g. thin ostracod valves and slender nodosariids. Thus the common absence even of fragments of the tests of planktonic foraminiferids in samples without complete specimens indicates that this damage is not an important factor. A countervailing phenomenon in harder calcareous rocks is that hollow fossils tend to become filled with solid calcite, which may protect the most fragile test. This factor operates in the samples inspected by the author from Yorkshire, and in many samples from the Lower Chalk. Not only does the infilling protect against damage in sample preparation, but also, it seems, against the effects of selective dissolution, as preparations from the Lower Chalk, like those mentioned from Yorkshire, typically contain high proportions of planktonics.

Selective survival of the Textulariina

The selective survival in some marls of members of the Textulariina in relation to the benthonic Rotaliina reported here is not readily explained. One observation may be relevant. In some cases when samples of very hard chalk are dissolved in dilute acid agglutinating foraminiferids (especially of the genus *Arenobulimina*) may be found in the residue. Such relics are typically very fragile and mostly fragmentary, but have survived in this very harsh environment when any Rotaliina have disappeared completely. The effect is apparently not due to slight silicification, which would probably have affected the Rotaliina also. It seems that some groups of agglutinating foraminiferids, at least, may have properties which ensure the preferential survival of their tests in an acidic environment.

Depth of the Chalk sea

A summary of recent opinions is provided by Hancock (1976), who suggests that a figure in the range 100–600 m is the most probable for the depth of the Chalk sea. This range of depths corresponds with that found by Grimsdale and Van Morkhoven (1955) to contain percentages of planktonic foraminiferids predominantly in the range 30–50% in Recent bottom deposits of the Gulf of Mexico (higher percentages, on average, were reported at these depths on the US Atlantic Slope). Contributions by the planktonic foraminiferids to the question of the depth of the Chalk sea have been controversial and conflicting because so many samples have yielded few or none of them, so suggesting very shallow depths and thus disagreeing with the other evidence. A recent example is provided by the work of Hart and Bailey (1979), which recorded percentages of planktonics ranging from nil to a maximum of 20% (with an average of less than 4%) in a series of seventy-four samples from the Coniacian and early Santonian of south-eastern England. Corresponding late Cenomanian and Turonian samples yielded much higher figures, with a mean of about 40%. The authors were using the proportion of keeled to unkeeled planktonic species as an indicator of seawater depth and found difficulty in reconciling the very sharp reduction in water depth indicated by the change in planktonic percentage at the Turonian–Coniacian boundary with the lack of obvious change in the keeled:unkeeled ratio at that level. However, many of the results for planktonic percentages in the Upper Chalk quoted by Barr (1962) lie within the range 30–50% and some which are notably lower (those from the *testudinarius* zone, for example) were probably based on chalk, not on flint meal.

The planktonic percentages reported here in relation to flint meal, phosphatised chalk faunas, the Chalk Rock and the Yorkshire chalks, lie in the range 18–70%. However, these figures are likely to understate the proportion of planktonics in the original sediments because of the probability of undetected removal before final fossilisation. The evidence from the planktonic foraminiferids is thus now consistent with that provided on depth by other groups of organisms, and by sedimentological and palaeogeographical considerations. The new figures for the Upper Chalk of the south of England now come in line with those for Yorkshire, so that the hypothesis (based on apparently much higher proportions of planktonics in the latter) that the Yorkshire Chalk was deposited in deeper water than that of southern England is no longer supported.

The figures now presented, joined to those published by Barr (1962), Jefferies (1962), Carter and Hart (1977) and Hart and Bailey (1979) demonstrate high planktonic percentages at least from mid-Cenomanian times, with some fall-off from the end of the Santonian onwards; a fall-off which hints at a gentle shallowing of the sea from that time.

The pattern of abundance of keeled to non-keeled trochospiral planktonics (the former held to occur preferentially in deeper water) which was noted in the present work provides some support for this. Though very variable, the percentage of keeled forms averages about 30% up to the lower part of the *mucronata* zone, after which it falls to less than 10%. In addition there is no indication in either case of any shallowing in the late Turonian. In these two respects the evidence of the planktonic foraminiferids conflicts with the pattern of transgressions and regressions presented by Hancock (1976, p. 512).

Interpretative use of fossil foraminiferid assemblages

In addition to their general use in stratigraphical correlation, widespread use has been made of foraminiferid assemblages in making determinations of salinity, water depth and other palaeoenvironmental factors. However, erroneous conclusions have been drawn from them in a number of cases because of a failure to recognise that the assemblages examined were not representative of the original fauna because of selective preservation. Examples of this latter include the preferential loss of robertinaceans and miliolaceans in relation to other groups, the preferential preservation of Textulariina in some non-calcareous rocks and the preferential loss of uncemented Textulariina in most rocks.

Preferential loss of planktonic foraminiferids such as is documented here is well known amongst Recent deep-water deposits and has been identified in ocean cores (e.g. Berger 1967, 1973; Schlanger & Douglas 1974). However, to the author's knowledge, it has not previously been proved to occur amongst rock sequences on land, though its probable presence may have been suspected on occasion. The new discovery thus adds one more hazard to the problems associated with making palaeoecological deductions from foraminiferal assemblages and emphasises the exceptional importance of considering the possibility of some kind of selective preservation in any samples used in such work.

Conclusions

(a) There has been heavy selective removal of the tests of planktonic foraminiferids from much of the Upper Chalk of southern England.
(b) There is some evidence for selective removal of benthonic foraminiferids also.
(c) Some samples of marl give evidence of even more severe selective removal, favouring the survival of agglutinating species. This probably took place in the immediate subsurface, and under acid conditions.
(d) The time-span of the removal process is very varied; in some cases it was almost contemporaneous, in others it took tens or hundreds of thousands of years. In one case it was still in progress after 15 Ma or more.
(e) Fossils preserved within flint meal or by phosphatisation are believed to represent fairly accurately the original death assemblage.
(f) Revised figures for the proportion of planktonic foraminiferids to the total population now reconcile depth indications (in the range 100–600 m) to those suggested by other data for the depth of the Chalk sea.
(g) Before foraminiferal assemblages are used for environmental analysis the possibility of selective destruction of the original fauna should be examined carefully.

Acknowledgements

The discerning reader will realise that this work owes much to the masterly synthesis by Hancock (1976), which nominally deals with the petrology of the English Chalk, but in fact ranges over many associated subjects in addition. I have a debt of gratitude to Jake Hancock in many other respects, not least in relation to his guidance in the field and in the exchange of ideas and opinions in the course of many conversations. I would also like to give thanks to those of my colleagues, past and present, at University College with whom I have discussed ideas contained herein, and in particular to Tom Barnard, who has maintained a life-long interest in the Chalk, and whose work is honoured in the present volume. Finally, I am grateful to Malcolm Hart for assistance with problems of nomenclature.

References

Alexandersson, E. T. 1978. Destructive diagenesis of carbonate sediments in the eastern Skagerrak, North Sea. *Geology* **6**, 324–7.
Barr, F. T. 1962. Upper Cretaceous planktonic foraminifera from the Isle of Wight, England. *Palaeontology* **4**, 552–80.
Barr, F. T. 1966. The foraminiferal genus *Bolivinoides* from the Upper Cretaceous of the British Isles. *Palaeontology* **9**, 220–43.
Berger, W. H. 1967. Foraminiferal ooze: solution at depths. *Science, N.Y.* **156**, 383–5.
Berger, W. H. 1973. Deep sea carbonates. Pleistocene dissolution cycles. *J. Foramin. Res.* **3**, 187–95.
Bromley, R. G. 1967. Some observations on burrows of thalassinidean Crustacea in chalk hardgrounds. *Q. J. Geol Soc. Lond.* **123**, 157–82.
Carter, D. J. and M. B. Hart 1977. Aspects of mid-Cretaceous stratigraphical micropalaeontology. *Bull. Br. Mus. Nat. Hist. (Geol.)* **29**, 1–135.
Cayeux, L. 1897. Contribution à l'étude micrographique des terrains sédimentaires. Mém. Soc. Géol. Nord IV (2).
Chapman, F. 1892. Microzoa from the phosphatic chalk of Taplow. *Q. J. Geol Soc. Lond.* **48**, 514-18.
Curry, D. 1965. The English Palaeogene pteropods. *Proc. Malac. Soc. Lond.* **36**, 357–71.
Eley, H. 1859. *Geology in the garden, or the fossils in the flint pebbles.* London: Bell & Daldy.
Fisher, R. A., A. S. Corbet and C. B. Williams 1943. The relation between the number of species and the number of individuals in a random sample of an animal population. *J. Anim. Ecol.* **12**, 42–58.
Grimsdale, T. F. and F. P. C. M. Van Morkhoven 1955. The relation between pelagic and benthonic foraminifera as a means of estimating depth of deposition of sedimentary rocks. *Proc. 4th World Petrol Congr.* Sect. 1/D, Paper 4, 473–91.
Håkansson, E., R. Bromley and K. Perch-Nielsen 1974. Maastrichtian chalk of north-west Europe – a pelagic shelf sediment. In *Pelagic sediments on land and under the sea*, K. J. Hsü & H. C. Jenkyns (eds), Sp. Publ. Int. Assoc. Sediment. 1, 211–33.
Hancock, J. M. 1976. The petrology of the Chalk. *Proc. Geol. Assoc.* **86**, 499–535.
Hart, M. B. and H. W. Bailey 1979. The distribution of planktonic Foraminiferida in the mid-Cretaceous of NW Europe. In *Aspekte der Kreide Europas*. IUGS Series A, no. 6, 527–42. Stuttgart: IUGS.
Hofker, J. 1957. Foraminiferen der Oberkreide von Nordwestdeutschland und Holland. *Beih. Geol. Jb.* **27**, 1–464.

Hofker, J. 1968. Studies of foraminifera, Part 1, General problems. *Natuurhist. Genoot. Limburg*, ser. 18, 5–135.

Jefferies, R. P. S. 1962. The palaeoecology of the *Actinocamax plenus* subzone (lowest Turonian) in the Anglo-Paris Basin. *Palaeontology* **4**, 609–47.

Jefferies, R, P, S, 1963. The stratigraphy of the *Actinocamax plenus* subzone (Turonian) in the Anglo-Paris Basin. *Proc. Geol. Assoc.* **74**, 1–34.

Murray, J. W. and C. A. Wright 1974. *Palaeogene foraminiferida and palaeoecology, Hampshire and Paris Basins and the English Channel.* Sp. Pap. Palaeont. no. 14.

Peake, N. B. and J. M. Hancock 1961. The Upper Cretaceous of Norfolk. *Trans Norfolk Norwich Nat. Soc.* **19**, 293–339.

Schlanger, S. O. and R. G. Douglas 1974. Pelagic ooze–chalk–limestone transition and its implications for marine stratigraphy. In *Pelagic sediments on land and under the sea*, K. J. Hsü & H. C. Jenkyns (eds), Sp. Publ. Int. Assoc. Sediment. 1, 117–48.

Scholle, P. A. 1974. Diagenesis of Upper Cretaceous chalks from England, Northern Ireland and the North Sea. In *Pelagic sediments on land and under the sea*, K. J. Hsü & H. C. Jenkyns (eds), Sp. Publ. Int. Assoc. Sediment. 1, 177–210.

White, H. J. O. 1921. *A short account of the geology of the Isle of Wight.* Mem. Geol Surv. UK.

White, H. J. O. and Ll. Treacher 1905. On the age and relations of the phosphatic chalk of Taplow. *Q. J. Geol Soc. Lond.* **61**, 461–94.

Wright, J. 1875. On the discovery of Microzoa in the Chalk-flints of the north of Ireland. *Rep. Br. Assoc.* (1874), 95–6.

7 Metacopine ostracods in the Lower Jurassic

A. R. Lord

The present state of knowledge of metacopine ostracods in the Jurassic is reviewed and their biostratigraphic significance analysed. Particular attention is paid to the disappearance of this group near the Pliensbachian/Toarcian boundary, which is an important but poorly known extinction event in the history of the Ostracoda.

Introduction

The Metacopina comprise a very important Palaeozoic group of families, the number depending on which classificatory concept is followed, with numerous representatives. When the suborder was erected (in Moore 1961) it was considered to be of essentially Palaeozoic age but with a final representative in *Robsoniella* Kuznetsova from the mid-Cretaceous (Aptian–Albian). Subsequently McKenzie (1967) discovered the living form *Saipanetta* in the Pacific which appeared to have certain features in common with Palaeozoic ostracods, in particular a large circular cluster of muscle-scars which caused that author to classify the genus in the metacopine Superfamily Healdiacea. The overall morphology of *Saipanetta* led McKenzie (1968) to suggest that the families Darwinulidae and Cyprididae were developed from metacopine rather than podocopine ancestors and that therefore the suborder Metacopina ranges from the Ordovician to Recent. Both *Robsoniella* and *Saipanetta* have since been re-investigated; in the case of the former genus Gramm and Kuznetsova (1970) conclude that it is a podocopine ostracod quite distinct from healdiacean types, and Schornikov and Gramm (1974) consider that *Saipanetta* is not a living example of fossil healdiaceans and further suggest that the Metacopa be restricted to the single superfamily Healdiacea and single family Healdiidae. Thus the last true metacopine ostracods are almost certainly *Ogmoconcha* Triebel and its allies in the Lower Jurassic.

An account of the genus *Ogmoconcha* and its possible synonymy with the Triassic *Hungarella* Méhes is given by Lord (1972). Despite considerable work on Triassic ostracods from various parts of the world, the matter of

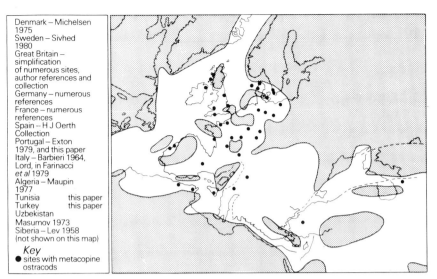

Figure 7.1 Distribution of metacopine ostracods in the Lower Jurassic. Siberia record of Lev (1958) not shown. Palaeogeographic reconstruction after Howarth (1981, Fig. 13.7).

Ogmoconcha's relationship with *Hungarella* is still unresolved, although Malz (1971, p. 435) maintains that the two are quite distinct. Kristan-Tollmann (1977a&b, and discussion of the latter paper) contends that the two are discrete genera based on the following features: (a) *Ogmoconcha* has fewer muscle scars, about six scars in two vertical rows inside a ring of smaller scars, in contrast to *Hungarella* which has two vertical rows of eight to ten scars inside a ring of scars more numerous than in *Ogmoconcha*; (b) left valves of *Hungarella* may differ greatly from the right morphologically, whereas *Ogmoconcha* left and right valves are symmetrical; (c) *Hungarella* may have a rim or spine on the anterior margin of the right valve but not the more numerous marginal spines found on both valves of *Ogmoconcha*. Valid as these distinctions may possibly be, the fact remains that the type-species of *Hungarella* (i.e. *Bairdia* (?) *problematica* Méhes) has yet to be re-illustrated; in particular the muscle-scar pattern must be accurately figured. Until this important preliminary step is carried out, the taxonomic status and relationships of species assigned to *Hungarella* will continue to be in doubt, as will the possible synonymy with *Ogmoconcha*. In general, much more information is now available about the later metacopine ostracods, although the occurrence and development of these ostracods in the Permian and Triassic is still somewhat obscure. Thus, for essentially the same reasons discussed in an earlier paper (Lord 1972, pp. 320–5), the name *Hungarella* is not applied to Jurassic species.

Muscle scars

A large volume of literature about healdiid muscle-scar patterns has appeared in recent years, stimulated by Gründel's recognition (Gründel 1964, see Fig. 6) of long term reduction in number of scars in *Healdia* and descended genera from the late Palaeozoic into the early Mesozoic. Work by Gramm (1970, 1973a&b, 1976) has been particularly valuable and has confirmed this evolutionary trend in healdiid ostracods from the Carboniferous to the Jurassic, with a relative strengthening of the central scars and a decrease in number or even disappearance of peripheral scars. Thus in the Pliensbachian, *Ledahia* Gründel, which has a simple muscle-scar pattern of biserial type reminiscent of cytherellid patterns, co-exists with *Ogmoconcha* with more complex and more typically metacopine scar patterns. Kristan-Tollmann, working on Alpine Triassic (Kristan-Tollmann 1971) and southern German Lower Jurassic material (Kristan-Tollmann 1977a), has documented further variety in healdiid muscle-scar patterns, differentiating the late Pliensbachian 'vallate-*Ogmoconcha*' of Malz (1975) as the genus *Hermiella*. In a phylogenetic diagram (Kristan-Tollmann 1977a, Fig. 10), she relates *Hermiella* to *Healdia* via Triassic *Torohealdia* and *Signohealdia*, rather than through a line including *Ogmoconcha/Hungarella*. At present I prefer to consider vallate forms as a final development of *Ogmoconcha* rather than a separate generic group with different antecedents.

Study of these muscle-scar patterns in detail must be carried out with caution, particularly in view of the variation possible as a result of preservation, as discussed by Herrig (1969a, pp. 447–52).

Marginal zone

Satisfactory investigations of the nature of the marginal zone of Jurassic healdiid ostracods have yet to be carried out, although marginal pore canals (and therefore a calcified inner lamella) have been illustrated (Herrig 1969a, Figs 5, 9 & 12).

Contact groove

Fine morphology of the contact groove has proved to be valuable for taxonomy, especially for the definition of *Ogmoconchella* (see Malz 1971, pp. 436, 437). The functional morphology of the contact groove is poorly understood. Adamczak (1976) has compared the contact grooves of Silurian *Kuresaaria gotlandica* Adamczak with *Ogmoconchella ellipsoidea* (Jones) from the Lower Jurassic; the two forms have dissimilar grooves along the dorsal margin, but both show an interrupted groove in the mid-ventral region of adults. Adamczak interprets this similarity of ventral contact groove as parallel evolution, which probably reflects a similar functional control.

External marginal features

Jurassic healdiids commonly exhibit anterior and posterior marginal spines (e.g, *Ogmoconcha amalthei* (Quenstedt)) or posteroventral spines (*Ogmoconchella bispinosa* (Gründel)). The presence or absence of these features may have a certain taxonomic value and Reyment *et al.* (1977) have suggested that they may be biostratigraphically useful, based on a study of *Cytheridea* from the Miocene.

Surface morphology and ornament

The rounded lateral and cross-section outlines of most forms of *Ogmoconcha* and its allies are well known and have been frequently illustrated. Certain forms, however, show other types of surface morphology, viz.

Ledahia septenaria (Gründel) – posterior angulation and posteroventral spine.
Ogmoconchella impressa Malz – longitudinal depression along centre of left valve.
Vallate-*Ogmoconcha* (Malz 1975) = *Hermiella* Kristan-Tollmann 1977a – with margin rim features, ranging from a ventral angulation plus dorsal rim in *O. circumvallata* Dreyer to strongly inflated rims along the free margins of *O. cincta* Malz and *O. ambo* Lord and Moorley. The vallate species are in most cases restricted in time to the late Pliensbachian and also, apparently, in geographical distribution (Fig. 7.2). Kristan-Tollmann

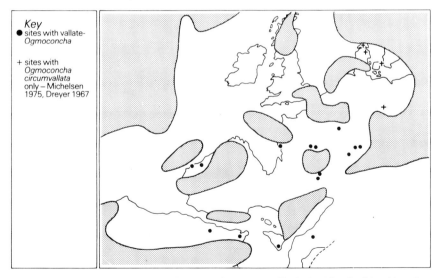

Figure 7.2 Distribution of vallate-*Ogmoconcha* (= *Hermiella*) in the Pliensbachian. Note that the northernmost records (+) are of *O. circumvallata* only; this species also occurs in south-western Germany and northeastern France.

(1977a, pp. 626–32) differentiates *Hermiella* from *Ogmoconcha* and related genera by the external (vallate) morphology and by muscle-scar pattern. At present I prefer to adopt a conservative view of these forms as late varieties of *Ogmoconcha*, particularly as differences in muscle-scar patterns can be unreliable. They represent the final, slightly exotic manifestation of metacopine ostracods.

Surface ornament has been figured in a few species. Malz (1971) described *Ogmoconchella propinqua* with 'fingerprint' ornament and *Ogmoconchella conversa* with surface reticulation. *Ogmoconchella adenticulata* (Pietrzenuk) also has a 'fingerprint' ornament of fine anastomosing ribs, as has *Ogmoconchella martini* (Anderson). In contrast, the curious form *Ogmoconcha nordvikensis* described by Lev (1958) from the USSR has anterior and posterior pustulose areas.

Zonal biostratigraphy

One of the fullest accounts of Lias metacopine ostracods is that of Michelsen (1975) on Danish assemblages. His systematic study of a number of borehole sequences allowed him to propose a zonal system based on metacopine, as well as podocopine ostracods. Sivhed (1980) has extended the scheme into southern Sweden. Elsewhere the ranges of metacopine species are still not fully known and any improvement of knowledge is hindered by some remaining taxonomic uncertainties.

In other respects metacopines have excellent biostratigraphic potential. They are widely distributed (Fig. 7.1) in assemblages from western Europe (numerous references), North Africa (Maupin 1977 – Algeria; author's collection – Tunisia), Turkey (author's collection, see Appendix), Uzbekistan (Masumov 1973) and the Lena Basin, Siberia (Lev 1958).

Marine Triassic metacopine ostracod records are even more widespread, from Israel (Sohn 1968), Iran (Kristan-Tollmann, in Kristan-Tollmann *et al.* 1980), Pakistan (Sohn 1970), China (Gou & Cao 1980), the USSR (e.g. Gramm 1970), Alaska (Sohn 1964; P. F. Sherrington, personal communication), Nevada (Sohn 1964) and Australia (Jones 1970), as well as Europe. I suspect that Lower Jurassic *Ogmoconcha* and allies are actually more widespread than Figure 7.1 indicates, but no records are known to me, for example, from the marine Lower Jurassic sediments of the Americas, that is from the Rockies, the Andes and Central America.

These ostracods are also unusually abundant and commonly exceed foraminifera in numbers of individuals, an unusual situation in marine benthic assemblages. This is the case in the Blue Lias (*angulata* and *bucklandi* zones, Hettangian–Sinemurian), Black Ven Marls (*raricostatum* zone, Upper Sinemurian) and most of the Belemnite Marls (*jamesoni* and *ibex* zones, Lower Pliensbachian) of the Dorset coast and particularly in an

Hettangian assemblage from North Humberside (Lord 1971, Fig. 4 – Hotham 'g') which is extremely rich and monospecific with *Ogmoconchella ellipsoidea* (Jones) (=*O. aspinata* Drexler of authors) together with a few foraminifera and holothurians. It is difficult to over-emphasise the numerical importance of these ostracods. They are abundant, occasionally to the exclusion of even cytheracean ostracods, which are otherwise the most successful post-Palaeozoic ostracod group. Absence of the Cytheracea characterises assemblages from the Apennines (Lord, in Farinacci *et al.* 1979), Bilecik (see Appendix) and Djebel Zaghouan, Tunisia (author's collection). From a stratigrapher's point of view the main difficulty with metacopine ostracods is ease of identification. Certain species are distinctive, but the range of inter- and intraspecific variation has yet to be fully explored.

Metacopine ostracods disappear from the fossil record at the base of the Toarcian, more precisely in the first zone of the Toarcian and not at the stage boundary. This extinction altered the whole character of Jurassic ostracod faunas and at a stroke cytheracean species were left dominating marine assemblages; a position unchanged to the present. If members of the Superfamily Cytheracea are the most successful later Mesozoic and Cenozoic marine ostracods, then this emphasises in contradistinction the Metacopina, which were so prominent in the early Jurassic that they could exclude rival forms. In many respects, then, the metacopine extinction in the early Toarcian was the most important single event in post-Palaeozoic ostracod history and thus worthy of special consideration.

The extinction of the Metacopina

The late Pliensbachian was a period of extensive marine regression and in northwestern Europe deposits of this age (*spinatum* zone) commonly reflect shallow-water conditions. The Toarcian, however, saw a substantial transgression with marine Jurassic sediments being deposited for the first time in many areas on the edges of continental masses. In north-western Europe conditions were not encouraging for marine benthic organisms, with the widespread deposition of bituminous paper shales such as the Posidonienschiefer of Germany, indicative of anoxic bottom conditions.

Denmark and Sweden

Marine sediments of late Pliensbachian and early Toarcian age are absent from southern Sweden (Sivhed 1980), while in the main Danish Embayment the late Pliensbachian is represented but not its upper boundary. The vallate form *Ogmoconcha circumvallata* Dreyer occurs (Michelsen 1975, p. 227), although it is not very common.

Great Britain

The *spinatum* zone is mostly represented by ironstones, from which few ostracods have been recovered, or in southern England by condensed carbonates. A poor ostracod fauna has been reported from Yorkshire (Lord, in Catt *et al.* 1971) and the sequence in this area merits a closer re-examination. However, the early Toarcian sediments have little or no bottom fauna preserved. Another argillaceous sequence is penetrated by the Mochras Borehole, western Wales, and some details of the ostracods are available (Sherrington, in Lord 1978). *Ogmoconcha contractula* Triebel and *Ogmoconchella bispinosa* (Gründel) both occur in the basal Toarcian, *tenuicostatum* zone and the important Toarcian cytheracean *Kinkelinella sermoisensis* (Apostolescu) appears in the succeeding *falciferum* zone. In contrast, Bate and Coleman (1975) have reported an assemblage from the late *tenuicostatum* zone of Empingham, East Midlands, where *Ogmoconcha* (a final metacopine representative) occurs with *K. sermoisensis*.

Samples from Tilton, Leicestershire, and Kirton-in-Lindsey, Lincolnshire (Lord, in Plumhoff 1967, p. 563), are shown in Figure 7.3, a composite diagram giving for the first time some detail of ostracods at the Pliensbachian/Toarcian boundary in Britain. Again, *Ogmoconcha* survives through the *tenuicostatum* zone.

No vallate – *Ogmoconcha* (=*Hermiella*) have yet been found in Britain. Figure 7.2 shows that *O. circumvallata* appears to be the boreal vallate form, as opposed to the *O. ambo* group which has a Tethyan distribution, with an area of overlap in southwestern Germany and northeastern France. There seems no obvious reason why *O. circumvallata* should not occur in Britain.

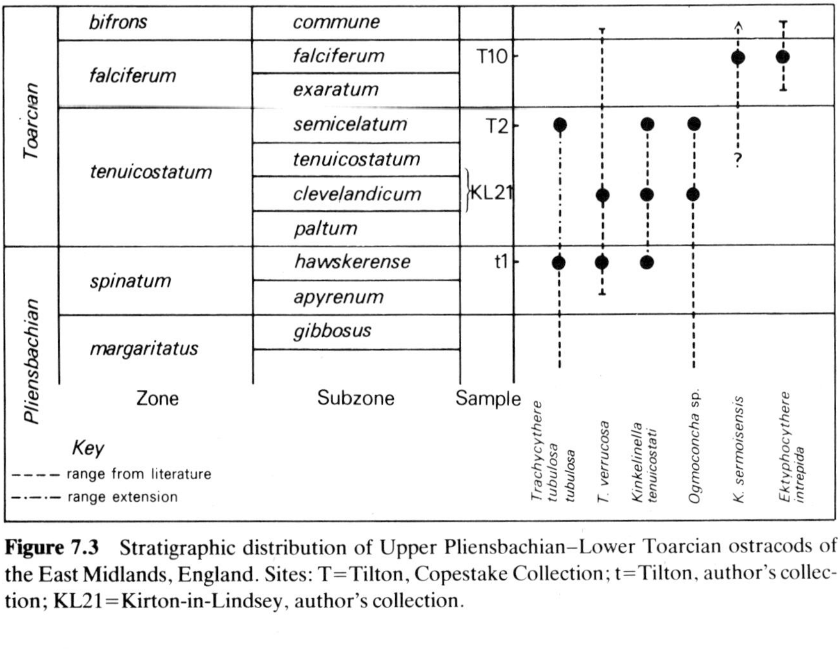

Figure 7.3 Stratigraphic distribution of Upper Pliensbachian–Lower Toarcian ostracods of the East Midlands, England. Sites: T=Tilton, Copestake Collection; t=Tilton, author's collection; KL21=Kirton-in-Lindsey, author's collection.

Germany

The first vallate-*Ogmoconcha* were figured from southwestern Germany by Buck (1954) and Klingler (1962), and described by Lord and Moorley (1974) and Malz (1975). Malz gave the fullest account, describing a sequence of species: *O. comes* Malz, *O. cista* Malz, *O. cincta* Malz and *O. ambo* Lord and Moorley, through the *spinatum* zone, accompanied by *O. circumvallata* Dreyer. Malz was concerned primarily with the vallate-*Ogmoconcha* and did not study the stage boundary or report on associated ostracods. The assemblages from the site examined by Malz at Aselfingen, and others from Reutlingen, have been studied by Moorley (1974). The assemblages are dominated by metacopines with species of *Bairdia, Isobythocypris, Polycope* and a few cytheraceans. Some samples are devoid of Cytheracea forms.

Ostracod and ammonite distributions in the type Pliensbachian of Pliensbach, southwestern Germany, have been recorded in commendable detail by Urlichs (1977). In this section *O. comes, O. cista, O. cincta* and *O. ambo* occur through the Upper Pliensbachian in the series demonstrated by Malz, accompanied by *O. circumvallata* and *O. intercedens* Dreyer. Urlichs does, however, observe that two of Malz's species are closely related and have the same range, so that the lineage is *O. comes (+O. cista) – O. cincta – O. ambo*. He also comments (Urlichs 1977, p. 6) that the vallate forms are common in marls (of the *spinatum* zone) as opposed to clays (of the *margaritatus* zone) where they are rare or absent. This facies distinction does not appear to hold true for all the known sites. At the stage boundary *O. ambo* and *Ogmoconchella impressa* survive into the *tenuicostatum* zone; *Ektyphocythere champeauae* (Bizon) appears at the base of the Toarcian.

Dreyer (1967) has described rimmed *Ogmoconcha (O. circumvallata* and *O. intercedens)* from the lower part of the Upper Pliensbachian of southwestern Brandenburg. These forms are slightly different from the *O. ambo* group of species. Herrig, in a series of papers (Herrig 1969a&b, 1979a–c), has documented Upper Pliensbachian material from the Baltic coast and Thüringen, East Germany, but not the metacopine extinction level; he records no vallate species.

France

A number of works record metacopines, including vallate-*Ogmoconcha*, from the late Pliensbachian (Apostolescu 1956 – Causses; Apostolescu 1959 – south Paris Basin; various authors in the *Colloque sur le Lias français* 1961; Oertli 1963) but with insufficient range data to be really useful. At Jouy aux Arches, south of Metz, Chester (personal communication) recorded vallate-*Ogmoconcha* ranging through the *spinatum* zone with other metacopines and all disappearing *at* the base of the Toarcian. The basal *tenuicostatum* zone sample was, however, in bituminous shales and barren. Higher in that zone *Kinkelinella tenuicostati* Martin, a characteristic form for the zone in Germany, and other small cytheraceans appear. In the late Pliensbachian *spinatum* zone assemblages the only cytheraceans were rare *Trachycythere*.

The most comprehensive account of Lower Jurassic ostracods in France is regrettably unpublished. Viaud (1963) examined numerous sections, documenting the almost total faunal change at the Pliensbachian/Toarcian boundary which is now well known for ostracods and other fossil groups. He recognised vallate-*Ogmoconcha* (under Oertli's (1963) notation 'Subgenus C') from a number of areas, and his data are included in Figure 7.2, but he did not record metacopines surviving into the basal Toarcian. One of Viaud's sites in the Vandée has since been re-examined and information published (Maupin 1975, 1978). The Toarcian *tenuicostatum* zone has *Ogmoconcha* species and *O. ambo* associated with *Ektyphocythere* species and *Kinkelinella sermoisensis*. A detailed range analysis of the ostracods at this locality would be very interesting.

Portugal

A series of samples across the Pliensbachian/Toarcian boundary at Zambujal have been analysed by Exton (1979). Forms referred to as '*Ogmoconcha* spp.' and a vallate form called *O. ambo* disappear at the stage boundary. The vallate species is restricted to the *spinatum* zone and it is accompanied by a *Kinkelinella* species. Basal Toarcian assemblages are small and dominated by '*Bairdiacypris*' with *Ektyphocythere* cf. *E. intrepida* Bate and Coleman and ?*Cytheropteron*.

A section on the western coast at Peniche has been studied by the present author. A rather poor series of ostracod assemblages is summarised in Figure 7.4, using the stratigraphic work of Mouterde (1955) as a reference. Lower samples contain *Ogmoconcha* and vallate-*Ogmoconcha*, but no

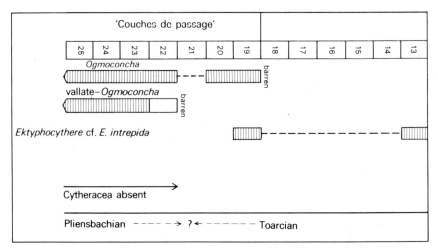

Figure 7.4 Stratigraphic distribution of Upper Pliensbachian–Lower Toarcian ostracods at Peniche, Portugal. Stratigraphy after Mouterde (1955).

cytheracean species. The normal *Ogmoconcha* survive longest and overlap the appearance of *Ektyphocythere* cf. *E. intrepida* Bate and Coleman. This faunal change occurs in the '*Couches de passage*' described by Mouterde (1955, p. 104) at the Pliensbachian/Toarcian stage boundary and it may be conjectured that the critical interval of samples 19 and 20 falls in the *tenuicostatum* zone.

Italy

A Pliensbachian–Toarcian sequence penetrated by a borehole at Ragusa, Sicily, has been described by Barbieri (1964). Domerian assemblages contain metacopine ostracods, especially *Ogmoconcha hyblea* (Barbieri) which appear vallate, but lack cytheraceans. Toarcian assemblages are poor but contain *Kinkelinella*. *Spinatum* zone assemblages from Strettura in Umbria, recorded by Lord (in Farinacci *et al.* 1979), contain vallate-*Ogmoconcha*, lack cytheraceans and have much in common with southwestern German material. At this site, unfortunately, the base of the Toarcian falls in a carbonate interval immediately above a slumped horizon.

North Africa

Ostracods from two sections in northern Algeria have been reported by Maupin (1977). The stratigraphy of these sections is based on some ammonite evidence and the microfaunas themselves. Assemblages with *Ogmoconcha* and vallate-*Ogmoconcha* are assigned to the Upper Pliensbachian *spinatum* zone, while those without metacopines but with *Kinkelinella sermoisensis* (Apostolescu) or *Ektyphocythere* cf. *E. bucki* (Bizon) are considered Toarcian in age. In the section from which ammonites were recorded the *tenuicostatum* and *falciferum* zones appear to be absent, as an Upper Pliensbachian ('Domerien', *spinatum* zone) microfauna occurs immediately below beds from which *bifrons* zone ammonites have been found. On the basis of other sites discussed above, the level with *Ogmoconcha* and vallate-*Ogmoconcha* could equally well be of early Toarcian, *tenuicostatum* zone age, and in fact at one site they occur with *Ektyphocythere* cf. *E. rugosa* (Bizon), a form described from the Toarcian of France.

Material from Djebel Zaghouan, northern Tunisia, collected by Dr R. H. Bate during the VIème Colloque Africain de Micropaleontologie from levels described as 'Domerien' and 'Toarcien', has been examined. Calcareous nannofossils suggest that the samples may be from the *margaritatus* and *spinatum* zones respectively. In both cases *Ogmoconcha* and vallate-*Ogmoconcha* occur with large *Triebelina (Ptychobairdia)* in assemblages lacking cytheracean ostracods. The material is more diverse than that described from Algeria and differs also in composition. From evidence at other localities these samples cannot be younger than *tenuicostatum* zone age.

Discussion

Metacopine ostracods in the Lower Jurassic merit detailed consideration because of their overwhelming abundance and also for their biostratigraphic potential. Their extinction in early Toarcian times is arguably one of the most important events in the post-Palaeozoic history of the ostracods, and for this reason particular attention has been paid in the present paper to sites from which late Pliensbachian–early Toarcian ostracods are known, to investigate the metacopine extinction event.

In the assemblages described above it is clear that *Ogmoconcha*, both vallate and rounded types, and some allied forms such as *Ogmoconchella* survived into the early Toarcian *tenuicostatum* zone, but not into the succeeding *falciferum* zone. The Domerian (Upper Pliensbachian) aspect of the *tenuicostatum* zone faunas was recognised by Hoffmann and Martin (1960) in their important study of the zone in Germany and this feature was emphasised by Plumhoff (1967) in a more broadly based survey. Many authors have described the faunal turnover at the Pliensbachian/Toarcian boundary, reflected as it is in a variety of fossil groups, but it is important to recognise that the change did not necessarily occur simultaneously in all groups or necessarily coincide with divisions of our ammonite-based stage and zone system. Indeed, in the records described above any total break in ostracod faunas coinciding exactly with the stage boundary now seems suspicious. Nonetheless, the disappearance of the metacopines in the *tenuicostatum* zone is a prominent biostratigraphic marker.

A marine regression during the late Pliensbachian, culminating during *spinatum* zone times, was followed by an important transgression in the early Toarcian, the effects of which can be recognised globally. Over wide areas of northern Europe, Lower Toarcian deposits are bituminous shales (Jet Rock of Yorkshire, Posidonienschiefer of Germany, etc.), usually explained by invoking a model of transgression, with poor marine circulation and consequent anoxic bottom conditions. Such conditions can be used to explain the absence or poor representation of benthic organisms, lack of bioturbation, etc., features discussed by Hallam and Bradshaw (1979, pp. 157–60). Such an adverse environment may explain the faunal changes. Bituminous shale deposition was initiated at different times and was of varying duration; some intervals occur in the *tenuicostatum* zone, but the shales are most widespread during the *falciferum* zone, apparently related to local subsidence (Hallam & Bradshaw 1979, p. 159, Fig. 2). It is noteworthy that *Ogmoconcha* and allies disappear approximately at the same time as the inception of bituminous shale deposition.

The early Toarcian anoxic events were clearly of great significance, but not perhaps enough to account for the extinction of the Metacopina:

(a) The metacopines disappeared more or less simultaneously from an area greater than that of bituminous shale deposition. Bituminous shales

were not deposited in southern England, for example, and marine conditions in the Iberian peninsula and Northern Africa during the Toarcian have been described as 'normal' (Ager 1981, p. 161).
(b) *Ogmoconcha* and *Ogmoconchella* had survived earlier transgressive phases with bituminous shale deposition, of possibly similar magnitude, during the Rhaetian and Hettangian when there were many extinctions of benthic organisms. Hallam and Bradshaw (1979, p. 163) suggest, however, that the Toarcian bituminous shale facies may be the most significant example to occur in the Jurassic, involving both shelf and oceanic depositional environments, although sediments of this age have yet to be found in the ocean basins.
(c) The Superfamily Cytheracea survived the early Toarcian and evolved rapidly in the Middle Jurassic to become the dominant group of marine ostracods, yet during the earlier part of the Lower Jurassic they were apparently less cosmopolitan and successful than the metacopines.
(d) A few species lived through Lower Toarcian time, occurring before and after the transgression, for example *Ektyphocythere champeauae* (Bizon), *Trachycythere verrucosa* Triebel and Klingler, *Liasina lanceolata* (Apostolescu) and *Polycope cerasia* Blake. The species *E. champeauae* is known from southwestern Germany at the base of the Toarcian, it disappears during the period of deposition of the Posidonienschiefer and reappears late in the Toarcian. If this species could survive, why not an *Ogmoconcha*?

For the reasons discussed, it seems unlikely that the anoxic events of the early Toarcian could have been solely responsible for the extinction of the Metacopina. Obvious oceanographic effects such as temperature change are also unconvincing as an explanation for the disappearance of a group that occupied a range of shelf environments over a range of latitude. The explanation must be sought in a more complex genetic–oceanographic interaction.

Acknowledgements

Material from Italy and northern France was collected with the aid of grants from the Central Research Fund of the University of London to Professor Barnard and the author, which are gratefully acknowledged. Dr Füsun Alkaya, Dr R. H. Bate and Dr P. Copestake provided sample material from Turkey, Tunisia and Leicestershire respectively. Professor D. T. Donovan and Dr R. Mason kindly discussed certain aspects of the work and Dr R. H. Bate read an early draft.

References

Adamczak, F. 1976. Morphology and carapace ultrastructure of some Healdiidae (Ostracoda). *Abh. Verh. Naturw. Ver. Hamburg (NF)* **18/19** (Suppl.), 315–18.

Ager, D. V. 1981. Major marine cycles in the Mesozoic. *J. Geol Soc. Lond.* **138**, 159–66.

Apostolescu, V. 1956. Ostracods in Corrélation dans le Lias marneux des Causses majeurs. *Rev. Inst. Fr. Pétrole* **XI**, 439–48.

Apostolescu, V. 1959. Ostracodes du Lias du bassin de Paris. *Rev. Inst. Fr. Pétrole* **XIV**, 795–826.

Barbieri, F. 1964. Micropaleontologia del Lias e Dogger del Pozzo Ragusa 1 (Sicilia). *Riv. Ital. Paleont.* **LXX**, 709–831.

Bate, R. H. and B. E. Coleman 1975. Upper Lias Ostracoda from Rutland and Huntingdonshire. *Bull. Geol Surv. Gt Br.* **55**, 1–42.

Buck, E. 1954. Stratigraphisch wichtige Ostracoden aus dem Lias und Dogger von S.W.-Deutschlands. Unpublished table.

Catt, J. A., M. A. Gad, H. H. LeRiche and A. R. Lord 1971. Geochemistry, micropalaeontology and origin of the Middle Lias Ironstones in north-east Yorkshire (Great Britain). *Chem. Geol.* **8**, 61–76.

Colloque sur le Lias français. Mem. Bur. Recherches Geol. Min. no. 4 (1961).

Dreyer, E. 1967. Mikrofossilien des Rät und Lias von SW-Brandenburg. *Jb. Geol. (Berlin)* **1**, 491–531 (for 1965).

Exton, J. 1979. *Pliensbachian and Toarcian microfauna of Zambujal, Portugal: systematic paleontology*. Carleton University, Ottawa, Geological Paper 79–1, 1–103.

Farinacci, A., A. R. Lord, G. Pallini and F. Schiavinotto 1979. The depositional environment of the Domerian–Toarcian sequence of Strettura (Umbria). *Geologica Romana* **XVII** (1978), 303–23.

Gou, Yun-sian and Cao, Mei-zhen 1980. Outline of Triassic Ostracoda in China. *Riv. Ital. Paleont.* **85**, 1227–9.

Gramm, M. N. 1970. Ostracodes of the family Healdiidae from Triassic deposits of southern Primorye. In *Triassic invertebrates and plants of east of USSR*, 41–93. Vladivostok: Academy of Sciences of the USSR. (In Russian, English summary.)

Gramm, M. N. 1973a. Neoteny and the directiveness of evolution of the adductor muscle scar of ostracodes. In *Scientific reports of the evolutionary seminar*, Vol. 1, 31–41. Vladivostok: Academy of Sciences of the USSR. (In Russian, English summary.)

Gramm, M. N. 1973b. Cases of neoteny in fossil ostracodes. *Paleont. J.* **1**, 3–12. (Translation from Russian, original pagination.)

Gramm, M. N. 1976. On two tendencies in the evolution of ostracod adductor muscle scar. *Abh. Verh. Naturw. Ver. Hamburg (NF)* **18/19** (Suppl.), 287–94.

Gramm, M. N. and Z. V. Kuznetsova 1970. Systematic position of the Lower Cretaceous ostracode genus *Robsoniella*. *Paleont. J.* **3**, 89–94. (Translation from Russian, original pagination.)

Gründel, J. 1964. Zur Gattung *Healdia* (Ostracoda) und zu einigen verwandten Formen aus dem unteren Jura. *Geologie* **13**, 456–77.

Hallam, A. and M. J. Bradshaw 1979. Bituminous shales and oolitic ironstones as indicators of transgressions and regressions. *J. Geol Soc. Lond.* **136**, 157–64.

Herrig, E. 1969a. Ostracoden aus dem Ober-Domérien von Grimmen westlich von Greifswald, Teil I. *Geologie* **18**, 446–71.

Herrig, E. 1969b. Ostracoden aus dem Ober-Domérien von Grimmen westlich von Greifswald, Teil II. *Geologie* **18**, 1072–101.

Herrig, E. 1979a. Die Gattung *Bairdia* (Ostracoda, Crustacea) im Lias von Thüringen, Teil I. *Z. Geol. Wiss. (Berlin)* **7**, 641–61.

Herrig, E. 1979b. Ostrakoden aus dem Lias von Thüringen: Die Gattungen *Bairdia* (Teil II) *Fabalicypris* und *Bairdiacypris*. *Z. Geol. Wiss. (Berlin)* **7**, 763–82.

Herrig, E. 1979c. Weitere glattschalige Ostrakoden aus dem Lias von Thüringen. *Z. Geol. Wiss. (Berlin)* **7**, 1343–61.
Hoffmann, K. and G. P. R. Martin 1960. Die Zone des *Dactylioceras tenuicostatum* (Toarcien, Lias) in NW- und SW-Deutschland. *Paläont. Z.* **34**, 103–49.
Howarth, M. K. 1981. Palaeogeography of the Mesozoic. In *The evolving earth*, L. R. M. Cocks (ed.), 197–220. London: British Museum (Natural History) and Cambridge University Press.
Jones, P. J. 1970. Marine Ostracoda (Palaeocopa, Podocopa) from the Lower Triassic of the Perth Basin, Western Australia. *Bull. Bur. Miner. Resources Aust.* **108**, 115–44.
Klingler, W. 1962. Lias Deutschlands. In *Leitfossilien der Mikropaläontologie*, W. Simon & H. Bartenstein (eds), 73–122. Berlin: Borntraeger.
Kristan-Tollmann, E. 1971. Zur phylogenetischen und stratigraphischen Stellung der triadischen Healdiiden (Ostracoda). *Erdöl-Erdgas-Z.* **87**, 428–38.
Kristan-Tollmann, E. 1977a. Zur Evolution des Schließmuskelfeldes bei Healdiidae und Cytherellidae (Ostracoda). *Neues Jb. Geol. Paläont. Mh.* **10**, 621–39.
Kristan-Tollmann, E. 1977b. On the development of the muscle-scar patterns in Triassic Ostracoda. In *Aspects of ecology and zoogeography of Recent and fossil Ostracoda*, H. Löffler & D. Danielopol (eds), 133–43. The Hague: Junk.
Kristan-Tollmann, E., A. Tollmann and A. Hamedani 1980. Beiträge zur Kenntnis der Trias von Persien, II. Zur Rhätfauna von Bagerabad bei Isfahan (Korallen, Ostracoden). *Mitt. Öst. Geol. Ges.* **73**, 163–235.
Lev, O. 1958. The Lower Jurassic Ostracoda of the Nordvik and Lena–Olenck areas. *Nauchno-issled. Inst. Geol. Arct.* **12**, 23–49. (In Russian.)
Lord, A. R. 1971. Revision of some Lower Lias Ostracoda from Yorkshire. *Palaeontology* **14**, 624–65.
Lord, A. R. 1972. The ostracod genera *Ogmoconcha* and *Procytheridea* in the Lower Jurassic. *Bull. Geol Soc. Denmark* **21**, 319–36.
Lord, A. R. 1978. The Jurassic Part 1 (Hettangian–Toarcian). In *A stratigraphical index of British Ostracoda*. R. H. Bate & J. E. Robinson (eds), 189–212. Liverpool: Seel House Press.
Lord, A. R. and A. Moorley 1974. On *Ogmoconcha ambo* sp. nov. Lord and Moorley. In *A stereo-atlas of ostracod shells* Vol. 2, P. C. Sylvester-Bradley & D. J Siveter (eds), 9–16. Leicester: University of Leicester Press.
Malz, H. 1971. Zur Taxonomie 'glattschaliger' Lias-Ostracoden. *Senckenberg. Leth.* **52**, 433–55.
Malz, H. 1975. Eine Entwicklungsreihe 'vallater' Ogmoconchen (Ostracoda) im S.-deutschen Lias. *Senckenberg. Leth.* **55**, 485–503.
Masumov, A. S. 1973. *Jurassic ostracods of Uzbekistan*. Tashkent (in Russian).
Maupin, C. 1975. Etude micropaléontologique de la zone à *Dactylioceras tenuicostatum* du Toarcien de l'anse Saint-Nicolas (Commune de Jard-Vandée). *C. r. Somm. Séanc. Soc. Géol. Fr.*, 11–13.
Maupin, C. 1977. Données micropaléontologiques nouvelles et précisions stratigraphiques sur le Lias du Kef Ben Chikr Bou Rouhou et du Kef Toumiette Nord (Chaine Calcaire Kabyle – nord du Constantinois – Algérie). *Rev. Micropaléont.* **20**, 91–9.
Maupin, C. 1978. Deux Ostracodes nouveaux du Toarcien de Vendée (France). *Geobios* **11**, 107–11.
McKenzie, K. G. 1967. Saipanellidae: a new family of podocopid Ostracoda. *Crustaceana* **13**, 103–13.
McKenzie, K. G. 1968. Contribution to the Ontogeny and Phylogeny of Ostracoda. *Proc. Int. Palaeont. Union, XXIII Int. Geol. Cong.*, 165–88.
Michelsen, O. 1975. Lower Jurassic biostratigraphy and ostracods of the Danish Embayment. *Danm. Geol. Unders. II Raekke* **104**, 1–287.
Moore, R. C. (ed.) 1961. *Treatise on invertebrate paleontology*, Part Q. Lawrence: University of Kansas Press and Geological Society of America.

Moorley, A. 1974. *Domerian ostracods from Baden Würtemberg, south-west Germany.* MSc dissertation, University of Wales.
Mouterde, R. 1955. Le Lias de Peniche. *Comunções Servs Geol. Port.* **36,** 87–115.
Oertli, H. J. 1963. *Faunes d'Ostracodes du Mésozoique de France,* 1–57. Leiden: Brill.
Plumhoff, F. 1967. Die Gattung *Aphelocythere* (Ostracoda) im NW-europäischen Jura und zur Entwicklung der Mikrofauna am Übergang Domerium/Toarcium. *Senckenberg. Leth.* **48,** 549–77.
Reyment, R. A., I. Hayami and G. Carbonnel 1977. Variation of discrete morphological characters in *Cytheridea* (Crustacea: Ostracoda). *Bull. Geol. Instn Univ. Uppsala* **7,** 23–36.
Schornikov, E. I. and M. N. Gramm 1974. *Saipanetta* McKenzie 1967 (Ostracoda) from the northern Pacific and some problems of classification. *Crustaceana* **27,** 92–102.
Sivhed, U. 1980. *Lower Jurassic ostracodes and stratigraphy of western Skåne, southern Sweden.* Sver. Geol. Unders. Afh. Ca 50, 1–85.
Sohn, I. G. 1964. *Significance of Triassic ostracodes from Alaska and Nevada.* Prof. Pap. US Geol Surv. 501-D, 40–2.
Sohn, I. G. 1968. Triassic ostracodes from Makhtesh Ramon, Israel. *Bull. Geol Surv. Israel* **44,** 1–71.
Sohn, I. G. 1970. Early Triassic Marine Ostracodes from the Salt Range and Surghar Range, West Pakistan. In *Stratigraphic boundary problems: Permian and Triassic of West Pakistan,* B. Kummel & C. Teichert (eds), 193–206. Lawrence: University of Kansas Press.
Urlichs, M. 1977. Stratigraphy, ammonite fauna and some ostracods of the Upper Pliensbachian at the type locality (Lias, SW-Germany). *Stuttg. Beitr. Naturk.* B **28,** 1–13.
Viaud, J. 1963. *Les Ostracodes des principaux bassins liasiques français.* Doctoral thesis, University of Bordeaux.

Appendix: Bilecik, Turkey

This locality, referred to in passing in the paper, is one of a number of Turkish sites from which Lower Jurassic samples were collected for the author by Dr Füsun Alkaya. Only one contained ostracods, from the top of the Bayirköy Formation in red marls of *jamesoni-ibex* zone, Lower Pliensbachian, age at Trafo, near Bilecik, western Turkey.

The microfauna, both ostracods and foraminifera, is identical with assemblages from northwestern Europe. The foraminifera could be from the Dorset coast, as described by Tom Barnard in his classic paper on the area. The ostracods, as abundant as the foraminifera, were: *Polycope* sp., *Paracypris* sp., *P. redcarensis* (Blake), *Isobythocypris* sp., *Bythocypris* sp., *Bairdia* cf. *B. fortis* Drexler, *B.* cf. *B. hilda* Jones and *Ogmoconcha* spp. including *O. amalthei* (Quenstedt). The apparent uncertainty of some of the identifications relates to taxa which are unornamented. The assemblage could be from the Lower Pliensbachian of northwestern Europe, except for the absence of cytheracean species. Samples from the *jamesoni* and *ibex* zones of the Dorset coast (Belemnite Marls) contain a variety of *Polycope* species, *Gammacythere ubiquita* Malz and Lord, *Monoceratina* species and other cytheraceans as well as metacopine species. The absence of cytheraceans is an important feature already commented upon in this paper.

Bilecik is just north of a major zone of tectonic disturbance and, in a palaeotectonic sense, can be thought of as part of Europe, as indicated by the benthic microfauna. Brachiopods from this site, examined by Professor D. V. Ager, are also European in character. Curiously, for palaeoceanographic reasons, the ammonites present are not species with northwestern European associations but are either indigenous or Tethyan. Thus, in the Lower Pliensbachian this site was on the northern edge of the Tethyan Ocean, with a benthic fauna very close to that of northwestern Europe, while the nektoplanktonic ammonites reflect Tethyan oceanic influence and faunal provincialism.

8 Palynofacies and salinity in the Purbeck and Wealden of southern England

D. J. Batten

Many associations between components of palynological preparations and lithofacies in the Purbeck Beds and Wealden (uppermost Jurassic–Lower Cretaceous) of southern England have been recognised and are progressively being employed to aid the identification of depositional environments. Detailed sampling has led, in particular, to the recovery of numerous assemblages of phytoplankton. All are taxonomically restricted and none has a fully marine aspect. Both numerical and compositional variations suggest near-marine to perhaps only slightly brackish and freshwater conditions of deposition. Selected aspects of these and other palynofacies studies are discussed and illustrated.

Introduction

I first became interested in palynology as a student enrolled in Professor Barnard's MSc course in micropalaeontology at University College London. In common with others on the course, I was given (by Dr, now Professor, W. G. Chaloner) a rock sample from an unspecified locality and asked to prepare it for palynological analysis, write a report on what I found and determine its age. The sample turned out to be from the early Cretaceous Hastings Beds Group of southeastern England (Fig. 8.1). My enthusiasm for this project caused me to follow it up with a study of the facies distribution of British Wealden palynomorphs under the supervision of Dr N. F. Hughes. My interest in palynofacies (palynological facies) has continued ever since and hence seemed an appropriate subject for my contribution to this volume.

The term palynofacies was coined by Combaz (1964). Although criticised by Sigal (1965), subsequent papers by Correia (1967, 1969, 1971), Kieser (1967) and other French workers showed that the identification of the main organic components of sedimentary rocks which comprise palynofacies were of value both in the analysis of source potential for petroleum and for palaeoecology. Hughes saw in palynological facies a means of refining local biostratigraphic correlation and an aid to palaeoenvironmental interpretation (Hughes & Moody-Stuart 1967a,b). As a result, my research on Wealden palynofacies was aimed mainly at finding a means of recognising different kinds of miospore assemblages and correlating these with depositional environments. Papers orientated towards both biostratigraphy and palaeoecology were the end-products (e.g. Batten 1968, 1969, 1973a&b, 1974).

When I began work on Wealden palynomorph distributions (in 1966), literature relating to the subject was very limited in scope. To be sure, palaeoenvironmental conclusions based on spore and pollen associations had been published for Pleistocene to Holocene and some Carboniferous sequences but little was directly relevant to my proposed analysis of the total organic recovery from rock samples as seen in transmitted light. I was, however, able to find useful comparative data in a number of papers on the distribution of phytoclasts in aqueous suspension and Recent sediments, including those by Muller (1959), Rossignol (1961), Cross *et al.* (1966), Groot (1966), Spackman *et al.* (1966), Traverse and Ginsburg (1966) and Williams and Sarjeant (1967), and on pre-Quaternary distributions, e.g. Neves (1958), Tschudy (1961), Smith (1962), de Jekhowsky (1963a&b), Upshaw (1964), Marshall and Smith (1965), Muir (1967), Hughes and Moody-Stuart (1967a), Chaloner (1968) and Chaloner and Muir (1968).

During the past decade, interest in the organic content of sedimentary rocks as seen in transmitted light has increased considerably, mainly in connection with organic maturation and petroleum source potential studies. References to 'visual kerogen analyses' are now commonplace in the petroleum geology literature. Articles in which consideration is given to the relationships between the distribution of organic matter and depositional environments are also beginning to appear more frequently, although only rarely (e.g. Habib 1979) does the subject constitute the main theme. In the light of current interest in the field I take this opportunity both to summarise past work and to discuss some aspects of Purbeck and Wealden (mostly Lower Cretaceous) palynofacies and their palaeoenvironmental significance in the light of new data.

Initially (in 1966–69) I spent much time trying to determine a satisfactory basis on which to distinguish Wealden palynological assemblages. Samples were collected from scattered localities in the Wealden district and selected from a few borehole cores. The majority came from the Hastings Beds Group, and the Wadhurst Clay was studied in detail (Fig. 8.1; Batten 1968). Subsequent work has included an examination of further material from the Hastings Beds, but has been concentrated on the older Purbeck Beds and the younger Weald Clay in the south-east of England and Wealden Marls and Shales of the Isle of Wight (Fig. 8.3). Recurrent associations between palynological entities and depositional environments have received particular attention. This has led to several interesting new discoveries, particularly concerning salinity variations in the Wessex–Weald Basin; some of these are discussed herein.

Previously I have relied largely on verbal description and illustration of selected components to describe palynofacies. I do, however, find that,

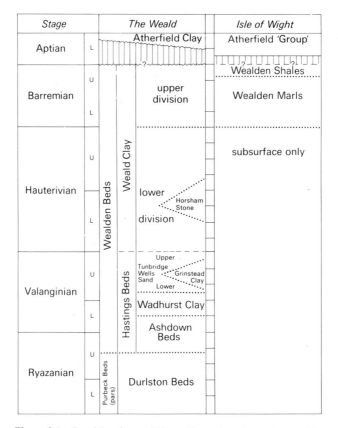

Figure 8.1 Provisional correlation with marine stages of part of the Purbeck Beds, the Wealden and the lowest division of the Lower Greensand of southeastern England and the Isle of Wight (adapted from Rawson *et al.* 1978, Fig. 3).

when accompanied by a few descriptive remarks, photographs showing the main characters have considerable practical value for reference. Although the accumulation of tabulated data is a necessary part of the procedure of organic facies analysis, annotated photographs are at least as convenient to use routinely as tables of relative abundance, if not more so. The discussion which follows is accompanied by a number of illustrations representing just a few of the many preparations which have been examined.

Geology

The uppermost Jurassic–Cretaceous rocks of the UK are mostly restricted in outcrop to the east and south of England, where they represent marginal extensions of three major offshore structures. One of these, the Wessex–

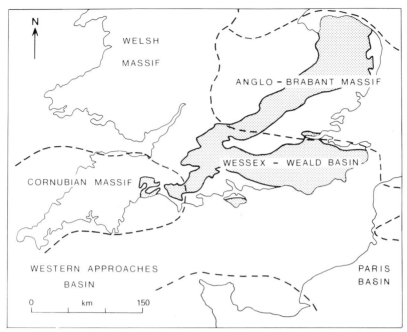

Figure 8.2 Structural setting of the Cretaceous deposits of southern England (adapted from Rawson *et al.* 1978, Figs 1, 2).

Weald Basin to the south of the Anglo-Brabant Massif (Fig. 8.2), was essentially non-marine from early Cretaceous Ryazanian/Berriasian to Barremian times. The base of the Cretaceous in southern England is taken to be the Cinder Beds of the Middle Purbeck (see Anderson & Bazley 1971, Casey 1973). The under- and overlying strata are termed the Lulworth and Durlston Beds respectively (Townson 1975). Whilst these divisions are appropriate for Dorset, they are less satisfactory for the Wealden district where, because the Cinder Beds horizon is not easy to identify (Anderson & Bazley 1971), their use is not accepted by all (see Lake & Holliday 1978). The boundary between the Upper Purbeck and the Wealden Beds in the south-east is gradational and thus also difficult to recognise. This led Allen (1975) to include the 30–70 m or so of Durlston Beds Formation in the Hastings Beds Group, although Rawson *et al.* (1978) maintain their separation (Fig. 8.1).

The overall structure of the Weald is a shallow dome, usually described as an anticlinorium. At its core approximately 70 m of the upper Ashdown Beds are exposed (Lake 1975), but faulting brings the Purbeck Beds to the surface in small inliers in the vicinity of Mountfield, Sussex (Fig. 8.3). The Ashdown Beds Formation comprises part of the Hastings Beds Group, a variable succession of argillaceous and arenaceous units which reaches a maximum thickness of approximately 430 m near the centre of the Weald. The overlying Weald Clay Group consists largely of argillaceous rocks and is

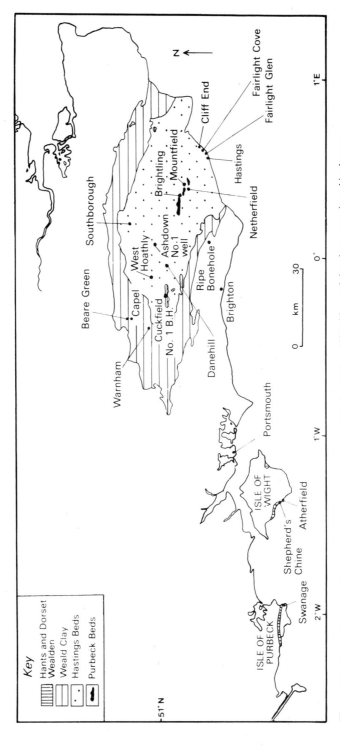

Figure 8.3 Map of southern England showing the areal extent of the Purbeck and Wealden and the places mentioned in the text.

thickest in the western Weald where it may attain 460 m (Thurrell *et al.* 1968, Lake 1975).

To the west of the Wealden district the pre-Aptian Lower Cretaceous succession on the Isle of Wight and in Swanage Bay, Dorset, is of similar thickness to that in the south-east. Although referred to as Wealden, the lithological subdivision of the Wealden district is not recognisable in this region. Further west still it rapidly becomes thinner and more arenaceous. On the Isle of Wight only the top 220 m or so of Wealden strata crop out, the succession being well exposed on the southern coast. The upper 50–60 m belong to the Wealden Shales, the remainder to the Wealden Marls (Fig. 8.1).

Purbeck Beds palynofacies

The Purbeck is made up of sediments which accumulated at a time of major regression following a phase of late Jurassic marine deposition. In Dorset gypsiferous evaporites formed low in the succession, but have since been largely replaced by carbonate and partly removed by solution (West 1975). Evaporites are, however, still present in the Mountfield Purbeck Beds of the Wealden district, where the basal part of the formation comprises some 15–20 m of mudstones, limestones, gypsum and anhydrite (Howitt 1964; Holliday & Shephard-Thorn 1974; Lake & Holliday 1978).

Palynological preparations from the Lower Purbeck in both the Weald and Dorset are usually dominated by *Classopollis* pollen grains, often occurring in tetrads (Pl. 8.1a). Associated palynomorphs tend to be mostly other gymnosperm pollen, particularly bisaccates and *Inaperturopollenites*. The abundance of *Classopollis* in samples adjacent to and within the evaporite beds (Norris 1969; M. A. Partington, in preparation) implies that the pollen grains are derived from a cheirolepidiaceous-dominated vegetation which was able to tolerate dry conditions. The climate may have been fairly arid–warm temperate, perhaps similar to that of present-day northern North Africa. The lowland Dorset vegetation was, at least for some of the time, dominated by trees whose wood has been referred to *Cupressinoxylon* and associated foliage compared with *Cupressinocladus* (Francis 1980) but whose male cones have yielded *Classopollis*. At least some of the plants which flooded these Purbeck lagoons with *Classopollis* seem, therefore, to have been arborescent, but at other times and in other places the Cheirolepidiaceae may also have been represented by shrubs (Batten 1976, Watson 1977).

This evaporite phase was followed by a period of less extreme conditions when a rhythmic series of shales, silts, current-bedded sandstones and limestones was deposited in waters of varying salinity. In the Mountfield area the Middle Purbeck horizon that is equivalent to the Dorset Cinder Beds consists of interbedded mudstones and limestones. These contain an invertebrate macrofauna that includes the bivalve *Praeexogyra* from which it is inferred that salinities were relatively high at this time (Lake & Holliday

1978). At other levels the occurrence of rootlet beds and sandstones which have erosive lower contacts (Allen 1975) implies a freshwater influence on the depositional environment. This is also suggested in palynological preparations by the lack of dinoflagellate cysts and the presence of algal masses of the *Botryococcus* type (Pl. 8.1b–d). Intermediate mildly saline conditions may be indicated by the occurrence of the palynomorph *Celyphus rallus* Batten (see below and Pl. 8.9a) and a few acritarchs and dinoflagellates (under investigation). Both *C. rallus* and *Botryococcus* masses also occur in, and are sometimes important components of, the Purbeck Beds in Dorset.

In addition to these microfossils, the palynomorph assemblages recovered from the Middle–Upper Purbeck in Sussex often contain large numbers of *Classopollis* pollen, though these grains are not quite as numerous as they are in the evaporite beds below. Other gymnosperm pollen, particularly bisaccate and inaperturate species, may also be very abundant (Pl. 8.2a). Pteridophyte spores are moderately varied in their morphology, but overall they are numerically subordinate to gymnospermous grains.

Hastings Beds palynofacies

The Hastings Beds consist mainly of light-coloured fine-grained sandstones and siltstones with subordinate shales and mudstones. The several formations into which the group has been divided vary in thickness and lithological content; whilst not recognisable throughout the district because of facies changes, they are nevertheless mappable over large parts of it. A variety of depositional environments is represented and these have received much attention during the past few decades, notably from Professor P. Allen. Originally thought to represent a deltaic pile, Allen is now of the opinion that the succession is dominated by braided alluvival sand plain and muddy alluvial and lagoonal facies (Allen 1959, 1967, 1975). However, this model has not yet wholly convinced all of those who have specialised in Wealden geology (see Lake 1976, Lake & Young 1978).

Hastings Beds palynofacies seem to show even more variation than the sedimentary facies. I now consider selected aspects of a few of these which are important from the general palaeoenvironmental standpoint.

Some of the best preserved miospore assemblages have been isolated from channel-fill deposits. These are most common in the Ashdown Beds and Tunbridge Wells Sand. Pteridophyte spores tend to dominate the assemblages. Cuticles (Pl. 8.2b), brown-black 'woody tissues' (vitrinite) and other 'humic' matter (Pls 8.2c, 8.6e&f), megaspores (Pl. 8.3a) and other 'large' plant microfossils (Pl. 8.4a–d) may also be abundant. A few channel deposits yield many hundreds of megaspores per 100 g of rock (Pl. 8.3a). The horizons concerned are light-medium to dark grey siltstones bearing plant debris dispersed at all angles. They readily disaggregate in water or in a solution of 'Regent' soft soap.

The composition and preservation of assemblages in channel deposits does, however, vary considerably according to the grain size and sorting of the sedimentary infill, and their general aspect may not be clearly distinct from palynofacies representing other depositional environments.

At outcrop, large parts of the Ashdown Beds and Tunbridge Wells Sand are commonly too coarse grained, well sorted and weathered to be of much palynological interest. However, where more argillaceous facies are developed locally, as in the Ashdown Beds along the coast east of Hastings (Fig. 8.3), miospores are usually recovered in large numbers (Batten 1973a, 1974). The state of preservation of the palynomorphs again varies widely. Although gymnosperm pollen grains are often abundant (Pls 8.3b&5a) the triradiate spore assemblages are more varied, reflecting what must have been a diverse lowland pteridophyte flora. Some are dominated by species or groups suggesting derivation from local vegetation. I have previously considered those in which *Trilobosporites* (Pl. 8.5b–d), *Pilosisporites* (Pl. 8.6a), *Concavissimisporites* (Pl. 8.6d), and *Cicatricosisporites* (Pl. 8.7a) are common to be key forms in the identification of assemblage types (Batten 1973a), but many others not selected for this purpose may also be abundant (e.g. Pl. 8.6b,c,e&f).

By contrast, the Wadhurst and Grinstead Clays yield miospore assemblages which are generally rather poorly preserved and taxonomically, though not numerically, impoverished (Pl. 8.7b&c). Usually gymnosperm pollen grains are dominant, and the triradiate spores are less varied than those of the Ashdown Beds and Tunbridge Wells Sand (Batten 1974). However, beds containing the 'horsetail' *Equisetites in situ* and adjacent horizons do yield rather different and distinctive assemblages (Pl. 8.8a; see also Batten 1968, 1973b), and there are also other preparations of note. A few samples have yielded large numbers of megaspores, mostly of the genus *Minerisporites* (Pl. 8.4k). Many more contain an abundance of *Classopollis* (Pl. 8.8b). By contrast with the Ashdown Beds and Tunbridge Wells Sand, which overall yield fewer pollen grains of this genus, assemblages from the transgressive Wadhurst and Grinstead Clays may be dominated by them. There seems to be a direct correlation between increased *Classopollis* representation and transgressive phases of deposition (Batten 1974; see also Médus & Pons 1967, Hughes & Moody-Stuart 1967a). *Classopollis* recovered from Ashdown and Tunbridge Wells Sand facies might have been produced by plants which coexisted with, but were not dominant in lowland communities. On the other hand, large quantities of *Classopollis* in association with an abundance of bisaccate and other gymnosperm pollen in Wadhurst and Grinstead assemblages may suggest a hinterland source for at least some of the components (Batten 1974).

Superabundance of *Classopollis* has been correlated with arid conditions (Vakhrameev 1970, and others). The overall lessening of the importance of this pollen during the earliest part of the Cretaceous is accompanied by an increase in the diversity of pteridophyte spores. Both phenomena suggest a climatic change to wetter conditions in the Wessex–Weald area which may well have been partly related to increased elevation of the London–Belgium ridge. As Allen (1975, 1976) has pointed out, the nature and extent of the fluvial deposits of the Hastings Beds Group suggest widely fluctuating rainfall.

In addition to the large numbers of degraded palynomorphs that occur in Wadhurst and Grinstead preparations, black wood fragments are also abundant. Some, perhaps much, of this debris may be charcoal produced by fire (Pl. 8.8d; compare with modern fragmented charcoal, Pl. 8.8c). I believe that all the palynological facts support the interpretation of the Wadhurst and Grinstead Clays as largely oxidised transgressive accumulations (Allen 1976). During these periods, the upland relief was low, the basin was subsiding, and the waters shallow and liable to fluctuate in salinity (Anderson et al. 1967, Allen et al. 1973). Although there is generally little positive evidence of marine influence on the depositional environment in the majority of the Wadhust palynological assemblages that I have examined, raised salinities are suggested by large quantities of dinoflagellate cysts at several horizons and, as noted earlier, it is possible that the occurrence of *Celyphus rallus* is also related to slightly saline phases.

C. rallus is rarely numerous in the Ashdown Beds and Tunbridge Wells Sand, whereas in the Purbeck Beds and in the Wadhurst, Grinstead and Weald Clays it is often present and sometimes abundant (Pl. 8.9a). It may be common in preparations which also contain dinoflagellate cysts. The affinity of this colonial organism remains uncertain (cf. Batten 1973a). I have compared it with various fungal remains and with modern and fossil loricae of presumed folliculinids (spirotrich ciliates). Any resemblance of individual specimens to the latter from offshore Belize (Pl. 8.8b&c) and from the Lower Cretaceous of Gabon (Pl. 8.8d) is probably superficial. Both fungal spores and folliculinids differ from *C. rallus* in the texture and colour of their walls. In transmitted light the outer surface of morphologically similar fungal spores is usually smoother. The Belize microfossils have thin smooth loricae. In thermally immature assemblages, fungal spores are generally darker in colour than associated pollen and spores of vascular plants. Likewise, the Gabonese Cretaceous ?folliculinids are significantly darker than the miospores, and the modern 'tests' have a blue-grey hue.

Indications of marine influence on the depositional environment in the form of dinoflagellate cysts, acritarchs, tasmanitids and foraminiferal linings do occur in the Ashdown Beds and Tunbridge Wells Sands but they are uncommon, and the dinoflagellates are often clearly reworked from older (Jurassic) strata. That the western Weald may have received transgressive saline pulses more frequently than further east (Allen 1976) is at least partly borne out by the recovery of several rich phytoplankton assemblages from the Grinstead Clay in the west. They comprise large numbers of only a few taxa and are not of typical marine aspect. Most of the species recorded to date from both the Wadhurst and Grinstead Clays are referable to the dinoflagellates *Muderongia* (Pls 8.8d, 8.10a), *Cyclonephelium*, and *Can-*

ningia, the acritarchs *Micrhystridium* and *Veryhachium* and to a 'simple sacs' category (cf. below).

Weald Clay palynofacies

The Weald Clay Group consists mainly of clay and silty clay, but also includes beds of sandstone, shelly limestone and clay ironstone. Water depths in the basin may occasionally have been somewhat greater during Weald Clay times than previously although Allen (1975) maintained that they seldom exceeded 2–3 m. The occurrence of mottled clays, suncrack and rootlet horizons, and beds with *Equisetites* and possibly other plants in position of growth (Allen 1959, 1975, Kennedy & MacDougall 1969, and others) lends some support to this suggestion, but they are less common than in the Hastings Beds.

Sedimentation was apparently governed by fluctuating fluvial and marine processes in a steadily subsidising basin. There is both invertebrate and palaeobotanical evidence to suggest that salinities varied more widely than during Hastings Beds deposition. A fossil series which includes *Equisetites, Viviparus, Filosina, Cassiope* and *Praeexogyra* is thought to indicate increasing salinity (see Allen *et al.* 1973, Morter in Worssam 1978). It is thus significant that *Equisetites lyellii* (Mantell) in position of growth is much more common in the Wadhurst than in the Grinstead and Weald Clays, and that *Neomiodon* is replaced by *Filosina* (Casey 1955, Allen 1975) in the lower Weald Clay. In addition most of the non-marine cyprid ostracods which characterise Hastings Beds assemblages are not found in the younger group (Anderson 1967, 1973). Although salinities fluctuated widely, these facts seem to indicate a change to generally more saline conditions of clay deposition, at least during the Hauterivian, in the western part of the Weald. It has long been known that a few horizons near the top of both the lower and upper divisions of the Weald Clay and the Wealden Shales of the Isle of Wight are brackish – marine (see Kilenyi & Allen 1968, Allen *et al.* 1973).

These general conclusions are broadly supported by the palynofloras. The 'marine bands' in the Weald Clay near Capel and Warnham (both Surrey, Fig. 8.3) have yielded large numbers of *Cribroperidinium* (Pl. 8.10b), *Cyclonephelium* cf. *C. distinctum* Deflandre and Cookson (Pl. 8.10d) and *Subtilisphaera terrula* (Davey) (Pl. 8.10e). Adjacent strata yield fewer of these 'marine' forms and varying quantities of an as yet unnamed peridiniacean cyst (Pl. 8.11d), a '*Cleistosphaeridium*' type (Pl. 8.11a), *Muderongia* (Pl. 8.11e), simple sacs (Pl. 8.11b) and *Celyphus rallus*, clearly indicating less saline conditions. Other strata yield only isolated microplankton or are entirely devoid of them.

The miospore assemblages associated with these restricted dinoflagellate 'populations' are normally dominated by bisaccate and other gymnospermous pollen. At best, their state of preservation is generally only fair, probably reflecting both deposition in oxidising environments and reworking. In the context of the latter, it is interesting that an abundance of Jurassic dinoflagellates has been recovered from samples adjacent to calcareous sandstones at Capel which contain burrows identified as *Ophiomorpha* and attributed to callianassid shrimps. This trace fossil has generally been considered to suggest marine conditions, at least during the period represented by the surface from which the burrows originate (Kennedy & MacDougall 1969, Dike 1972, Allen 1975), although its acceptance as a marine indicator has been questioned recently (Stewart 1978).

Other Weald Clay sections of different ages have also yielded numerous dinoflagellates at certain horizons, some of which are new (Pl. 8.11c&f). All will be described formally elsewhere.

Wealden Marls and Shales palynofacies, Isle of Wight

The Wealden Marls of the Isle of Wight (Figs 8.1&3) consist mainly of mottled reddish-purplish brown mudstones with subordinate beds of grey mudstone and siltstone, sandstone, limestone and conglomerate. The bulk is thought to be of fluvial origin (Stewart 1978, Daley & Stewart 1979). The mottled muds and silts are usually devoid of palynomorphs, or yield impoverished assemblages. However the grey argillaceous units are often rich in miospores (Pl. 8.12a) and, as in the Hastings Beds, some of the fine-grained channel deposits contain not only diverse well preserved assemblages but also large quantities of megaspores.

I have seen little palynological evidence of marine influence on the depositional environment except near the top of the formation. The interbedded sandstones and mudstones which make up the highest beds show a variety of sedimentary structures including trace fossils of the *Ophiomorpha* type. On the basis of the ostracod fauna, the presence of charophyte oogonia and sedimentological studies, Stewart (1978) suggested that the burrows were constructed by animals tolerant of low or fluctuating salinities. The recovery of assemblages containing numerous acritarchs and some dinoflagellates from horizons within these beds (Pls 8.11g, 12b&13a) does, however, suggest that at times conditions were more saline than hitherto.

The Wealden Shales are grey to greenish-grey mudstones with subordinate beds of clay ironstone, sandstone and shelly limestone. The grey colour and even bedding contrasts strongly with the variegated and massive marls below. Above the prominent Barnes High Sandstone (White 1921) the mudstones are often finely laminated with silt and rich in ostracods. The occurrence of *Viviparus, Unio, Filosina,* cyprid ostracods, remains of the fish *Hybodus* (Patterson 1966), and ferns (Alvin 1974, D. J. Batten, in preparation) suggests that deposition took place in a lagoon of varying salinity (Stewart 1978) not so very far from which plants were able to maintain a foothold. The Barnes High Sandstone is thought to be a river mouth bar (Daley & Stewart 1979). Near the top of the Wealden Shales, salinities apparently increased because oysters and the gastropod *Cassiope* are found, heralding the onset of the early Aptian marine trangression and deposition of the Lower Greensand.

The palynological assemblages isolated from the Wealden Shales are of varied composition reflecting the fluctuating salinities. As in the Weald Clay, relative abundances of *Cribroperidinium* and *Cyclonephelium* are associated with brackish-marine phases. Occurring with these marine genera and on their own are other plankton which were presumably able to tolerate less saline conditions (e.g. Pl. 8.13a).

Conclusion

The determination of sedimentary environments forms an important part of the research of many sedimentologists, but attention is generally concentrated on sandstones and limestones. I believe that the study of the organic contents of not only shales and siltstones, but also of limestones and the finer sandstones, can significantly aid the interpretation of sedimentary facies. Certainly the examples selected for discussion and illustration in this paper provide a basis for more reliable assessments of Purbeck–Wealden depositional conditions. Although I have not discussed palynofacies relationships in detail, bed by bed, the generalisations made are based on such data.

An important outcome of this work from the palynological viewpoint, on which I have placed some emphasis herein, has been the recognition of non-marine dinoflagellates. Although modern dinoflagellates are widely distributed in both freshwater and marine environments, with a few Tertiary exceptions, virtually all known fossil forms are considered to be marine or brackish-marine. Some Wealden cysts are clearly reworked from older strata (Pl. 8.13d&e). Some are identified as marine species (e.g. *Cyclonephelium distinctum, Subtilisphaera terrula*) which were presumably able to tolerate lower salinities. Others are referable to marine genera (e.g. *Muderongia*) but cannot be identified with any known species. Yet others are unknown in marine assemblages and are probably the cysts of dinoflagellates which thrived in brackish and/or fresh water. Not until the Perna Bed of the Lower Greensand was deposited do the assemblages take on a fully marine aspect (Pl. 8.13b,c&f).

Acknowledgements

Some of the dinoflagellate data on which the conclusions presented here are based were first discussed at the Fifth International Palynological Conference, Cambridge (Batten & Eaton 1980). I thank Dr J. D. Burgess (Gulf Research and Development Company, Houston, Texas) for the loan of slides of palynological material from offshore Belize and the Cretaceous of Gabon, and Dr C. Sladen for his comments on the manuscript of this paper. I am grateful to the Institute of Geological Sciences, British Petroleum Company Ltd, and Dr Sladen for making samples available for study, and to a number of companies and individuals for access to the quarries and clay pits under their management; these include London Brick Company Ltd, Redland Bricks Ltd, Ibstock Brick Hudsons, Freshfield Lane Brickworks Ltd, British Gypsum Ltd, and Mr L. A. Hannah, Philpots Quarry.

References

Allen, P. 1959. The Wealden environment: Anglo-Paris basin. *Phil Trans R. Soc. B* **242**, 283–346.

Allen, P. 1967. Origin of the Hastings facies in northwestern Europe. *Proc. Geol. Assoc.* **78**, 27–105.

Allen, P. 1975. Wealden of the Weald: a new model. *Proc. Geol. Assoc.* **86**, 389–437.

Allen, P. 1976. Wealden of the Weald – a new model. Reply *Proc. Geol. Assoc.* **87**, 433–42.

Allen, P., M. L. Keith, F. C. Tan and P. Deines, 1973. Isotopic ratios and Wealden environments. *Palaeontology* **16**, 607–21.

Alvin, K. L. 1974. Leaf anatomy of *Weichselia* based on fusainized material. *Palaeontology* **17**, 587–98.

Anderson, F. W. 1967. Ostracods from the Weald Clay of England. *Bull. Geol Surv. Gt Br.* **27**, 237–69.

Anderson, F. W. 1973. The Jurassic–Cretaceous transition: the non-marine ostracod faunas. In *The boreal Lower Cretaceous*, R. Casey & P. F. Rawson (eds), 101–10. Liverpool: Seel House Press.

Anderson, F. W. and R. A. B. Bazley 1971. The Purbeck Beds of the Weald (England). *Bull. Geol Surv. Gt Br.* **34**.

Anderson, F. W., R. A. B. Bazley and E. R. Shephard-Thorn 1967. The sedimentary and faunal sequence of the Wadhurst Clay (Wealden) in boreholes at Wadhurst Park, Sussex. *Bull. Geol Surv. Gt Br.* **27**, 171–235.

Batten, D. J. 1968. Probable dispersed spores of Cretaceous *Equisetites*. *Palaeontology* **11**, 633–42.

Batten, D. J. 1969. Some British Wealden megaspores and their facies distribution. *Palaeontology* **12**, 333–50.

Batten, D. J. 1973a. Use of palynologic assemblage-types in Wealden correlation. *Palaeontology* **16**, 1–40.

Batten, D. J. 1973b. Palynology of early Cretaceous soil beds and associated strata. *Palaeontology* **16**, 399–424.

Batten, D. J. 1974. Wealden palaeoecology from the distribution of plant fossils. *Proc. Geol. Assoc.* **85**, 433–58.

Batten, D. J. 1976. Wealden of the Weald – a new model. Written discussion of a paper previously published. *Proc. Geol. Assoc.* **87**, 431–3.

Batten, D. J. and G. L. Eaton 1980. Dinoflagellates and salinity variations in the Wealden (Lower Cretaceous) of southern England. *5th int. Palynol. Conf., Cambridge*, p. 32 (abstract).

Casey, R. 1955. The pelecypod family Corbiculidae in the Mesozoic of Europe and the Near East. *J. Wash. Acad. Sci.* **45**, 366–72.

Casey, R. 1973. The ammonite succession at the Jurassic–Cretaceous boundary in eastern England. In *The boreal Lower Cretaceous*, R. Casey & P. F. Rawson (eds), 193–266. Liverpool: Seel House Press.

Chaloner, W. G. 1968. The paleoecology of fossil spores. In *Evolution and environment*, 125–38. New Haven, Conn.: Yale University Press.

Chaloner, W. G. and M. D. Muir 1968. Spores and floras. In *Coal and coal-bearing strata*, D. G. Murchison & T. S. Westoll (eds), 127–46. Edinburgh: Oliver & Boyd.

Combaz, A. 1964. Les palynofaciès. *Rev. Micropaléont.* **7**, 205–18.

Correia, M. 1967. Relations possible entre l'état de conservation des éléments figurés de la matière organique (microfossiles palynoplanctologiques) et l'existence de gisements d'hydrocarbures. *Rev. Inst. Fr. Pétrole* **22**, 1285–306.

Correia, M. 1969. Contribution à la recherche de zones favorables à la genèse du pétrole par l'observation microscopique de la matière organique figurée. *Rev. Inst. Fr. Pétrole* **24**, 1417–54.

Correia, M. 1971. Diagenesis of sporopollenin and other comparable organic substances: application to hydrocarbon research. In *Sporopollenin*, J. Brooks, P. R. Grant, M. D. Muir, P. van Gijzel & G. Shaw (eds), 569–620. London: Academic Press.

Cross, A. T., G. G. Thompson and J. B. Zaitzeff 1966. Source and distribution of palynomorphs in bottom sediments, southern part of Gulf of California. *Marine Geol.* **4**, 467–524.

Daley, B. and D. J. Stewart 1979. Week-end field meeting: the Wealden Group in the Isle of Wight. *Proc. Geol. Assoc.* **90**, 51–4.

de Jekhowsky, B. 1963a. Répartition quantitative des grandes groups de "microorganontes" (spores, hystrichosphères, etc.) dans les sédiments marins du plateau continental. *C. R. Somm. Séanc. Soc. Biogéogr.* **349**, 29–47.

de Jekhowsky, B. 1963b. Variations latérales en palynologie quantitative et passage du continental au marin: le Dogger Supérieur du sud-ouest de Madagascar. *Rev. Inst. Fr. Pétrole* **18**, 977–95.

Dike, E. F. 1972. *Ophiomorpha nodosa* Lundgren: environmental implications in the Lower Greensand of the Isle of Wight. *Proc. Geol. Assoc.* **83**, 165–77.

Francis, J. 1980. The Purbeck Forest of the Great Dirt Bed of Dorset. *Int. Palaeobot. Conf. Reading*, p. 19 (abstract).

Groot, J. J. 1966. Some observations on pollen grains in suspension in the estuary of the Delaware River. *Marine Geol.* **4**, 409–16.

Habib, D. 1979. Sedimentary origin of North Atlantic Cretaceous palynofacies. In *Deep drilling results in the Atlantic Ocean: continental margins and palaeoenvironment*, M. Talwani, W. Hay & W. B. F. Ryan (eds), 420–37. Washington, D.C., American Geophysical Union.

Holliday, D. W. and E. R. Shephard-Thorn 1974. *Basal Purbeck evaporites of the Fairlight Borehole, Sussex*. Rep. Inst. Geol Sci. 74/4.

Howitt, F. 1964. Stratigraphy and structure of the Purbeck inliers of Sussex (England). *Q. J. Geol Soc. Lond.* **120**, 77–113.

Hughes, N. F. and J. C. Moody-Stuart 1967a. Palynological facies and correlation in the English Wealden. *Rev. Palaeobot. Palynol.* **1**, 259–68.

Hughes, N. F. and J. C. Moody-Stuart 1967b. Proposed method of recording pre-Quaternary palynological data. *Rev. Palaeobot. Palynol.* **3**, 347–58.

Kennedy, W. J. and J. D. S. MacDougall 1969. Crustacean burrows in the Weald Clay (Lower Cretaceous) of southeastern England and their environmental significance. *Palaeontology* **12**, 459–71.

Kieser, G. 1967. Quelques aspects particuliers de la palynologie du Supérieur du Crétacé Sénegal. *Rev. Palaeobot. Palynol.* **5**, 199–210.

Kilenyi, T. J. and N. W. Allen 1968. Marine-brackish bands and their microfauna from the lower part of the Weald Clay of Sussex and Surrey. *Palaeontology* **11**, 141–62.

Lake, R. D. 1975. The structure of the Weald – a review. *Proc. Geol. Assoc.* **86**, 549–57.

Lake, R. D. 1976. Wealden of the Weald – a new model. Written discussion of a paper previously published. *Proc. Geol. Assoc.* **87**, 827–9.

Lake, R. D. and D. W. Holliday 1978. *Purbeck Beds of the Broadoak Borehole, Sussex*. Rep. Inst. Geol Sci. 78/3, 1–12.

Lake, R. D. and R. G. Thurrell 1974. *The sedimentary sequence of the Wealden beds in boreholes near Cuckfield, Sussex*. Rep. Inst. Geol Sci. 74/2.

Lake, R. D. and B. Young 1978. *Boreholes in the Wealden Beds of the Hailsham area, Sussex*. Rep. Inst. Geol Sci. 78/23.

Marshall, A. E. and A. H. V. Smith 1965. Assemblages of miospores from some Upper Carboniferous coals and their associated sediments in the Yorkshire Coalfield. *Palaeontology* **7**, 656–73.

Médus, J. and A. Pons 1967. Étude palynologique du Crétacé Pyrénéo-Provençal. *Rev. Palaeobot. Palynol.* **2**, 111–17.

Muir, M. D. 1967. Reworking in Jurassic and Cretaceous spore assemblages. *Rev. Palaeobot. Palynol.* **5**, 145–54.

Muller, J. 1959. Palynology of Recent Orinoco delta and shelf sediments. *Micropaleontology* **5**, 1–32.

Neves, R. 1958. Upper Carboniferous plant spore assemblages from the *Gastrioceras subcrenatum* horizon, north Staffordshire. *Geol Mag.* **95**, 1–19.

Norris, G. 1969. Miospores from the Purbeck Beds and marine Upper Jurassic of southern England. *Palaeontology* **12**, 574–620.

Patterson, C. 1966. British Wealden sharks. *Bull. Br. Mus. Nat. Hist. (Geol.)* **11**, 281–350.

Rawson, P. F., D. Curry, F. C. Dilley, J. M. Hancock, W. J. Kennedy, J. W. Neale, C. J. Wood and B. C. Worssam 1978. *A correlation of Cretaceous rocks in the British Isles*. Geol Soc. Lond. Sp. Rep. no. 9.

Rossignol, M. 1961. Analyse pollinique de sédiments marins quaternaires en Israël. I. Sédiments Récents. *Pollen Spores* **3**, 303–24.

Sigal, J. 1965. Palynofaciès: un néologisme illogique et superflu. *Rev. Micropaléont* **8**, 59–60.

Smith, A. H. V. 1962. The palaeoecology of Carboniferous peats based on the miospores and petrography of bituminous coals. *Proc. Yorks Geol Soc.* **33**, 423–74.

Spackman, W., C. P. Dolsen and W. Riegel 1966. Phytogenetic organic sediments and sedimentary environments in the Everglades-mangrove complex. Part I: Evidence of a trangressing sea and its effects on environments of the Shark River area of southwestern Florida. *Palaeontographica B* **117**, 135–52.

Stewart, D. J. 1978. *Ophiomorpha*: a marine indicator? *Proc. Geol. Assoc.* **89**, 33–41.

Thurrell, R. G., B. C. Worssam and E. A. Edmonds 1968. *Geology of the country around Haslemere*. Mem. Geol Surv. Gt Br.

Townson, W. G. 1975. Lithostratigraphy and deposition of the type Portlandian. *J. Geol Soc. Lond.* **131**, 619–38.

Traverse, A. and R. N. Ginsburg 1966. Palynology of the surface sediments of Great Bahama Bank, as related to water movement and sedimentation. *Marine Geol.* **4**, 417–59.

Tschudy, R. H. 1961. Palynomorphs as indicators of facies environments in Upper Cretaceous and Lower Tertiary strata, Colorado and Wyoming. In *Guidebook, 13th Annual Field Conference*, 53–9 Geological Association of Wyoming.

Upshaw, C. F. 1964. Palynological zonation of the Upper Cretaceous Frontier Formation near Dubois, Wyoming. In *Palynology in oil exploration. A symposium*, Sp. Publs Soc. Econ. Palaeont. Miner., Tulsa, no. 11, 153–68.

Vakhrameev, V. A. 1970. Range and palaeoecology of Mesozoic conifers, the Cheirolepidiaceae. *Paleont. Zh.* **1**, 19–34. (In Russian.)

Watson, J. 1977. Some Lower Cretaceous conifers of the Cheirolepidiaceae from the USA and England. *Palaeontology* **20**, 715–49.

West, I. M. 1975. Evaporites and associated sediments of the basal Purbeck Formation (Upper Jurassic) of Dorset. *Proc. Geol. Assoc.* **86**, 205–25.

White, H. J. O. 1921. *A short account of the geology of the Isle of Wight*. Mem. Geol Surv. UK.

Williams, D. B. and W. A. S. Sarjeant 1967. Organic-walled microfossils as depth and shore-line indicators. *Marine Geol.* **5**, 389–412.

Worssam, B. C. 1978. *The stratigraphy of the Weald Clay*. Rep. Inst. Geol Sci. 78/11.

Plates

In this and subsequent plate captions in this chapter, stage co-ordinates refer to Leitz Dialux (LD) microscope number 322, Department of Geology and Mineralogy, Aberdeen University.

Plate 8.1 (a) Palynofacies in which degraded *Classopollis* pollen grains are dominant. These occur singly and in tetrads. Pale background detritus is undigested gypsum. Sample DJB 80/M1, a calcareous mudstone containing gypsum from British Gypsum Brightling Mine, Stonehouse ventilation shaft, Lower Purbeck Beds. Preparation MCP 1504.3, LD 34.3 120.9, ×225.

(b) Amorphous organic matter of non-marine (?algal) origin with algae of the *Botryococcus* type. Sample, a shaley mudstone from the Purbeck Beds, depth 219.2 m in Ashdown No. 1 well, drilled during 1954–55 by D'Arcy Exploration Company Ltd. Preparation DB 1197.2, LD 31.9 133.1, ×450.

(c) Alga of the *Botryococcus* type. Sample and preparation as above, LD 32.1 134.0, ×450.

(d) Alga of the *Botryococcus* type. Sample and preparation as above, LD 18.1 122.2, ×450.

Plate 8.2 (a) Palynofacies in which *Classopollis* and other gymnosperm pollen are abundant. Numerous tetrads of *Classopollis*. Much of the black detritus is probably fusinite (inertinite). Sample DJB 80/42, a calcareous shale from above the 'Cinder Bed' horizon, River Line stream section, Netherfield, Sussex. Preparation MCP 903.2, LD 43.3 122.2, ×135.

(b) Cuticle-rich palynofacies recovered from a siltstone in Fairlight Clays facies of the Ashdown Beds Formation on the coast east of the Fairlight anticline, Fairlight Cove. Sample DJB/CE 26, preparation MCP 618.2, LD 33.2 128.0, ×90.

(c) Plant debris/'humic'-rich palynofacies from a siltstone in Fairlight Clays facies, Fairlight Cove as above. Sample DJB/CE 30, preparation MCP 622.2, LD 31.1 124.9, ×90.

Plate 8.3 (a) Palynofacies in which small triradiate spores and megaspores (mostly *Minerisporites alius* Batten) are exceptionally abundant. *Botryococcus* s.l. and other possible freshwater algae referable to *Schizosporis reticulatus* Cookson and Dettmann (and another undescribed species) occur in association. Sample DJB/CE 12 from a channel-fill siltstone in the Ashdown Beds coastal section near Cliff End. Preparation MCP 527.2, LD 37.1 132.9, ×90.

(b) Gymnosperm (bisaccate) pollen-dominated palynofacies from a siltstone in Fairlight Clays facies of the Ashdown Beds coastal section east of the Fairlight anticline, Fairlight Cove. Sample DJB/CE 20, preparation MCP 535.2, LD 36.6 129.2, ×135.

Plate 8.4 (a),(d) Spore mass of *Trilobosporites* sp. Unregistered silty mudstone sample labelled CUC 971 from depth of 296 m in IGS Cuckfield No. 1 borehole, Ashdown Beds (see Lake & Thurrell 1974). (a) ×90; (d) detail, ×180.

(b),(c) Part of a sporangium-bearing smooth-walled triradiate miospores. Sample DJB 174, a siltstone from the Fairlight Clays facies of the Ashdown Beds near Fairlight Glen. (b) ×90; (c) detail, ×225.

(e)–(k) Megaspores.

(e) *Trileites* sp., ×90. Sample DJB 170, a siltstone from the Fairlight Clays facies of the Ashdown Beds near Fairlight Glen as above.

(f),(j) *Arcellites pyriformis* (Hughes), ×45. Sample CUC 971, see explanation of (a),(d) above.

(g),(h) *Hughesisporites galericulatus* (Dijkstra emend. Hughes). (g) ×90; (h) detail, ×225. Sample DJB 170, see explanation of (e).

(i) *Paxillitriletes alatus* (Batten), ×45. Unregistered silty mudstone sample labelled CUC 442 from depth of 134.7 m in IGS Cuckfield No. 1 borehole, Upper Tunbridge Wells Sand (see Lake & Thurrell 1974).

(k) *Minerisporites marginatus* (Dijkstra), ×135, from sample CUC 442 as above.

Plate 8.5 (a) Gymnosperm pollen-dominated palynofacies with algal mass of the *Botryococcus* type. Sample DJB/CE 6, a mudstone from the Ashdown Beds coastal section near Cliff End. Preparation MCP 514.2, LD 41.8 133.8, ×135.

(b),(c) *Trilobosporites* spp. Sample DJB/CE 17, a siltstone from the Ashdown Beds coastal section, Goldbury Point, Fairlight Cove. Preparation MCP 532.2: (b) LD 38.1 126.3; (c) LD 37.7 126.2, both ×450.

(d) Palynofacies in which species of *Trilobosporites* are unusually common. Sample and preparation as above, LD 37.3 126.1, ×135.

Plate 8.6 (a) *Pilosisporites trichopapillosus* (Thiergart) s.l. Sample DJB/CE 51, a siltstone from the Ashdown Beds coastal section near Cliff End. Preparation MCP 651.2, LD 34.5, 133.3, ×450.

(b) *Couperisporites* cf.*C. complexus* (Pocock). Sample and preparation as above, LD 21.9 132.2, ×450.

(c) *Triproroletes* sp. Sample and preparation as above, LD 35.0 137.1, ×450.

(d) *Concavissimisporites* sp. Sample and preparation as above, LD 24.0 130.1, ×450.

(e) Miospore assemblage dominated by *Gleicheniidites*. Sample DJB/CE 25, a siltstone from Fairlight Clays facies of the Ashdown Beds coastal section east of the Fairlight anticline, Fairlight Cove. Preparation MCP 617.2, LD 29.7 125.2, ×135.

(f) Miospore assemblage dominated by smooth-walled triradiate spores. Sample DJB/CE 50, a siltstone from the Ashdown Beds coastal exposure northeast of Haddock's reversed fault near Cliff End. Preparation MCP 650.2, LD 23.2 122.2, ×135.

Plate 8.7 (a) Miospore assemblage dominated by species of *Cicatricosisporites*. Sample DJB/CE 33, a mudstone containing plant debris from the Fairlight Clays facies of the Ashdown Beds coastal section east of the Fairlight anticline, Fairlight Cove. Preparation MCP 627.2, LD 38.3 132.2, ×135.

(b) Typical Wadhurst Clay palynofacies. Miospores abundant but poorly preserved. Sample DJB 348A, a mudstone from the upper Wadhurst Clay, High Brooms pit, Southborough. Preparation A581.2, LD 36.9 126.8, ×225.

(c) Wadhurst Clay palynofacies with abundant miospores, the majority in a very degraded state. Sample DJB 80/47, a laminated siltstone containing vertebrate debris, upper Wadhurst Clay, Freshfield Lane Brickworks clay pit, Danehill. Preparation MCP 874.2, LD 29.8 121.0, ×135.

Plate 8.8 (a) Palynofacies in which cuticles and other tissues, and presumed spores of *Equisetites*, dominate; recovered from a plant fragment parting in mudstone associated with *Equisetites* in position of growth. Unregistered IGS sample CUC 792D from a depth of approximately 241 m in Cuckfield No. 1 borehole, Wadhurst Clay (see Lake & Thurrell 1974). Preparation T244.3, LD 49.1 129.1, ×135.

(b) *Classopollis*-rich palynofacies. Sample DJB 292, a mudstone from the upper Wadhurst Clay, High Brooms pit, Southborough. Preparation T224/7, LD 36.9 124.1, ×225.

(c) Fragmented modern charcoal. Sample DJB 80/M2, preparation MCP 1430.1, LD 40.2 126.9, ×135.

(d) Grinstead Clay palynofacies in which dinoflagellate cysts are abundant. Sample DJB 79/P3, a silty mudstone from the Lower Grinstead Clay of Philpots Quarry, West Hoathly. Preparation MCP 676.2, LD 35.1 124.4, ×135.

Plate 8.9 (a) *Celyphus rallus* Batten palynofacies. Unregistered IGS sample CUC 869, a mudstone from a depth of 264.8 m in Cuckfield No. 1 borehole, Wadhurst Clay (see Lake & Thurrell 1974). Preparation A281.1, LD 22.9 130.4, ×135.

(b),(c) Presumed folliculinids from bottom sediments, southern end of Tobacco Cay, offshore Belize. Sample/preparation JR 16: (b) LD 32.1 131.0; (c) LD 34.0 119.1; both ×135.

(d) Presumed fossil folliculinids from the Lower Cretaceous of Gabon. Sample/preparation A0757, LD 17.1 125.7, ×135.

Plate 8.10 (a) A ceratiacean cyst of the *Muderongia* type. Sample CSL/P12, a calcareous mudstone from the Lower Grinstead Clay of Philpots Quarry, West Hoathly. Preparation MCP 358.2, LD 29.9, 120.1, interference contrast, ×900.

(b) *Cribroperidinium* sp. Mudstone sample DJB 388, part of a 'marine band' containing the bivalve *Filosina* in the upper part of the lower division of the Weald Clay exposed in the Clockhouse Brickworks pit, Capel. Preparation A699.4, LD 24.1 135.1, ×900.

(c) *Cribroperidinium*-dominated palynofacies from 'marine band' sample DJB 158, a calcareous mudstone containing *Cassiope* from the lower division of the Weald Clay, Warnham Brick pit. Preparation T139.4, LD 11.0 124.9, ×135.

(d) *Cyclonephelium* cf. *C. distinctum* Deflandre and Cookson. Sample DJB 388, details as for (b) above. Preparation A699.4, LD 22.9 134.1, ×900.

(e) Peridiniacean cyst referable to *Subtilisphaera terrula* (Davey). Sample DJB 390, a mudstone from a 'marine band' in the upper part of the lower division of the Weald Clay exposed in the Clockhouse Brickworks pit, Capel. Preparation A701.3, LD 22.3 122.1, interference contrast, ×900.

Plate 8.11 (a) *'Cleistosphaeridium'*. Sample DJB 379, a calcareous shale from the upper part of the lower division of the Weald Clay exposed in Clockhouse Brickworks pit, Capel. Preparation A677.4, LD 46.7 128.6, interference contrast, ×900.

(b) Cyst type A, a 'simple sac' from the same sample and preparation as above, LD 57.8 124.8, ×900.

(c) Cyst type B from a calcareous mudstone sample, IGS Ripe Borehole, depth 170.4–171 m, Weald Clay (see Lake & Young 1978). Preparation MCP 1083.2, LD 28.1 121.7, interference contrast, ×900.

(d) Peridiniacean cyst type A from the same sample and preparation as (a) and (b), LD 43.1 130.2, interference contrast, ×900.

(e) *Muderongia* sp. from calcareous mudstone sample DJB 382, Weald Clay, Capel, see (a). Preparation A680.4, LD 117.7 52.9, interference contrast, ×900.

(f) Peridiniacean cyst type B from a Weald Clay mudstone with silty laminae, Beare Green Brickworks pit. Sample CSL/BG 19, preparation MCP 494.1, L D20.9 129.9, interference contrast, ×900.

(g) *Veryhachium* sp. from a mudstone at the top of the Wealden Marls on the coast northwest of Shepherd's Chine, Isle of Wight. Sample DJB 80/27, preparation MCP 888.2, LD 42.5 122.8, interference contrast, ×450.

Plate 8.12 (a) Diverse assemblage of miospores in varying states of preservation from a siltstone at the top of the Wealden Marls coastal exposure northwest of Shepherd's Chine, Isle of Wight. Sample DJB 80/28, preparation MCP 889.2, LD 41.6 127.2, ×135.

(b) General aspect of the palynofacies in which the *Veryhachium* on Plate 8.11g and Cyst type C (Pl. 8.13a) occur. Preparation MCP 888.2, LD 27.1 126.3, ×135.

Plate 8.13 (a) Cyst type C from sample DJB 80/27 (see explanation of Plate 8.11g for details of section). Preparation MCP 888.2, LD 28.9 128.0, ×450.

(b) *Spiniferites ramosus* (Ehrenberg), Perna Bed (sandstone), basal Lower Greensand, Atherfield Point, Isle of Wight. Sample DJB 80/25, preparation MCP 887.2, LD 21.9 130.1, ×450.

(c) *Protoellipsodinium spinosum* Davey and Verdier, Perna Bed, sample and preparation as above, LD 17.9 131.4, ×450.

(d) *Wanaea fimbriata* Sarjeant reworked from the late Jurassic (probably early Oxfordian) in a laminated muddy siltstone, upper part of the lower division of the Weald Clay, Clockhouse Brickworks pit, Capel. Sample DJB 370, preparation A583.4, LD 38.8 136.9, ×450.

(e) *Gonyaulacysta areolata* Sarjeant reworked from the Jurassic (late Callovian to early Oxfordian). Sample DJB 370 as above, preparation A583.4, LD 57.9 134.6, ×450.

(f) Perna Bed palynofacies; scattered marine dinoflagellate cysts, foraminiferal linings, miospores, brown-black woody and amorphous detritus. Sample DJB 80/25; for locality details see (b) above. Preparation MCP 887.2, LD 30.3 132.9, ×135.

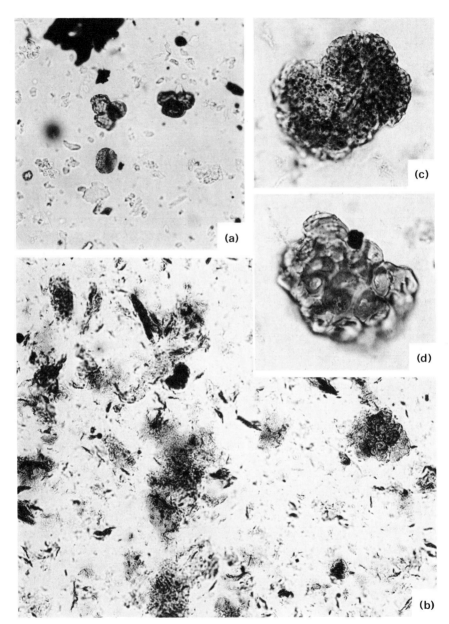

Plate 8.1

Plate 8.2

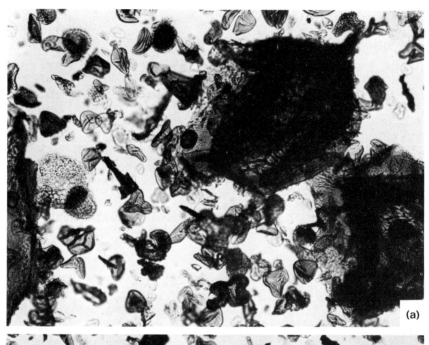
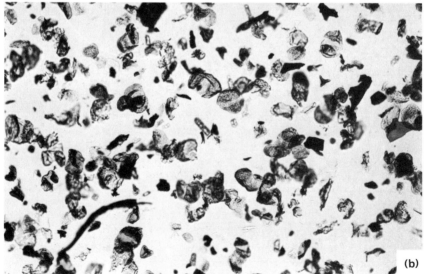

Plate 8.3

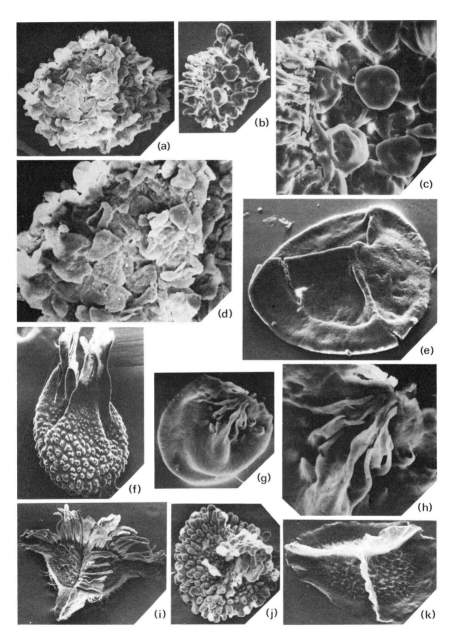

Plate 8.4

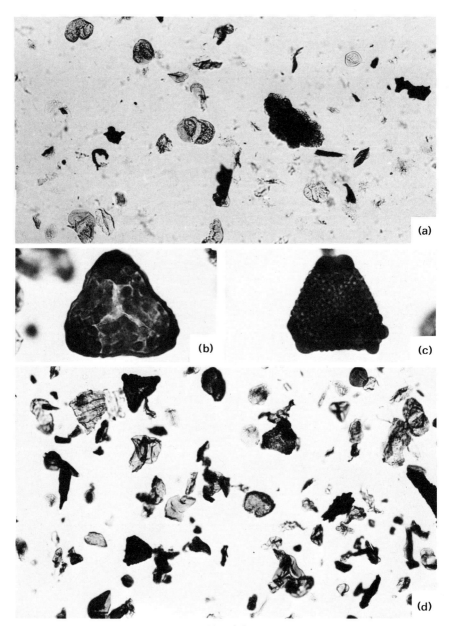

Plate 8.5

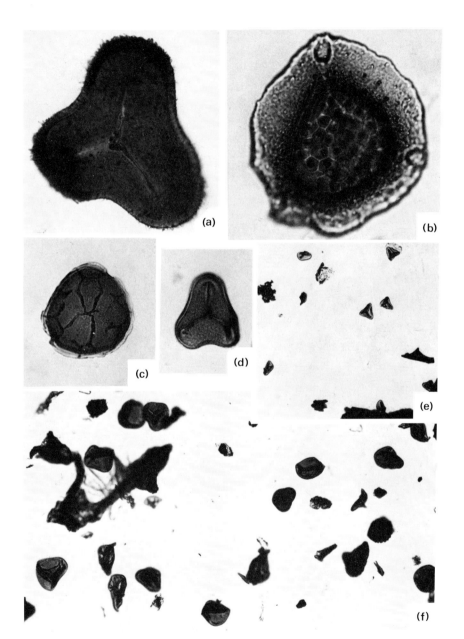

Plate 8.6

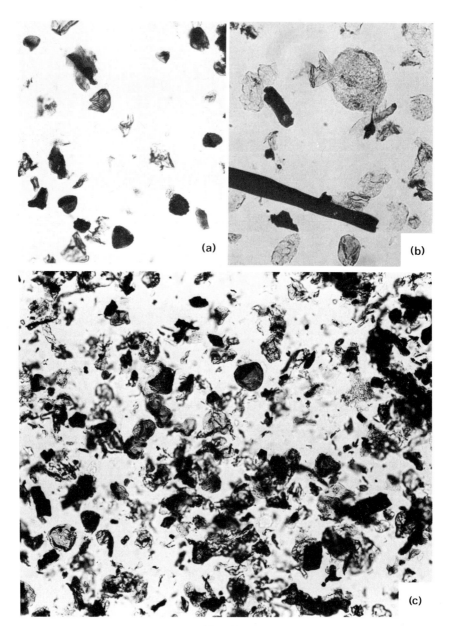

Plate 8.7

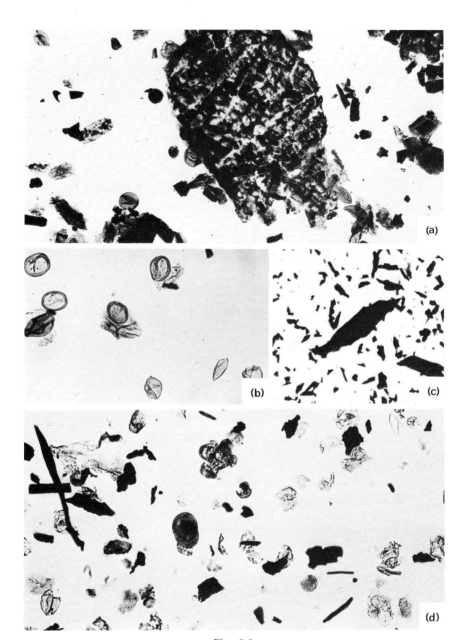

Plate 8.8

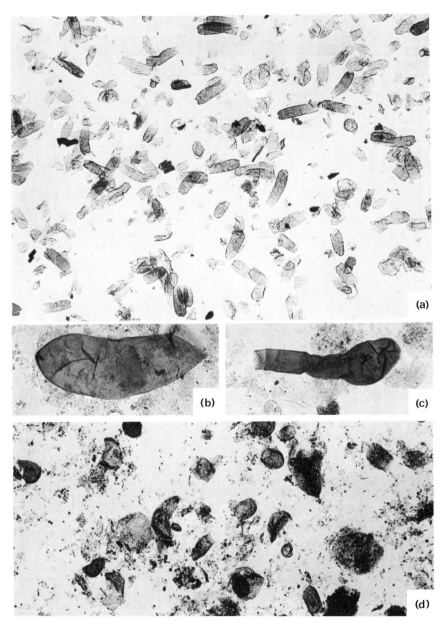

Plate 8.9

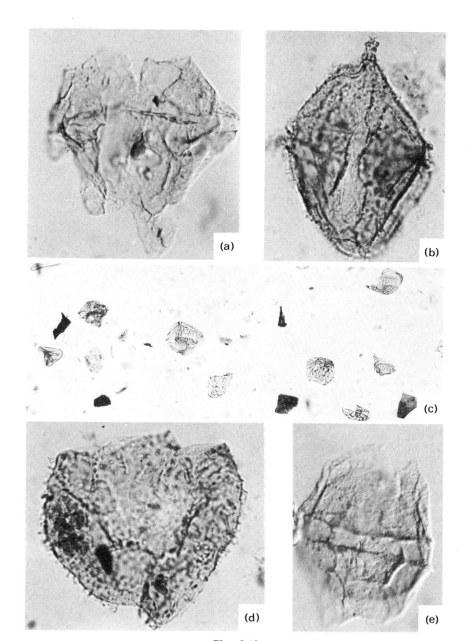

Plate 8.10

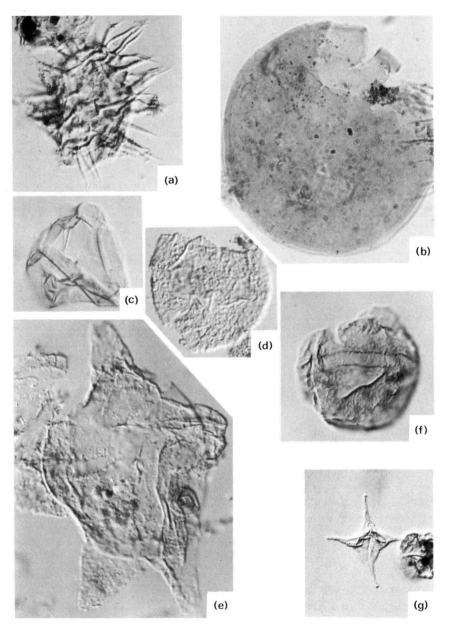

Plate 8.11

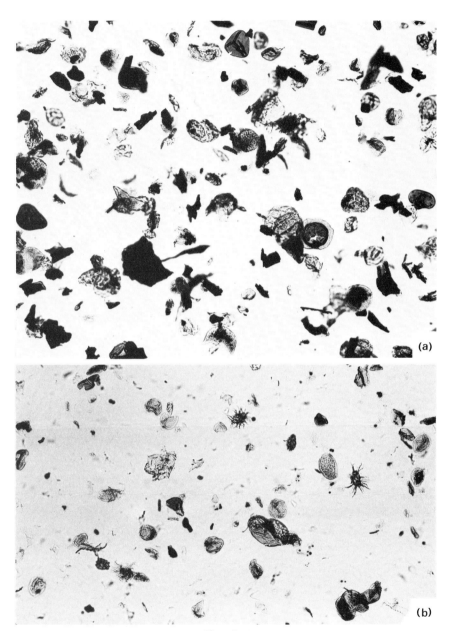

Plate 8.12

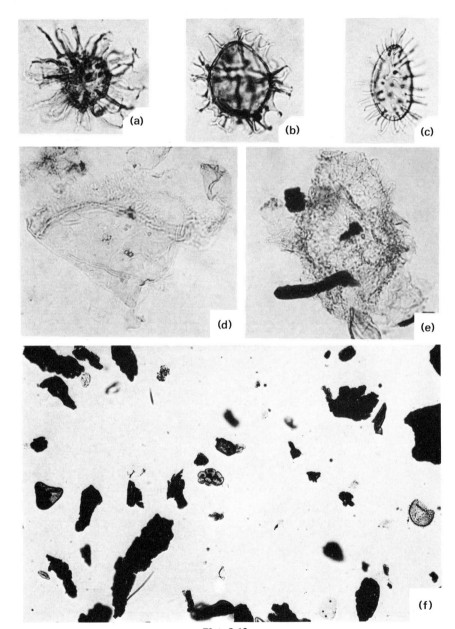

Plate 8.13

Appendix: Past and present students of micropalaeontology at University College London

The following list shows the length of registration of each student and the degree(s) for which they registered. If a person is employed in geology then the most recent address, if known, is given. In other cases the person no longer practises geology or their whereabouts is unknown. In some cases the addresses are old and we are certain that this list is also imperfect in other respects. Where a person's name is prefixed by an asterisk, that person is known to be deceased.

Name	Years	Degree	Position
Al-Ansary, S. E.	1949–52	PhD	Professor of Petroleum Geology, Cairo, Egypt
Banner, F. T.	1951–3	PhD	Professor of Oceanography, University College, Swansea
Bowen, R. N. C.	1951–2	PhD	Professor of Geology, Fourah Bay College, University of Sierra Leone
Dilley, F. C.	1952–3	MSc	British Petroleum (retired)
Lloyd, A. J.	1952–4	PhD	Senior Lecturer in Geology, University College, London
Gordon, W. A.	1954–6	PhD	Professor of Geology, University of Saskatchewan, Regina, Canada
Hughes Clarke, M. W.	1954–7	PhD	Shell International Petroleum Co. Ltd
McNicol, S. J.	1954–7	PhD	Wolverhampton Polytechnic
Omara, S. M.	1954–5		Dean of Science, University of Sohag, Egypt
Sizer, C. A.	1954–5		
Rasheed, D. A.	1955–7	PhD	Professor of Geology, University of Madras, India
Bell, S. V.	1957–60	PhD	Broken Hill Pty Co. Ltd, Whyalla, South Australia, Australia
Brown, B. R.	1957–8	DUC, MSc	British Petroleum
Jones, P. C.	1957–8		
	1961–2	MSc	Portsmouth Polytechnic
Pallott, Jane M.	1957–60	MSc	
Khan, M. H.	1958–62	PhD	Geological Survey of Ghana
Cordey, W. G.	1959–63	DUC, PhD	Shell International Petroleum Co. Ltd
Edmonds, Barbara E.	1959–60	DUC	Broken Hill Pty Co. Ltd, Whyalla, South Australia, Australia
Muir, Marjorie D.	1959–62	DUC, PhD	Bureau of Mineral Resources, Canberra, ACT, Australia
Williams, B. G.	1959–60	DUC	
Barr, F. T.	1959–61	PhD	Houston Oil and Minerals, Houston, Texas, USA
Gurr, P. R.	1959–60	MSc	
*Clarke, R. F. A.	1960–3	PhD	Shell International Petroleum Co. Ltd

APPENDIX

Jenkins, W. A. M.	1960–1	DUC
Najdi, A. M.	1960–4	DUC, MSc, PhD — Professor of Geology, University of Baghdad, Iraq
Royall, J. J.	1960–1	DUC
Church, J. W.	1961–4	DUC, PhD — Robertson Research International Ltd
Hallet, D.	1961–6	DUC, PhD — British National Oil Corporation
Gollesstaneh, A.	1961–5	DUC, PhD — (formerly) National Iranian Oil Company
Fisher, M. J.	1962–3	DUC, MSc — British National Oil Corporation
Field, R. A.	1962–7	DUC, MSc, PhD — Chief Palaeontologist, British Petroleum
Rowlands, P. H.	1962–3	DUC
Thewlis, C. R.	1962–3	DUC — National Coal Board
King, C. B.	1963–7	PhD
Bush, J. S.	1963–4	DUC, MSc
Parmenter, J. C.	1963–4	DUC, MSc — Esso (Exxon) Petroleum Company
Wright, R. M.	1963–6	DUC, MSc — Director, Geological Survey of Jamaica
Mills, S. J.	1964–5	DUC, MSc — Anglo-Ecuadorian Oilfields Ltd
Williams, A. J.	1964–5	DUC, MSc — Queensland Geological Survey, Brisbane, Australia
Mortimer, Margaret G.	1964–5	MSc — Consultant
Simpson, Marjorie	1965–70	MPhil — British Petroleum
Battan, D. J.	1965–6	DUC, MSc — Lecturer in Geology, University of Aberdeen
Futyan, A. I. H.	1965–9	DUC, MSc, PhD — Robertson Research International Ltd
Horn, R. A.	1965–6	DUC, MSc — Consolidated Gold Fields Ltd
James, R. E. M.	1965–6	DUC, MSc
Parry, G.	1965–6	DUC, MSc
Kalantari, A.	1966–9	PhD — National Iranian Oil Company
Kasal, Judith A.	1966–7	DUC, MSc — Bureau of Mineral Resources, Canberra, ACT, Australia
Owen, M.	1966–9	PhD — Bureau of Mineral Resources, Canberra, ACT, Australia
Backhouse, J.	1967–8	DUC, MSc — Geological Survey of Western Australia, Perth, Western Australia, Australia
Hoskin, I. R.	1967–71	DUC, MSc, PhD — British Petroleum
Khalili, Maniyeh	1967–71	DUC, MSc, PhD — National Iranian Oil Company
Ajina, T. M.	1968–71	PhD
Bigg, P. J.	1968–72	PhD — Gearhart Geodata Services Ltd
Basha, S. H. S.	1968–9, 1973–5	DUC, MSc, PhD — Lecturer in Geology, University of Amman, Jordan
Chandra, A.	1968–9	DUC, MSc — Consultant
Croxton, Catherine A.	1968–9	DUC, MSc — Robertson Research International Ltd
Dempsey, M.	1968–9	DUC, MSc — Marathon International Petroleum (GB) Ltd
Rahman, M.	1968–9	DUC, MSC

Brasier, M. D.	1969–72	PhD	Lecturer in Geology, University of Hull
Radford, Sally S.	1969–72	PhD	(formerly) Shell UK Exploration and Production Ltd
Wallace, D. J.	1969–70	MPhil	
Allen, Lynn O.	1969–70	DUC, MSc	Paleoservices Ltd
Attewell, R. A. K.	1969–70	DUC, MSc	Robertson Research International Ltd
Meyrick, R. W.	1969–70	DUC, MSc	Paleoservices Ltd
Sezer, S.	1969–70	MPhil	
Ogbe, F. G. A.	1970–4	PhD	Geo-consults, Benin City, Nigeria
Russell, Valerie S.	1970–3	MPhil	Department of Earth Sciences, The Open University
Causebrook, R. M.	1970–1	MPhil	Exploration Logging Ltd
Gosling, Carolyn A.	1970–1	DUC, MSc	
Henderson, I. M.	1970–1	DUC, MSc	
Sarmah, K. N.	1970–2	DUC, MSc	Soil Geologist, Indonesia
Shipp, D. J.	1970–4	DUC, MSc, PhD	Robertson Research International Ltd
Walker, Carol A.	1970–1	DUC, MSc	
Tmalla, A. F. A.	1970–1	DUC, MSc	Occidental of Libya Inc., Libya
Scantlebury, Penny J.	1971–2	MPhil	
Clarke, D. R. D.	1971–2	DUC, MSc	
Evans, A. M.	1971–2	DUC, MSc	Shell International Petroleum Co. Ltd
Marshall, P. R.	1971–5	DUC, MSc, PhD	Robertson Research International Ltd
Rutherford, K. J.	1971–2	DUC, MSc	Cities Service Europe-Africa Petroleum Corporation
Abdalla, Y. H.	1971–2	DUC	Oasis Oil Company Inc., Tripoli, Libya
Price, R. J.	1972–5	PhD	Amoco Canada Petroleum Co. Ltd, Calgary, Alberta, Canada
Roberts, J. D.	1972–3	DUC, MSc	
Tzen, Maria I.	1972–3	DUC, MSc	
Wright, G. F.	1972–3	DUC, MSc	Strand Oil and Gas Ltd, Calgary, Alberta, Canada
Chamney, T. P.	1973–5	PhD	Petro-Canada, Calgary, Alberta, Canada
El-Demerdash Mohamed, Gihad	1973–7	MPhil	Agip (UK) Ltd
Burnhill, T. J.	1973–4	DUC, MSc	British Petroleum
Croad, Jennifer M.	1973–4	DUC, MSc	Petroleum Exploration Division, Department of Energy
Fernandez, P.	1973–4	DUC, MSc	
Ghosh, K. K.	1973–4	DUC, MSc	Consultant
Hojjatzadeh, Minoo	1973–8	DUC, MSc, PhD	Marathon International Petroleum (GB) Ltd
Jutson, D. J.	1973–8	DUC, MSc, PhD	Deutsche Texaco AG, Celle, Germany
Kyriacou, K.	1973–4	DUC, MSc	
Nellis, G. A.	1973–4	DUC, MSc	Exploration Logging Canada Ltd, Calgary, Alberta, Canada
Padley, C.	1973–4	DUC, MSc	

APPENDIX

Name	Years	Degree	Affiliation
Szegedi, Verona E.	1974–80	PhD	Gulf Oil (UK) Ltd
Afshar, Simin A.	1974–6	DUC, MSc	National Iranian Oil Company
Djahromy, Guity D.	1974–5	DUC, MSc	
Edwards, P. G.	1974–9	DUC, MSc, PhD	Robertson Research International Ltd
Robertson, Gillian B.	1974–9	DUC, MSc, PhD	Consultant
Salmon, D. A.	1974–5	DUC, MSc	Marine Geoscience Unit, Geological Survey of South Africa
Walker, I.	1974–5	DUC, MSc	
West, C. C.	1974–5	DUC, MSc	
Afejuku, A.	1975–8	PhD	
Mojab, F.	1975–9	PhD	(formerly) Department of Geology, University of Shiraz, Iran
Taylor, Rosanna J.	1975–8	PhD	
Capsey, M.	1975–6	DUC, MSc	Robertson Research International Ltd
Chance, Sarah J.	1975–6	DUC, MSc	
Davies, S. C.	1975–6	DUC, MSc	British Petroleum
Edwards, Vivienne A.	1975–6	DUC, MSc	
Horne, D. J.	1975–6	DUC, MSc	Research Fellow, City of London Polytechnic
Kay, A. M.	1975–6	DUC, MSc	
Mountford, Patricia A.	1975–6	DUC, MSc	Seismograph Services Ltd
Bahadori, A. A.	1976–7	MPhil	
Ashburn, A. J.	1976–7	DUC, MSc	National Coal Board
Chow, Y. C.	1976–7	DUC, MSc	Robertson Research Company, Singapore
Comet, P. A.	1976–7	DUC, MSc	Organic Geochemistry Unit, University of Bristol
Craig, D. L.	1976–7	DUC, MSc	
Esat, B.	1976–8	DUC, MSc	
Mabillard, J. E.	1976–7	DUC, MSc	Shell International Petroleum Co. Ltd
Stuart, I. A.	1976–7	DUC, Msc	Texaco Production Services Ltd
Weighell, A. J.	1976–7	DUC, MSc	British National Oil Corporation
Memon, A. A.	1976–80	DUC, MSc, MPhil	Lecturer in Geology, University of Sind, Pakistan
Crux, J. A.	1977–80	PhD	British Petroleum
Varol, O.	1977–81	PhD	Robertson Research Company, Singapore
Chester, Rosemary	1977–8	DUC, MSc	
Corbett, P. W. M.	1977–8	DUC, MSc	Union Oil Co. of GB Ltd
Dublin-Green, Caroline	1977–8	DUC, MSc	
Fakrai, A.	1977–8	DUC, MSc	
Fuller, N. G.	1977–	DUC, MSc, PhD	
Moghadam, Azam M.	1977–8	DUC, MSc	National Iranian Oil Comapny
Partington, M. A.	1977–8	DUC, MSc	Robertson Research International Ltd
Ali, M. H.	1978–	PhD	

Clarke, D. G.	1978–9	DUC, MSc	Seismograph Services Ltd
Dewey, C. P.	1978–9	DUC, MSc	Department of Geology, Memorial University of Newfoundland, St Johns, Newfoundland, Canada
Girgis, M. H.	1978–	DUC, MSc, PhD	
*Hargreaves, R. H.	1978–9	DUC, MSc	
Holmes, N. A.	1978–9	DUC, MSc	Department of Geology, University College, Aberystwyth
Irvine, R. M.	1978–9	DUC, MSc	Anglo-American Corporation, Botswana
Massoudkhan, B.	1978–9	DUC, MSc	
Samadian, Fatemeh	1978–81	PhD	Geological Survey of Iran
Forsey, G.	1979–80 1981–	DUC, MSc, MPhil	Lecturer, Nene College
Lee, H.-Y.	1979–80	DUC, MSc	Korea Research Institute of Geoscience and Mineral Resources, Seoul, Korea
Park, Se-Moon	1979–	DUC, MSc, PhD	
Riddick, A.	1979–80	DUC, MSc	North-East London Polytechnic
Thomson, A.	1979–80	DUC, MSc	Institute of Geological Sciences
Wade, J. M.	1979–80	DUC, MSc	Gearhart Geodata Services Ltd
Farhan, Anba	1979–	MPhil	
Cooper, M. K. E.	1980–	MPhil	
Kerim, Bariwan A.	1980–	MPhil	
Shakib, S.	1980–	MPhil	
Bou-Dagher, Marcelle	1980–	DUC, MSc	
Evans, I. P.	1980–1	DUC, MSc	
Fitz-gerald, B.	1980–	DUC, MSc, PhD	
Mortimer, C. P.	1980–1	DUC, MSc	Robertson Research International Ltd
Vakil, P. P.	1980–	DUC, MSc	
Okosun, E. A.	1981–	M.Phil	Geological Survey of Nigeria
Aguilar, Paloma	1981–	DUC, MSc	
Barlow, N. D.	1981–	DUC, MSc	
Douglas, P. K.	1981–	DUC, MSc	
Herguera, J. C.	1981–	DUC	
Jakubowski, M.	1981–	DUC, MSc	
Lowry, Florence M. D.	1981–	DUC, MSc	
Mostofi, A.	1981–	DUC, MSc	
Ogundiran, O. A.	1981–	DUC, MSc	Exploration and Mining Division, National Steel Council, Nigeria
Okpiabhele, K. I.	1981–	DUC, MSc	
Parsons, D. G.	1981–	DUC, MSc	
Read, Helen F.	1981–	DUC, MSc	

Visitors have included: Dr J. J. Bizon (France), Dr R. Cifelli (USA), Professor R. V. Dingle (South Africa), Professor A. Dizer (Turkey), Dr L. Fahraeus (Canada), Dr W. W. Hay (USA), Professor D. J. Jones (USA), Dr A. Q. Rathur (Australia), Professor R. Said (Egypt), Dr S. Yoshida (Japan).

Index

Numbers in bold type indicate tables within the text; numbers in italics indicate text figures and plates.

The following key words for each chapter have not been indexed:

Chapter 1
  Foraminifera
    architecture
    ecology
    evolution
    growth
    morphology

Chapter 2
  Albian–Cenomanian
  Foraminifera
    biostratigraphy
    taxonomy
  late Cretaceous
  northern Europe
  USSR

Chapter 3
  Eocene
  Foraminifera
    morphometry
    taxonomy
  Iran

Chapter 4
  Foraminifera
    benthos
    distribution patterns
    ecology
  northwestern Scotland
  Recent

Chapter 5
  family-group
  Foraminifera
    classification

  phylogeny
  plankton
  genus-group
  index
  key

Chapter 6
  chalk
  England
  flints
  Foraminifera
    preservation
  Upper Cretaceous

Chapter 7
  Lower Jurassic
  Ostracods
    biostratigraphy
    distribution patterns
    extinction
    Metacopina
  Pliensbachian–Toarcian

Chapter 8
  depositional environments
  England
  Lower Cretaceous
  palynofacies
  phytoplankton
  Purbeck/Wealden
  salinity
  Upper Jurassic

N.B. Localities and species names have not been indexed for Chapter 5.

General index

alpha (α) index 247, 250–3, *6.1, 6.2*
Alum Bay, Isle of Wight 244, **6.2**
Anglo-Brabant Massif 281
Apennines, Italy 267
Aselfingen, Germany 269
Ashdown Beds 281, 284–6
Ashtian, Iran 82

Bakhtiyari area, Iran 81
Bangestan Group 83
Barnes High Sandstone 288
Belemnite Marls 276

Belize 286
Bemerode, Germany 77
Bileçik, Turkey 267, 276–7
biocoenosis, foraminiferal 121–2, 135
biofacies
  Bulimina-dominated 132, 135, **4.1**
  Cibicides-dominated 132–3, **4.1**
  Eponides-dominated 132, 135, **4.1**
  Gaudryina-dominated 132, 135, **4.1**
  Hyalinea-dominated 132, 135, **4.1**
  Textularia-dominated 132, 133, **4.1**
Black Ven Marls 266

GENERAL INDEX   315

Blue Lias 266
Bochum, Germany 48, 54, 77
Brochterbeck, Germany 64, 77
'bulliform' chambers, foraminifera 219
Burujen, Iran 81–100

calcispheres 252
Cap Blanc Nez, France 54
Capel, Surrey 287–8
Cardigan Bay 113
Carnaby, Yorkshire 251, **6.2**
Chalk Rock 246, 251, 258, **6.3**
Chalk sea, depth 257–9
Châlons-sur-Vesle, France 252
Chalton Down, Hampshire 246, **6.1, 6.2**
Cinder Bed 281, 283
correlation, by planktonic foraminifera 142
cryptogenic appearances, planktonic foraminifera 147, 224
Culver Cliff, Isle of Wight 240–1, 246, **6.2**

Deep Sea Drilling Project 143
Djebel Zaghouan, Tunisia 267, 271
Dorset Coast 266, 276
Durleston Beds 281

East Harnham, Wiltshire 244, **6.2**
Empingham, East Midlands 268
English Channel 241
evolution, iterative 147, 220

Fisher's alpha index 247, 250–3, *6.1*, **6.2**
Flammenmergel 64
flaserkalk 253, 255–6
Flaunden, Hertfordshire 246, 250, **6.2**
flint, formation of 254–5
'flint meal' 240, 246–7, 250–5, 258, 259, **6.1**
Folkestone, Kent 64, 69, 77
foraminifera
 apertures 3–7, 8–10, 14–18, 20–9, 34–6
 axopodia 2
 buoyancy 2
 chamber shape 15, 21–3, 30, 35
 chamber volume 15, 20, 22, 30, 35–6
 cytoplasm 3, 8, 24–5, 30, 34
 depth habitat 12–13, 18–19, 21, 24–5, 28, 32, 34, 37
 ectoplasm 2
 foramen (pl. foramina) 3, 8, 14, 21–2
 genetics 225
 Max LOC, definition 3
 megalospheric tests 15, 35, Chapter 3, 206
 microspheric tests 15, 35, Chapter 3, 205
 Min LOC, definition 3
 nucleus (pl. nuclei) 2, 5, 8, 14
 photosynthesis 34
 proluculus 3, 8, 13, 15, 24, 35
 protoplasm 2, 5, 6
 pseudopodia 3
 salinity 2, 6–8, 12–13, 18–19, 21, 24–5, 28, 32, 38, 115
 test form and substrate 125, **4.1**
 test thickness 125
 ultrastructure and composition in classification 13
Formations
 Aleg 158
 Aragon 149
 Bayirköy 276
 Chapapote 156, 182
 Cipero 154
 Colon 190
 Cuche 150
 Duck Creek 154
 Greenhorn 149, 194
 Guayabal 184
 Guayaguayare 148, 194
 Jackson 156
 Jahrum 81, *3.2*
 Laki 94
 Lengua 162, 164, 166
 Lodo 148–9
 Main Street 180
 Monte Rey 184
 Musa 160
 Navet 154, 160, 162, 195
 Oficina 164
 Pozon 158, 184, 192
 Rio Yauco 174
 San Fernando 170
 Shahbazan 81
 Siju 95–6
 South Bosque 196
 Taleh Zang 81–3
 Tantoyuca 192
 Taylor 149
 Yegua 149
Forstal, Aylesford, Kent 77
Frettenham, Norfolk 244, **6.2**

Gabon 286
Gault Clay 69
Grinstead Clay 285–7

Hamzeh-Ali Mountain, Iran 81–3, 100, *3.2*
hard grounds, in chalk 254–5
Hastings Beds 278–81, 284, 286–8
Hastings, Sussex 285
Haymana, Turkey 82, 86
Hildesheim, Poland 48
homeomorphy 219–25

interstitial habitats, foraminifera 122, 135
Isle of Wight 251, 257, 279, 283, 287–8

Jaz-Mourian, Iran 82
Jet Rock 272
Jouy aux Arches, Metz, France 269

Katsamonu, Turkey 82

Kensworth, Bedfordshire 246, 251, 253, **6.2, 6.3**
kerogen 279
Kirthar Series 82
Kirton-in-Lindsey, Lincolnshire 268, *7.3*
'kummerform' chambers, foraminifera 219

live:dead ratios
   see also biocoenosis, thanatocoenosis 112
Lower Greensand 289
Lulworth Beds 281
Lunata chalk 244

Minch Fault 116
Mochras Borehole, Wales 268
Montereau-faut-Yonne, France 77
Mountfield, Sussex 281–4

Newhaven, Sussex 246, Table 6.2
North Minch Channel
   bathymetry 116–18, *4.4*
   currents 118–20, *4.5–6*
   hydrography 113–*22*
   oxygen & nitrogen 115
   pH 115
   phosphates 115–16
   salinity 115
   sediment types 122–4, *4.7–9*
   substrate 120–2
   temperature 114

Orphan Knoll, north-west Atlantic 66

Paramoudra chalk 244
Peniche, Portugal 270, *7.4*
phosphatic chalks and phosphatised foraminifera 247–51, 254, 258–9
Pincent's Kiln, Reading, Berkshire 252, **6.2**
Pläner Marls 53
Plenus Marl 255
Pliensbach, Germany 269
polyphyletic genera, foraminifera 147
Portsmouth, Hampshire 241
Posidonienschiefer 267, 272–3
pteropods 251

radiolaria 1–2, 34, 242
Ragusa, Sicily 271
Reading Beds 252, 254

Reutlingen, Germany 269
rhizopods 5–8
Rickmansworth, Hertfordshire 246, **6.2**
Rügen Island, Germany 54, 57, 67, 77

sarcodines 1–2
scanning electron microscopy 145
Schoonebeek, The Netherlands 77
Seaford, Sussex 253, **6.2, 6.3**
Sens, France 56, 67
Shiant Bank, north-west Scotland 112, 122–3, 125–9, 135
Shiant Isles, north-west Scotland 116, 124
Sidestrand, Norfolk 244, **6.2**
silicification, in chalk 242
similarity index 244
Sinser Mähre, Germany 64
Sistan, Iran 81
Soleymaniyeh, Turkey 82
Strettura, Italy 271
Strouanne, France 77
sulphides and foraminifera 115
Swanage Bay, Dorset 283

Taplow, Buckinghamshire 247–51, **6.2**
Tethyan Ocean 277
thanatocoenoses 122
Tilton, Leicestershire 268, *7.3*
Trimingham chalk 241
Trimingham, Norfolk 244, **6.2**
Tunbridge Wells Sand 284–7

Vandée, France 270
Villemoyenne, France 77

Wadhurst Clay 279, 285–7
Warnham, Surrey 287
Wayrham, Yorkshire 251, **6.2**
Weald, The 279–83
Weald Clay 279, 281, 286 8
Wealden Marls 283, 288
Wealden Shales 283, 287–9
Wessex–Weald Basin 279–81, 286
Weybourne chalk 244
Weybourne, Norfolk 244, **6.2**
Wissant, France 77
Wünnenberg 2 Borehole, Germany 77

Zagros Mountains, Iran 81
zonation, planktonic foraminifera 144

Taxonomic index

Abathomphalidae 148
Abathomphalinae 211
Abathomphalus 148, 212, *5.117*
Acantharia 2
Acarinina 148, 203, 222, *5.38*
Acarininae 149

Acarinininae 149, 203
Acervulina inhaerens 125, 137
Actinomma 34
Adercotryma 24
Agathammina 32
Allogromiina 3, 7–8

TAXONOMIC INDEX

Alveolinidae 26
Ammobaculites 19, 25–6, 29, 36
Ammodiscacea 8, 32
Ammodiscidae 10, 13, 22
Ammodiscoides 10, 13
  A. turbinatus 12
Ammodiscus 10–13, 30
Ammolagena 13
Ammomarginula 19, 25
Ammonia 28
  A. beccarii 28, 137
    A. beccarii batava 28
    A. beccarii tepida 28
Ammoscalaria 19, 25, 37, 137
Ammospirata 36
Ammotium 19, 25
Ammovertella 10, 13
Amoebida 6
Amphicoryna cf. *A. scalaris* 137
  A. sp. 137
Amphistegina 28
Anaticinella 149, 186, 210, *5.103*
Applinella 149, 207, *5.74*
Aragonella 149, 207, *5.75*
Arcellinida 3–7
Arcellites pyriformis Plate 8.4f,j
Archaediscidae 13–14
Archaeoglobigerina 149, 210, 222, *5.104*
Arenobulimina 43, 46, 48–51, 54, 63–4, 66–7, 72, 247, 257, **6.2**
  '*A.*' *advena* 60
  A. convexocamerata 66, 72
  A. d'orbignyi 70
  A. flandrini 62, 71
  '*A.*' *frankei* 47–8
  A. presli **6.1**
  A. pseudodorbignyi 70
  '*A. sabulosa*' 64, 66
  A. sabulosa 66
  '*A. truncata*' 58
  A. cf. *truncata* 63
  A. vialovi 66, 72
Arenobulimina (Arenobulimina) 42, 46, 49–51, 53, 67
  A. (A.) macfadyeni 62
    A. (A.) macfadyeni elongata 59, 62, 71
  A. (A.) obliqua 53
  A. (A.) cf. *obliqua* 42, 51, 53–4, 71–2, *Plate 2.1i*
  A. (A.) presli 42, 52–4, 72, *Plate 2.1d–h*
  A. (A.) pseudalbiana 54
Arenobulimina (Columnella) 51, 66–7
  A. (C.) sabulosa 66, 72
Arenobulimina (Hagenowella) 42, 54, 57
  A. (H.) courta 43, 55–6, *Plate 2.1k*
  A. (H.) elevata 43, 55–6, *Plates 2.1l;2.2a,b*
  A. (H.) obesa 43, 56–7, *Plate 2.2c–e*
Arenobulimina (Hagenowina) 46
  see also *Hagenowina*
Arenobulimina (Harena) 42, 49–51, 57

  A. (H.) improcera 57–8, *Plate 2.1j*
Arenobulimina (Novatrix) 42, 51, 54, 67
Arenobulimina (Pasternakia) 42, 51, 57–8, 64, 70–1
  A. (P.) barnardi 42, 58–9, 63, 71–2, *Plate 2.2f*
  A. (P.) bochumensis 42, 59, 62, 71, *Plate 2.2g,h*
  A. (P.) chapmani 42, 51–4, 59–61, 68–9, 71–2, *Plate 2.2i–k,n*
  A. (P.) macfadyeni 42, 59–62, 68–9, 71, *Plate 2.2l,m,o*
  A. (P.) minima 42, 62–3, 72, *Plate 2.3a*
  A. (P.) truncata 42, 63, 71, *Plate 2.3b,c*
Arenobulimina (Sabulina) 42, 51, 64
  A. (S.) gaworbiedowae 42, 64–5, 72, *Plate 2.3d*
  A. (S.) sabulosa 42, 48, 63, 65–6, 71–2, *Plate 2.3e,f*
Arenobulimina (Voloshinoides) 42, 46, 48, 51, 66–7
  A. (V.) bulletta 70–1
  A. (V.) postchapmani 60–1
  A. (V.) postchapmani praecursor 60–1
Arenonina 18, 36
Arenoparella 27
  A. mexicana 28
Articulina 19, 36
Assilina 81, 84, 88
  A. aspera 83–7, *3.3*, *Plate 3.1a–i*
  A. aff. *aspera* 86–7, 98, *Plate 3.2b,h,i*
  A. assamica 95
  A. burujenensis 81, 84, 87–8, *3.3*, *Plate 3.2a,c,d*
  A. daviesi 94
    A. daviesi nammalensis 94
  A. exponens 82–3, 88–90, *3.3*, *Plates 3.2e–g,j,k; 3.3a–c; 3.7a–d*
    A. exponens tenuimarginata 81, 90–1, *3.3 Plate 3.5a–g*
    A. exponens var. b 89–90, *3.3*, *Plate 3.4a–c,e,f*
  A. hamzehi 81–2, 84, 91–2, *3.3*, *Plate 3.6a–n*
  A. laminosa 92
  A. leymerie 88
  A. orientalis 82, 93
    A. orientalis gargoensis 93
    A. orientalis iranica 81, 88, 93, *3.3*, *Plate 3.7e–h*
  A. persica 81, 84, 94–5, 97, *3.3*, *Plate 3.8a,b,d–h*
  A. spira 82–3, 95–6, *3.3*, *Plate 3.9a,c–e,h*
    A. spira corrugata 96, 98, *3.3*, *Plate 3.9b,f,g,i,j*
  A. aff. *spira* 97, *Plate 3.8c,i,j*
  A. subspira corrugata 96
  A. sp. 1 97, *Plate 3.4d*
  A. sp. 2 98, *Plate 3.4g*
Astacolus, sp. 137
Asterigerina 28

Asterigerinata mamilla 137
·*Asterohedbergella* 149, 209, 221, *5.91*
Astrorhiza 9, 12
Astrorotalia 149, 203, *5.37*
Ataxogyroidina 67
Ataxophragmiidae 42–3, 45, 48, 58, 68
Ataxophragmiinae 42–3, 45–6, 49, 51, 67, 72
Ataxophragmium 42, 46, 49, 72, **6.2**, **2.2**
   A. variabile **6.1**

Bairdia 269
   B. cf. *B. fortis* 276
   B. cf. *B. hilda* 276
   B. (?) *problematica* 263
'*Bairdiacypris*' 270
Bathysiphon 8–9, 12–13
Beella 149, 200, *5.13*
Berggrenia 150, 206, 224, *5.69*
Bifarina 37
Bigenerina 22, 35
Biglobigerinella 146, 150, 207, 224, *5.79*
Biloculinidea 13
Biorbulina 150, 205, *5.60*
Biticinella 150, 208, 224, *5.81*
Blowiella 150, 207, 224, *5.78*
Bolivinitidae 22
Bolivinoides 240, 247, **6.2**
Bolivinopsis 35
Bolliella 152, 206, 224, *5.70*
Botellina 9
Botryococcus 284, Plates *8.1b–d; 8.3a; 8.5a*
Bradyina 24
Brizalina 21
   B. pseudopunctata 137
   B. spathulata 137
   B. variabilis 137
Bucherina 152, 211, *5.110*
Bulimina 28, 57, 63, 135, **4.1**
   B. gibba/elongata 137
   B. marginata 119, 125, 131–2, 137, *4.16*, Plate *4.2c*
   B. presli 49
Buliminacea 29, **6.2**, **6.3**
Bythocypris 276

Calcivertellinae 12
Cancris 115
   C. auricula 137
Candeina 152, 201, *5.26*
Candeininae 152, 201
Candorbulina 152, 205, *5.61*
Canningia 286–7
Cassidulina 35
   C. carinata 125, 137
   C. obtusa 137
Cassidulinacea 29, 196
Cassigerinella 196
Cassigerinellidae 196
Cassigerinellinae 196

Cassigerinelloita 197
Cassiope 287, 289
Catapsydracinae 152
Catapsydrax 152, 200, *5.15*
Caucasella 154, 168, 199, 219–20, 225, *5.3*
Caucasellidae 154
Caucasellinae 199, 212–21, *5.1–11*, *5.122*
Celyphus rallus 284, 286–7, Plate *8.9a*
Ceratobulimina 27
Cheirolepidiaceae 283
Chitinodendron 13
Cibicides 28, 112, 125, **4.1**
   C. fletcheri 137
   C. lobatulus 122, 126–9, 137, *4.10*, Plate *4.1a–c*
   C. pseudoungerianus 137
Cicatricosisporites 285, Plate *8.7a*
Classopollis 283–5, Plates *8.1a; 8.2a; 8.8b*
Clavatorella 154, 203, 220–1, *5.44*
Clavigerinella 154, 207, 220, *5.71*
Clavihedbergella 154, 208, 221, *5.90*
Claviticinella 154, 210, *5.99*
Clavulina 19, 36
'*Cleistosphaeridium*' 287, Plate *8.11a*
Climacammina 35
Colianella 26
Colomia 15
Colonammina 5
Concavissimisporites 285, Plate *8.6d*
Conicospirillina 22
Conoglobigerina 154, 199, 219, *5.1*
Coscinosphaera 198
Coskinolina 36
Couperisporites cf. *C. complexus* Plate *8.6b*
Crenaverneuilina 48, 64, **2.2**
   C. frankei 42, 45, 48
   C. intermedia 43, 45, 48, 66, 71–2
   C. mariae 43, 45, 72
Cribrogoesella 35
Cribrohantkenina 154, 207, *5.76*
Cribroperidinium 287, 289, Plate *8.10b,c*
Cribrostomoides jeffreysii 137
Cuneolina 21
Cupressinocladus 283
Cupressinoxylon 283
Cyclammina 24–5
   C. cancellata 137
Cyclogyra 12
   C. involvens 137
Cyclonephelium 286, 289
   C. distinctum 289
   C. cf. *C. distinctum* 287, Plate *8.10d*
Cymbalopora 28
Cymbaloporetta 28
cyprid ostracods 287–8
Cyprididae 262
Cytheracea 267, 269, 273
cytherellid ostracods 264
Cytheridea 265
?*Cytheropteron* 270

Darwinulidae 262
Daucinoides 34
Daxia 24
Dendrophyra 12
Dentalina subarcuata 137
  D. sp. 137
Dentoglobigerina 156, 200, 222, *5.8*
Dicarinella 156, 209, 223, *5.94*
Dicyclinidae 18, 34
Discorbacea 197
Discorbidae 197
Discorbis 28
  D. sp. 137
Discorinopsis aguayoi 28
Dorothia 21
Dorothiinae 45–6
Duostominacea 29

Earlandia 22
Echinocorys 252
Echinogromia 5
Eggerella scabra 137
Ehrenbergina 35
Ektyphocythere 270
  E. cf. *E. bucki* 271
  E. champeauae 269, 273
  E. cf. *E. intrepida* 270–1
  E. cf. *E. rugosa* 271
Elphidium 25, 125
  E. articulatum 137
  E. crispum 137
  E. excavatum 137
  E. gerthi 137
  '*E.*' *incertum* 25
  E. magellanicum 25
  E. williamsoni 137
  E. sp. 137
Endothyra 26
Endothyranopsis 24
Endothyridae 25–6
Eoglobigerina 156, 202, 220–2, *5.34*
Eoglobigerinidae 158
Eoglobigerininae 202, *5.34*, *5.35*, *5.122*
Eogloborotalia 196
Eohastigerinella 158, 208, 220, 224, *5.85*
Eosigmoilina 32
Epistominella vitrea 137
  E. sp. 137
Eponides 111, **4.1**
  E. repandus 125, 127–9, 137, *4.12*, Plate *4.1g–i*
Equisetites 285, 287, Plate 8.8a
  E. lyellii 287

Falsotruncana 158, 184, 209, *5.118*
Favusella 158, 188, 199
Filosina 287–8, Plate 8.10b
Fissoarchaeoglobigerina 158, 210, 222, *5.105*
Fissurina 3

F. lucida 137
F. marginata 137
F. orbignyana 137
F. sp. 137
Fohsella 147, 158, 204, *5.47*
Foraminiferida 3, 5–6
Flabellamina 16
Frondicularia 16, 25
Fursenkoina fusiformis 137
Fusarchaias 26
Fusulina 26
Fusulinina 8, 14, 26

Gammacythere ubiquita 276
Gaudryina 111, **4.1**
  G. parallela **6.1**
  G. rudis 127, 129, 137, *4.13*, Plate *4.2d–f*
  G. rugosa **6.1**
Gaudryinella 47
Gavelinella 247, **6.2**
  G. costata **6.1**
  G. pertusa **6.1**
  G. stelligera **6.1**
  G. thalmanni **6.1**
Gavelinopsis praeggeri 137
Glabratella 28
Gleicheniidites Plates 8.6e
Globalternina 196
Globanomalina 158, 203, 220–2, *5.35*
Globastica 160, 199, 219, *5.6*
Globicuniculus 160, 201, *5.20*
Globigerapsidae 160
Globigerapsinae 204, 222–3, 225, *5.53*, *5.54*, *5.123*
Globigerapsis 160, 205, 222, *5.54*
Globigerina 28, 54, 147, 160, 200, 219–20, 222, *5.14*
  '*G.*' *elevata* 67
Globigerinacea 146, **6.2, 6.3**
Globigerinanus 160, 201, *5.23*
Globigerinatella 160, 205, *5.59*
Globigerinatheka 162, 205, 222, *5.56*
Globigerinella 162, 206, 224, *5.68*
Globigerinelloides 162, 174, 207, 224, 246, **6.1, 6.2**, *5.77*
Globigerinelloididae 162
Globigerinelloidinae 207
Globigerinida 146, 160
Globigerinidae 143, 199, 212, 219–23, 225, *5.1–5*, 250, *5.122–3*
Globigerininae 200, 219–20, 222–5, *5.7–25*, *5.122–3*
Globigerinita 162, 200, *5.16*
Globigerinoides 28, 164, 201, 222, *5.21*
Globigerinoidesella 164, 201, *5.22*
Globigerinoita 164, 201, *5.25*
Globigerinopsis 164, 204, 223, *5.51*
Globigerinopsoides 164, 206, *5.63*
Globobulimina 29
Globocassidulina subglobosa 137

Globoconusa 197
Globoquadrina 164, 202, 222, *5.31*
Globoquadrinidae 164, 202
Globorotalia 147, 166, 198, 204, 220–1, 223, *5.46*
Globorotalidae 143
Globorotaliidae 166
Globorotaliinae 203, 221, 223–4, *5.42–50*, *5.123*
Globorotalites 247, **6.2**
 G. michelinianus **6.1**
Globorotaloides 166, 202, *5.30*
Globorotaloidinae 166, 202, *5.30–1*
Globotextulariinae 42, 46, 51
Globotruncana 166, 211, 246–7, 254, **6.2**, *5.112*
 G. bulloides **6.1**
 G. pseudolinneiana **6.1**
Globotruncanella 166, 209, 221, *5.95*
Globotruncanellinae 166, 209, 221, 223, *5.92–5*, *5.118*, *5.125*
Globotruncanidae 166, 210, 222, *5.104–6*, *5.126*
Globotruncaninae 211, *5.112–17*, *5.126*
Globotruncanita 166, 211, *5.115*
Globoturborotalita 168, 200, 219, *5.12*
Globuligerina 168, 182, 199, 219, *5.2*
Globulina gibba 137
 G. sp. 137
Glomospira 12, 13, 37
Glomospirella 13
Gonyaulacysta areolata Plate 8.13e
Gromiida 3–8
Gubkinella 197
Guembelitria 28
Guembelitriidae 197
Guembelitriinae 197
Guembelitrioides 168, 201, 222, *5.19*
Guttulina 29
 G. lucteu 137
Gyroidinoides 247, **6.2**
 G. nitidus **6.1**

Hagenowella 46, 51, 54, 66–7, 247
 H. elevata **6.1**
 H. gibbosa 67
Hagenowina 42, 46, 49–51, 54, 56, 58, 60, 64, 66–9, 71–2, **2.2**
 H. advena 42, 60, 68–9, 71–2, *Plate 2.3g,h*
  H. advena praeadvena 68, 71
 H. anglica 42, 69–70, *Plate 2.3i–m*
 H. bulletta 71
 H. d'orbignyi 42, 70–2, *Plate 2.3n,o*
 '*H. postchapmani*' 60
 H. postchapmani 68–9, 71
  H. postchapmani praecursor 68–9, 71
 H. voloshinae 68, 71
  H. voloshinae praevoloshinae 68, 71
Hantkenina 168, 192, 207, *5.73*
Hantkeninella 170, 207, *5.72*

Hantkeninidae 143, 168
Hantkenininae 206, *5.71–6*, *5.124*
Haplophragmoides 19, 24–6
 H. bradyi 137
Harena 57
 see also *Arenobulimina (Harena)*
Hastigerina 170, 206, 224, *5.65*
Hastigerinella 170, 172, 206, 220, 224, *5.67*
Hastigerininae 170, 206, *5.64–70*, *5.124*
Hastigerinoides 24, 170, 208, 224, *5.84*
Hastigerinopsis 170, 172, 206, *5.67*
Haynesina 24–5
 H. albiumbilicata 25
 H. depressula 25
 H. germanica 25, 137
 H. orbiculare 25
Healdia 264
Healdiacea 262
Healdiidae 262
Hedbergella 146, 172, 198, 208, 220–2, 224, 246, **6.1**, **6.2**, *5.87*
 H. brittonensis **6.1**
Hedbergellidae 208, 219, 225, *5.125–6*
Hedbergellinae 172, 208, *5.86–91*, *5.125*
Hedbergelloidea 172
Hedbergina 198
Heliozoa 2
Helvetoglobotruncana 172, 209, *5.96*
Helvetoglobotruncaninae 172, 209, 220, 222, *5.96–7*, *5.126*
Hemigordiopsis 32
Hemigordius 32
Hemisphaeramminae 5, 8
Hermiella 264–6, 268, **7.2**
Heterohelicacea 21, 144, 147, 197, 212
Heterohelicidae 197
Heterohelix 21, 35, 246, 250, **6.2**
 H. globulosa **6.1**
Hippocrepina 10, 12–13
Hirsutellu 172, 178, 198, 204
Hoeglundina 27
Hormosina 15, 32
 H. globulifera 3
Hughesisporites galericulatus Plate 8.4g,h
Hungarella 263–4
Hyalinea 135 **4.1**
 H. balthica 111, 115, 131, 137, *4.15*, *Plate 4.2a,b*
Hybodus 288
Hyderia 197
Hyperammina 13, 19

Inaperturopollenites 283
Indicola 197
Indicolinae 197
Inoceramus 242, 253
Inordinatosphaera 172, 205, *5.57*
Involutinidae 13–14
Iridia 5
Isobythocypris 269, 276

TAXONOMIC INDEX   321

Jadammina macrescens 28
  J. polystoma 28
Jurassorotalia 197

Kinkelinella 270–1
  K. sermoisensis 268, 270–1
  K. tenuicostati 269
Kuglerina 172, 211, *5.107*
Kuresaaria gotlandica 264

Labrobiglobigerinella 174
Labroglobigerina 174
Labroglobigerinella 174, 207
Lagena 3, 125
  L. clavata 137
  L. interrupta 137
  L. laevis 137
  L. perlucida 137
  L. semistriata 137
  L. striata 137
  L. substriata 137
  L. sp. 137
Lagenammina 5
  L. arenulata 137
Lagenina 19, 26, 37
Lasiodiscidae 13–14
Ledahia 264
  L. septenaria 265
Lenticulina 25
  L. rotulata **6.1**
Leupoldina 174, 208, 225, *5.82*
Liasina lanceolata 273
Lituolidae 26, 34
Lituotuba 13, 36
Loeblichella 174, 208, 222, *5.89*
Loeblichellinae 174, 208
Loeblichia 23
Loftusiinae 26

Marginopora 33–4
Marginotruncana 176, 211, *5.114*
Marginotruncanidae 176
Marginotruncaninae 211
Mariannenina 197
Marsipella 9
  M. elongata 137
Marssonella **6.2**
  M. trochus **6.1**
Massilina 30, 32
  M. secans 137
Melonis 24, 115
  M. pompilioides 137
Menardella 147, 176, 198, 204, *5.45*
Metacopa 262
Micrhystridium 287
Miliammina 32
  M. fusca 137
Miliolidae 26
Miliolina 8, 13, 26, 29, 32, 37
Miliolinella 32

  M. circularis elongata 137
  M. subrotunda 137
Minerisporites 285, Plate *8.4k*
  M. alius Plate *8.3a*
  M. marginatus Plate *8.4k*
Monalysidium 19
Monoceratina 276
Moravammina 22
Morozovella 176, 203, 221, 223, *5.41*
Muderongia 286–7, 289, Plates *8.10a; 8.11e*
Muricoglobigerina 176, 200, 220, 222, *5.9*

Nanicella 23, 25
Nemogullmia 12
Neoacarinina 178, 200, *5.11*
Neogloboquadrina 178, 203, 222, *5.43*
Neomiodon 287
Nodobaculariella 19
Nodosarella 14
Nodosariacea 19, 25–6, **6.2, 6.3**
Nodosariidae 29, 34, 36
Nodosinellidae 25–6
Nonionella turgida 137
Novatrix 67
  see *Arenobulimina (Novatrix)*
Nummulites 81
  N. aturicus 83

Obandyella 147, 172, 178, 204, 223, *5.48*
Oberhauserellidae 197, 212
Ogmoconcha 262–6, 268, 270–3, 276
  O. amalthei 265, 276
  O. ambo 265, 268–70
  O. cincta 265, 269
  O. circumvallata 265, 267–9, *7.2*
  O. cista 269
  O. comes 269
  O. contractula 268
  O. hyblea 271
  O. intercedens 269
  O. nordvikensis 266
Ogmoconchella 264, 272
  O. adenticulata 266
  O. aspinata 267
  O. bispinosa 265, 268
  O. conversa 266
  O. ellipsoidea 264, 267
  O. impressa 265, 269
  O. martini 266
  O. propinqua 266
Oolina 3
  O. hexagona 137
  O. squamosa 137
  O. williamsoni 137
  O. sp. 137
Ophiomorpha 288
Ophthalmidium 32
Orbignya 72
Orbitoidacea 34
Orbitolina 15, 19

Orbitolinidae 18–19
Orbitopsella 33–4
Orbulina 2, 34, 152, 178, 198, 205–6, 222, *5.62*
'Orbulinacea' 146
Orbulinida 146
Orbulininae 205, 222–3, *5.58–63*, *5.123*
Orbulinoides 5, 180, 205, *5.53*
Osangularia **6.2**
 O. whitei **6.1**
Ostrea bellovacina 252
Oxinoxis 19, 26
 O. ligula 29
Ozawainella 24

Pachyphloia 16
Palaeobigenerina 35
Palaeotextulariina 22, 29
Paracypris 276
 P. redcarensis 276
Paradoxiella 34
Paraendothyra 22
Pararotalia 149
Parastafella 24
Paratextularia 21
Paratharamminacea 8
Paratikhinella 19, 22
Parvularugoglobigerina 180, 202, *5.32*
Patellina 22
Pateoris hauerinoides 137
Pavonina 18, 36
Pavonitinae 18
Paxillitriletes alatus Plate *8.4i*
Peneroplis 18, 34, 36
Periloculina 30, 32
Pilosisporites 285
 P. trichopapillosus Plate *8.6a*
Planogyrina 146, 180, 208, 224, *5.88*
Planomalina 180, 208, 224, *5.80*
Planomalinidae 146, 207, *5.77–81*, *5.125*
Planomalininae 146, 207, *5.77 81*, *5.125*
Planorbulina mediterranensis 115, 125, 137
Planorotalia 198
Planorotalites 149, 182, 198, 203, 220–1, 223, *5.36*
Planorotalitinae 203, 223, *5.36–7*, *5.122*
Platysolenites 13
Plectina 47
 P. cenomana 47
 P. ruthemia 48
Pleurostomella sp. 137
Pleurostomellidae 19
Plummerella 182
Plummerita 182, 186, 211, *5.111*
Polskanella 182, 199
Polycope 269, 276
 P. cerasia 273
Polyperibola 182, 201, *5.18*
Porticulasphaera 182, 205, *5.55*
Porticulasphaerinae 205, 222–3, *5.55–7*, *5.123*
Praebulimina reussi **6.1**

Praeexogyra 283, 287
Praeglobotruncana 182, 209, 221, 223, *5.92*
Praegubkinella 197, 219
Praehedbergella 184, 208, 221, 224, *5.86*
Praeindicola 198
Praeorbulina 184, 205, *5.58*
Prosphaeroidinella 184, 202, *5.27*
Protentella 184, 206, 224, *5.66*
Protoellipsodinium spinosum Plate *8.13c*
Psamminopelta 32
Psammosphaera 12
Pseudoclavulina 19
Pseudoeponides 28
Pseudogloboquadrina 184, 200, *5.10*
Pseudoglomospira 22
Pseudohastigerina 184, 206, 224, *5.64*
Pseudopalmula 21
Pseudopolymorphina cf. *novanglae* 137
 P. sp. 137
Pseudospiroplectinata 35
Pseudotextularia 212
Pseudothalmanninella 186, 210, *5.100*
Pseudoticinella 149, 186, 210, *5.103*
Pseudotruncorotalia 196
Pulleniatina 186, 204, 223, *5.62*
Pulleniatininae 186, 204, 223, *5.51–2*, *5.123*
Pylodexia 198–9
Pyrgo 30, 32
 P. constricta 137
 P. depressa 137
 P. williamsoni 137
 P. sp. 137

Quinqueloculina 30, 32
 Q. aspera 137
 Q. bicornis 137
  Q. bicornis angulata 137
 Q. dimidiata 137
 Q. duthiersi 137
 Q. lamarckiana 32
 Q. lata 137
 Q. oblonga 137
 Q. rugosa 137
 Q. seminulum 137
 Q. subpoeyana 32

Racemiguembelina 212
Radiolaria 2, 8
Radotruncana 186, 212, *5.116*
Rectobolivina 36
Reophax 14, 18–19, 26
 R. arctica 18
 R. fusiformis 138
 R. scorpiurus 138
Reticuloglobigerina 158, 188, 199, 219–20, *5.5*
Reussella **6.2**
 R. kelleri **6.1**
Rhabdammina 9
Rhapydionininae 18, 19
Rhizopoda 3

Rhynchospira 199
Riveroinella 197
Robsoniella 262
Robulus sp. 138
Rosalina 28, 133
  R. globularis 122, 125, 127-8, 138, *4.11*, Plate *4.1d-f*
  R. sp. 138
Rosalinella 188, 211, *5.113*
Rotaliacea 197
Rotaliidae 197
Rotaliina 8, 26, 28-9, 34, 37, 143, 253, 257, **6.3**
Rotalina 197
Rotalipora 188, 210, *5.102*
Rotaliporidae 188, 209, 221, 223, *5.98-100*, *5.126*
Rotaliporinae 210, *5.101-3*
Rotorbinella 28
Rotundina 190, 209, *5.93*
Rugoglobigerina 190, 210, *5.106*
Rugoglobigerininae 190, 210, *5.104-11*, *5.126*
Rugotruncana 190, 211, *5.109*
Rzehakinidae 29, 32

Saccammina 5, 19
Saccamminidae 22
Saccamminopsis 8, 19, 22
Saipanetta 262
Schackoina 190, 208, *5.83*
Schackoinidae 190
Schackoininae 208, *5.82-5*, *5.125*
Schizammina 13
Schizosporis reticulatus Plate *8.3a*
Semitextularia 18, 21, 36
Semitextulariida 22
Semitextulariidae 22, 37
Semivulvulina 21
Shepheardella 12
Sigmorphina sp. 138
Signohealdia 264
Siphonides 36
Siphotrochammina 27
Soritidae 18, 26, 34
Sorosphaera 5
Sphaeroidinella 190, 202, *5.29*
Sphaeroidinellinae 192, 201, *5.27-9*, *5.123*
Sphaeroidinellopsis 192, 202, *5.28*
Spiniferites ramosus Plate *8.13b*
Spirillina 8, 12
  S. vivipara 138
Spirillinacea 13-14
Spirillinina 14, 22, 29, 37
Spirocylininae 18
Spirolina 19, 36
Spiroloculina 30
  S. excavata 138
  S. cf. *S. rotunda* 138
Spirophthalmidium acutimargo 138
Spiroplectammina 22

S. wrightii 130, 138, Plate *4.2i,j*
Spiroplectinata 36
Spirosolenites 13
Sporohantkenina 192, 207, *5.73*
Stainforthia sp. 138
Stensioina 247, **6.2**
Subbotina 192, 200, 220, 222, *5.7*
Subtilisphaera terrula 287, 289, Plate *8.10e*

Technitella 5
Tectoglobigerina 197
Tenuitella 192, 202, *5.33*
Tenuitellinae 202, *5.32-3*, *5.122*
Testacarinata 194, 203, 222, *5.39*
Tetrataxidae 29
Tetrataxis 29
Textularia 21-2, **4.1**
  T. baudouiniana 250
  T. conica 21
  T. earlandi 21, 138
  T. sagittula 130-1, *4.14*, Plate *4.2g,h*
Textulariidae 250
Textulariina 5, 7, 13, 22, 25-6, 37, 247, 250, 253-4, 257, 259, **6.2, 6.3**
Thalmanninella 194, 210, *5.101*
Thurammina 5
Ticinella 194, 209, 224, *5.98*
Ticinellidae 194
Ticinellinae 209, *5.98-100*
Tinophodella 194, 200-1, *5.17*
Tiphotrocha 28
  T. comprimata 28
Tolypammina 10, 13
Tolypammininae 12
Torohealdia 264
Tournayellidae 13-14, 25, 37
Trachycythere 270
  T. verrucosa 273
Trepeilopsis 12
Trichosida 5
Triebelina (Ptychobairdia) 271
Trifarina angulosa 138
  T. bradyi 138
  T. sp. 138
Trileites Plate *8.4e*
Trilobosporites 285, Plates *8.4a,d; 8.5b-d*
Trilocularena 32
Triloculina 30, 32, 138
  T. oblonga 32
Trinitella 194, 211, *5.108*
Triporoletes Plate *8.6c*
? *Tritaxia* sp. 250
Trochammina 26-7, 29, 54
  T. inflata 28
  T. rotaliformis 138
Trochamminidae 28
Trochamminita 23, 25
Trochamminoides 23
Truncorotalia 194, 204, *5.49*
Truncorotaliinae 194, 203

Truncorotaloides 195, 203, 222–3, *5.40*
  T. rohri 83
Truncorotaloidinae 195, 203, 221–3, 225, *5.38–41*, *5.123*
Tubinella 18, 37
Turborotalia 147, 195, 203, 220, 222–4, *5.42*
Turborotalita 196, 204, 224, *5.50*
Turborotalitidae 196
Turborotalitinae 203
Turitellella 12, 13, 37

Unio 288
Uvigerina peregrina 138

'vallate-*Ogmoconcha*' 264–6, 268–73, *7.2*
Valvulina 54
  V. gibbosa 54, 66–7
  V. quadribullata 54–6, 66–8
Valvulineria **6.2**
  V. lenticula **6.1**
Valvulininae 42, 46

Vanhoeffenella 8
Velapertina 196, 201, *5.24*
Verneuilina 47
Verneuilininae 43, 45, 47–8, 72
Verneuilinoides 47
Veryhachium 287, *Plates 8.11g; 8.12b*
Vialovella 42, 47–8, 64, *2.2*
  V. oblonga 47–8
  V. praefrankei 42–3, 45, 47–9, 71–2, *Plate 2.1a–c*
Virgulinella 28
Viviparus 287–8
Voloshinoides 67
  see also *Arenobulimina (Voloshinoides)*
Voloshinovella 72

Wanaea fimbriata Plate 8.13d
'Wetheredella' 13
Whiteinella 196, 209, 220, 222, *5.97*
Woletzina 196, 199, *5.4*